CORROSION-RESISTANT PIPING SYSTEMS

CORROSION TECHNOLOGY

Editor
Philip A. Schwietzer, P.E.
Consultant
Fallston, Maryland

CORROSION-RESISTANT PIPING SYSTEMS

PHILIP A. SCHWEITZER, P.E.

Consultant
Fallston, Maryland

Marcel Dekker, Inc.   New York•Basel•Hong Kong

Library of Congress Cataloging-in-Publication Data

Schweitzer, Philip A.
    Corrosion-resistant piping systems / Philip A. Schweitzer.
        p. cm. — (Corrosion technology : 5)
    Includes index.
    ISBN 0-8247-9023-5
    1. Piping—Corrosion.   I. Title.   II. Series: Corrosion
technology (New York, N.Y.) : 5.
    TJ930.S348   1994
    621.8'672—dc20   93-38083
                                                                              CIP

The publisher offers discounts on this book when ordered in bulk quantities. For more information, write to Special Sales/Professional Marketing at the address below.

This book is printed on acid-free paper.

Marcel Dekker, Inc.
270 Madison Avenue, New York, New York 10016

Current printing (last digit):
10 9 8 7 6 5 4 3 2 1

PRINTED IN THE UNITED STATES OF AMERICA

Preface

Corrosion was not recognized or faced as a problem by the first people installing piping systems. The earliest pipelines were natural tubes, such as hollow logs or bamboo used by the Chinese (who are credited with having invented the pipeline) to transport water. As early as 500 B.C. the Greeks used clay pipes and the Romans used lead pipes to convey water from aqueducts made of hollowed stone. In the fourth century B.C. the Egyptians at Kom Ombo hollowed out stones to transport water to their temple from the Nile River.

The first material-of-construction problem arose about 400 B.C. when the Chinese decided to convey marsh gas for lighting. They discovered that the gas leaked from the bamboo pipes. The problem was overcome by wrapping the bamboo with waxed cloth.

As early as A.D. 600 large quantities of manufactured clay pipe were used by the Japanese, but large-scale commercial production of pipe only began approximately 200 years ago in Europe. At that time, cast iron pipe was produced for water and sewer systems and later was used to transport gas. The first steel water pipes were laid in the United States in 1860. For many years cast iron and steel were satisfactory as materials of construction for the transport of fluids.

During the industrial revolution, particularly in the chemical process industry, problems of corrosion became more pronounced. In order to combat these problems, various metallic alloys were produced to resist the chemical attack. Since that time a myriad of synthetic materials have been developed to provide resistance to the ever-growing list of aggressive corrodents, which include manufactured chemicals and their by-products and/or waste streams resulting from manufacturing operations.

In addition to the growing list of metallic alloys are plastic materials used to manufacture piping systems, or as liners for metallic piping systems, and various materials such as glass, carbon, and even wood. Wooden pipe is still made and sold in wood-producing regions. In fact, the chemical-resistant properties of wood make it more suitable than metal for some applications.

Today's piping designer is faced with the problem of selecting the correct material of construction to resist the corrosive attack of the fluids being conveyed. The problem is aggravated by the large variety of materials available and their varying corrosion-resistant, physical, and mechanical properties. As with all construction projects, economics also play a part in the selection process.

This book has been prepared to provide a single source with which to compare the corrosion-resistant, physical, and mechanical properties of various piping materials. To this end, solid plastic piping systems, metallic piping systems, and miscellaneous piping systems such as glass, carbon, and clay, etc. are included. A section covering the materials of construction available for double-containment piping systems, as well as their design, use, and installation, is also included. Sections are also included on good piping design practice, which is necessary to maintain the corrosion-resistant properties of the piping material. The compatibility of each piping system is compared to a list of 180 of the most common corrodents.

This book will be a useful tool for all persons involved with piping systems, whether it be the engineer, designer, installer, maintenance worker, or user. Here in one place may be found the answers to most questions regarding corrosion-resistant piping systems.

Philip A. Schweitzer, P. E.

Contents

1

Piping Design Considerations

Selecting the proper material of construction for a piping system need not be a burdensome chore, however, it does take some thought and knowledge of the process. Consideration must be given not only to the material of construction from a corrosion-resistance viewpoint, but also from physical, mechanical, and safety aspects. Location of the pipeline, whether indoors or outdoors, is another factor that enters into the evaluation. Some synthetic materials are affected by ultraviolet light, therefore a material of construction which may be compatible with the corrodent may not be suitable for an outside installation. External atmospheric conditions must also be considered. At times, external corrosion of the pipe can be as serious a problem as internal corrosion. These and other factors must be taken into account when selecting a material of construction.

1.1   EVALUATION OF MATERIALS OF CONSTRUCTION

In order to make an evaluation of the best material of construction suitable for a particular application, certain basic information must be obtained. The questions to ask include:

1. What material is being handled and at what concentration? Does the material have hazardous properties (explosive, toxic, combustible, etc.)?
2. What are the operating temperatures and pressures? Will there be temperature cycling?
3. Is static grounding required?
4. What is the installation location? (buried, inside of building, outside of building, etc.) What are the corrosive conditions of the surrounding atmosphere? In some instances external corrosion can be a problem. If piping is to be outside, what are the ambient

conditions throughout the year (maximum and minimum temperatures)? Will heat tracing be required to prevent freeze-up?

With the answers to these questions available, the screening process can begin. The first step is to list all of the piping materials compatible with the fluid being handled. This will usually narrow the selection appreciably. The next step is to eliminate those piping systems with operating temperatures and/or pressures not within the operating conditions of the system being installed. After these steps have been taken, in all probability a representative list of potential satisfactory piping systems will remain, including some metallic, solid plastic, and lined systems. As would be expected, each category of system will have certain advantages and disadvantages. An analysis of these features will help to zero in on the most appropriate system.

1.2 ADVANTAGES AND DISADVANTAGES OF EACH CATEGORY OF PIPING SYSTEM

Basically we are concerned with four types of piping systems, namely, plastic, lined, metallic, and miscellaneous. Each general type of system has its own specific set of advantages and disadvantages.

1.2.1 Plastic Piping Systems

Plastic piping systems should be broken down into two types: thermoplastic and thermoset. The following advantages are applicable to both types.

1. Usually resistant to external corrosions
2. Not susceptible to galvanic corrosion
3. Very resistant to nonoxidizing acids and chlorides
4. Can withstand full vacuum
5. Relatively inexpensive

The disadvantages of plastic piping systems vary between the thermoplastic and thermoset series. Thermoplastic types have the following disadvantages:

1. Relatively low maximum operating temperatures
2. Limited operating pressures
3. More complex system support than metallic piping system
4. May require protection from ultraviolet rays
5. Steam tracing problems

The thermoset materials have higher operating temperatures and pressures than the thermoplastics to such a degree that this cannot be considered a serious disadvantage. Their mechanical strength is such that support for these systems is no more complicated than for metallic systems. These systems are more expensive than the thermoplastics.

1.2.2 Lined Piping Systems

The advantages of these systems include:

1. Operating pressures usually the same as for metallic pipe
2. Support of system simpler than for thermoplastic pipe
3. Protection from ultraviolet rays not required

Disadvantages of these systems include:

1. Sometimes limited vacuum service
2. External corrosion of metallic pipe possible
3. Possible premature failure of liners due to thermal cycling
4. Steam tracing possibly a problem
5. Static grounding possibly a problem
6. Special fixtures possibly required to produce odd lengths of pipe

1.2.3 Metallic Piping Systems

The advantages of these systems include:

1. Withstands higher and lower operating temperatures than plastic or plastic-lined pipe (is only affected by an extremely wide range of temperature cycling)
2. Can withstand full vacuum
3. Can be steam jacketed and/or readily steam traced
4. Simple support system
5. Thermal expansion less than that of plastic piping systems
6. Protection from ultraviolet rays not required
7. Easy to static ground

Disadvantages include:

1. System must be designed to prevent galvanic corrosion
2. External corrosion possible
3. Alloy piping tends to be more expensive than plastic piping

1.2.4 Miscellaneous Piping Systems

These systems are usually designed to handle mixtures of corrosive materials, which will cause problems in plastic and/or metallic systems. Their main advantage is a wider range of corrosion resistance. The major disadvantage is that these systems are somewhat more expensive than metallic or plastic systems.

One further step must be undertaken. A piping system contains, besides the basic pipe, accessory items such as shut-off valves, check valves, relief valves, control valves, expansion joints, etc. Not all of these accessory items may be available in the same material of construction as the piping systems, therefore, it may be necessary to use one material of construction for the piping system and an alternate material of construction for some of the accessory items. Care must be taken in the selection of the alternate materials so that they will be compatible with the material from which the basic piping system is made and also so that galvanic corrosion will not be introduced into the system.

1.3 OTHER CONCERNS

Once the determination has been made as to the proper material of construction, other aspects of the installation must be considered in order to get the maximum benefit from the selected material of construction. General concerns will be discussed here, while specific concerns for each piping material will be discussed in the section covering that material.

1.3.1 Water Hammer

When the velocity of a fluid flowing in a pipeline is suddenly reduced and/or stopped, a pounding of the line is experienced, as though someone were hitting the line with a hammer. The pounding in the line is caused by a pressure rise due to the deceleration of a truly incompressible fluid in a nonexpandable pipe. Theoretically the pressure rise would be infinite since the fluid in the line would act like a plug and the pressure rise would correspond to the inertial effects of the plug. However, there is a finite pressure since part of the kinetic energy of the moving fluid is expended in stretching the pipe walls and compressing the fluid.

Taking into account the fact that the pipe walls are elastic the speed of the pressure wave can be calculated by:

$$S = \sqrt{\frac{EE^1}{(E^1 + Ed/T)}}$$

where:

S = speed of the pressure wave (ft/sec)
E = bulk modulus of the fluid
ρ = density of the fluid
E^1 = modulus of elasticity of the pipe material
d = inside diameter of the pipe
T = pipe wall thickness

and the change in pressure can be calculated by:

$$\Delta P = -\frac{\rho}{g} S \Delta V$$

where:

ΔP = change in pressure lb/ft^2
ρ = density of the fluid
S = speed of the pressure wave (ft/sec)
ΔV = change in velocity (ft/sec)

The pressure surge developed can be illustrated by the example of reducing the velocity of water in a pipeline by 8 ft/sec when the pressure wave travels at 4720 ft/sec:

$$P = \frac{-62.4(4720)(-8)}{32.17} = 73243 \; PSF = 509 \text{ psi}$$

This sudden pressure rise can be extremely harmful to plastic piping systems whose allowable operating pressures are relatively low.

When a valve at the end of a pipe connected to a reservoir is suddenly closed, a compression wave travels up the pipe to the reservoir. When the compression wave reaches the reservoir, it is reflected as an expansion wave, which travels down the pipe to the closed valve, where it is reflected as an expansion wave. Compression and expansion waves would travel up and down the pipe indefinitely were it not for friction. The pressure wave makes a round trip from the valve to the reservoir and back to the valve in:

$$t = 2L/S$$

where:

t = time (sec)
L = pipe length (ft)
S = speed of the wave (ft/sec)

In the preceding example, if the pipe length were 100 ft,

$$t = \frac{2(100)}{4720} = 0.04 \text{ sec}$$

In order to prevent the full pressure, rise the valve should close in a time greater than 2L/S.

If it is not feasible to close a valve slowly, air chambers or surge tanks can be installed on the pipe line to absorb the excess pressure (see Fig. 1-1). Alternatively, vacuum breakers and relief valves can be installed in the line to admit air when the fluid separates and to release air and some fluid when the fluid rejoins. However, this approach is not recommended on lines handling corrosive or toxic materials.

Water hammer can also occur as the result of a loose valve seat or washers. This is a common occurrence in homes. Regardless of what approach is followed, some means of preventing or controlling water hammer should be considered, particularly in solid plastic piping systems.

1.3.2   Piping Support

In order to prevent sagging and damage to the pipe, correct support, both in design and spacing, is required. The distance between piping supports is dependent upon the pipe diameter, pipe wall thickness, specific gravity of the fluid being transported, maximum ambient temperature, and maximum working temperature. The determination of piping support centers is based on the permissible amount of deflection between two brackets of 0.01 inch (0.25 cm) and may be calculated from:

$$L^2 = \frac{12ZS}{W}$$

where:

L = span between supports (in.)
Z = section modulus (in.^{3})
S = allowable design stress divided by 4 (psi)
W = weight per inch of length (lb)

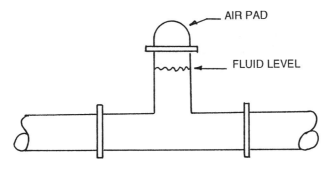

Figure 1–1   A typical air chamber.

The section modulus Z may be calculated by:

$$Z = 0.0982 \, \frac{OD^4 - ID^4}{OD}$$

where:

OD = outside diameter of pipe (in.)
 ID = inside diameter of pipe (in.)

The weight referred to in the support equation is a combination of the weight of the pipe plus the weight of the fluid being transported, which occupies one lineal inch of pipe.

The equation is based on a uniformly distributed load and does not take into account local concentrated loads such as flanges, valves, meters, controls, or other items that may be installed in a line. When handling highly corrosive fluids it is desirable to support such accessories in the pipe line independently. If not done, these accessories must be considered as a concentrated load and the piping support spacing adjusted accordingly.

Proper spacing of supports is essential. Improper support spacing and/or inadequate support will result in excessive spans between supports. This will lead to inducing stress in the pipe. Induced stress in a pipe, if severe enough, can cause the pipe to fail structurally. This rarely occurs in metallic pipe but is more likely to occur in plastic piping systems.

Induced stress in a piping system will also tend to make that piping system susceptible to corrosive attack by a corrodent to which the piping material is normally resistant. This is particularly prevalent in metallic piping systems.

The foregoing equations permit the design of a piping support system that will not induce stresses in the piping system. However, the pipe will tend to sag between the supports, which will prevent the line from draining completely. In order to overcome this problem, each successive support must be positioned lower than the maximum sag of the pipe between supports. The vertical distance that the span must be pitched so that the outlet is lower than the maximum sag of the pipe can be calculated from the following equation:

$$h = \frac{144 \, S^2 y}{36 \, S^2 - y^2}$$

where:

h = difference in elevation of span ends (in.)
S = distance between supports (ft)
y = deflection between supports (in.)

The term y^2, being inconsequential, may be eliminated, reducing the equation to

$$h = 4y$$

The amount of deflection, or sag between supports, is calculated from the following formula:

$$d = \frac{WL^4}{384EI}$$

where:

d = deflection or sag (in.)
w = weight of the pipe, covering, fluid, valves, etc. (lb/lineal in.)
L = distance between supports (in.)
E = modulus of elasticity of the pipe material
I = moment of inertia

The recommended maximum support spacing for each specific piping system, without inducing stress in the pipe line, is tabulated separately under each system. The spacing is provided for handling fluids of various specific gravities. If it is desired to have a line drain completely, it will be necessary to pitch the line in accordance with the above calculations. The degree of pitch required may be reduced by reducing the spacing between supports. This will reduce the sag between supports, thereby permitting a lesser pitch to have the line drain completely.

The type of support used is also critical. Most horizontal overhead lines are supported by means of clevis-type hangers or pipe racks. When used to support plastic pipe and some of the miscellaneous pipe materials, special thought must be given to these supports. They must have a larger bearing area than for metallic pipe, and care must be exercized that they do not score or otherwise damage the pipe. These considerations are discussed under the appropriate piping system. It is also essential that piping supports do not force the pipe out of alignment, which would tend to impart a stress to the pipe.

1.3.3   Thermal Expansion

Changes in temperature will produce a lengthening or shortening of pipework, which must be compensated for. Any piping system must be designed so that it will not fail because of excessive stresses, will not produce excessive thrusts or movements at connected equipment, and will not leak at joints because of expansion or contraction of the pipe. These problems are greater in plastic piping systems than in metallic or metallic-lined systems because of the larger coefficient of thermal expansion of plastic materials than metallic materials. When considering thermal expansion and/or contraction of piping, the general approach is to think of the temperature changes within the pipe resulting from the fluid being transported. However, with plastic piping an additional factor must be taken into account which can be very critical because of their relatively large coefficients of thermal expansion.

Piping systems are usually installed at a temperature other than that at which they will operate. In addition, there are wide temperature changes between summer and winter in many sections of the country, and this becomes a significant factor in outside installations. It is also possible that a pipe line is designed to handle a fluid at ambient temperature but external temperatures may be considerably higher or lower than the "ambient" temperature of the fluid.

Length changes can be accommodated by the inclusion of flexible sections and/or expansion joints in the system. For plastic piping systems, a temperature change of 30°F (17°C) or more requires that compensation for expansion or contraction be made. The inclusion of a flexible section in the piping system is the simplest and most economical procedure to follow. Details of calculations for these flexible sections will be found in Chapter 2.

Expansion loops can also be incorporated into a piping system. However, in metallic piping systems caution should be used when corrosive materials particularly chlorides, are handled, since the possibility of inducing stress corrosion exists. Properly designed expansion joints are the better solution.

Expansion joints consisting of two tubes, one telescoping within the other, are available. The outer tube is firmly anchored, while the inner tube is permitted to move freely, compensating for the contraction or expansion of the pipe. Proper installation of these joints particularly their alignment, is critical. If the pipe is misaligned, binding will take place. Guides should be installed approximately one foot from each end of the expansion joint to insure proper alignment.

The correct position of the inner tube at time of installation can be calculated by:

$$E_x = \frac{[T_m - T_i] L}{T_c}$$

where:

E_x = extension of inner tube out of the joint (in.)
T_m = maximum temperature (°F)
T_i = installation temperature (°F)
L = length of allowable travel in expansion joint (in.)
T_c = total temperature change (°F)

The total amount of expansion and contraction must be calculated for each piping run. Since each expansion joint will compensate for a specific amount of movement, it may be necessary to install more than one per piping run.

Bellows-type expansion joints are also available. In addition to compensating for thermal expansion, these also allow for angular, lateral, or axial displacement. Preferably these should be installed in a vertical section of line to permit drainage of the corrodent from the convolutions.

1.3.4   Gaskets

Flanged piping systems require gaskets between the flange faces in order to effect a seal. Materials of construction for gaskets vary from soft metals through a wide variety of elastomeric materials. In addition to being resistant to the corrodent being handled, the gasket material must also be sufficiently compressed so that it will "flow" into all of the voids in the mating surfaces. Unless this "flow" occurs, a seal cannot be effected. The amount of force required by a specific gasket material to effect a seal is known as the yield stress (psi) of the gasket. This is a constant for each thickness of each gasket material. Gasket manufacturers will supply this information.

The required yield stress (psi) of a gasket can be calculated using the following equation:

$$y = \frac{2L}{A}$$

where:

y = yield stress of gasket material (psi)
L = bolt load (lb)
A = gasket area (in.^{2})

The bolt load in pounds may be calculated from the following equation:

$$L = \frac{T}{KD}$$

where:

L = bolt load (lb)
T = tightening torque (in.-lb)
K = torque friction coefficient
      0.20 for tightening torque, dry
      0.15 for tightening torque, lubricated
D = nominal bolt diameter (in.)

If the tightening torque to be used is known, the bolt load may be gotten from standard bolt tables.

Gaskets for use with plastic flanges should be of the full-face variety. For dimensions of full-face gaskets, see Table 3-7. For bolt torques to be used when assembling plastic flanges, see Table 2-5.

As previously mentioned, selection of a gasket material resistant to the corrodent being handled in both concentration and temperature is critical. Table 1-1 provides the corrosion resistance of the more common elastomeric materials used as gaskets when in contact with selected corrodents.

Table 1-1 Corrosion Resistance of Selected Elastomers. The chemicals listed are in the pure state or in a saturated solution unless otherwise indicated. Compatibility is shown to the maximum allowable temperature for which data are available. Incompatibility is indicated by an x. A blank space indicates that data are unavailable.

Chemical	Butyl °F	Butyl °C	Hypalon °F	Hypalon °C	EPDM °F	EPDM °C	EPT °F	EPT °C	Viton A °F	Viton A °C	Kalrez °F	Kalrez °C	Natural rubber °F	Natural rubber °C	Neoprene °F	Neoprene °C	Buna N °F	Buna N °C
Acetaldehyde	80	27	60	16	200	93	210	99	x	x	x	x	x	x	200	93	x	x
Acetamide	x	x	x	x	200	93	200	93	210	99	x	x	x	x	200	93	180	82
Acetic acid 10%	150	66	200	93	140	60	x	x	190	88	200	93	150	66	160	71	200	93
Acetic acid 50%	110	43	200	93	140	60	x	x	180	82	200	93	x	x	160	71	200	93
Acetic acid 80%	110	43	200	93	140	60	x	x	180	82	90	32	x	x	160	71	210	99
Acetic acid, glacial	90	32	x	x	140	60	x	x	x	x	80	27	x	x	x	x	100	38
Acetic anhydride	150	66	200	93	x	x	x	x	x	x	210	99	x	x	90	32	200	93
Acetone	160	71	x	x	300	148	x	x	x	x	210	99	x	x	x	x	x	x
Acetyl chloride			x	x	x	x	x	x	190	88	210	99	x	x	x	x	x	x
Acrylic acid									x	x	210	99	x	x	x	x	x	x
Acrylonitrile	x	x	140	60	140	60	100	38	x	x	110	43	90	32	160	71	x	x
Adipic acid	x	x	140	60	200	93	140	60	180	82	210	99	80	27	160	71	180	82
Allyl alcohol	190	88	200	93	300	148	80	27	190	88			80	27	120	49	180	82
Allyl chloride	x	x	x	x	x	x	x	x	100	38			x	x	x	x	x	x
Alum	190	88	200	93	200	93	140	60	190	88	210	99	150	66	200	93	200	93
Aluminum acetate			x	x	200	93	180	82	180	82	210	99	x	x	x	x	200	93
Aluminum chloride, aqueous	150	66	250	121	210	99	180	82	190	88	210	99	140	60	200	93	200	93
Aluminum chloride, dry											190	88						
Aluminum fluoride	180	82	200	93	210	99	180	82	180	82	210	99	150	66	200	93	190	88
Aluminum hydroxide	100	38	250	121	210	99	140	60	190	88	210	99	x	x	180	82	180	82
Aluminum nitrate	190	88	250	121	210	99	180	82	190	88	210	99	150	66	200	93	200	93
Aluminum oxychloride									x	x								
Aluminum sulfate	190	88	200	93	210	99	210	99	190	88	210	99	160	71	200	93	210	99
Ammonia gas			140	60	140	60	140	60	x	x	210	99	x	x	140	60	190	88
Ammonium bifluoride	x	x	x	x	300	148	140	60	140	60	210	99	x	x	x	x	180	82
Ammonium carbonate	190	88	140	60	300	148	180	82	190	88	210	99	150	66	200	93	200	93

	P1 °F	P1 °C	P2 °F	P2 °C	P3 °F	P3 °C	P4 °F	P4 °C	P5 °F	P5 °C	P6 °F	P6 °C	P7 °F	P7 °C	P8 °F	P8 °C	P9 °F	P9 °C
Ammonium chloride 10%	190	88	200	93	210	99	180	82	190	88	210	99	150	66	200	93	200	93
Ammonium chloride 50%	190	88	200	93	210	99	180	82	190	88	210	99	150	66	190	88	200	93
Ammonium chloride, sat.	190	88	200	93	300	148	180	82	190	88	210	99	150	66	200	93	200	93
Ammonium fluoride 10%	150	66	200	93	210	99	210	99	140	60	210	99	160	71	100	38	200	93
Ammonium fluoride 25%	150	66			300	148	140	60	140	60	210	99	80	27	200	93	120	49
Ammonium hydroxide 25%	190	88	250	121	100	38	140	60	190	88	210	99			200	93	200	93
Ammonium hydroxide, sat.	190	88	250	121	100	38	140	60	190	88	210	99	90	32	210	99	200	93
Ammonium nitrate	180	82	200	93	250	121	180	82	x	x	300	148	170	77	200	93	180	82
Ammonium persulfate	190	88	80	27	300	148	210	99	140	60	210	99	150	66	200	93	200	93
Ammonium phosphate	180	82	140	60	300	148	180	82	180	82	210	99	150	66	200	93	200	93
Ammonium sulfate 10–40%	190	88	200	93	300	148	180	82	180	82	210	99	150	66	200	93	200	93
Ammonium sulfide			200	93	300	148	210	99	x	x	210	99			160	71	180	82
Ammonium sulfite											210	99					160	71
Amyl acetate	x	x	60	16	210	99	x	x	x	x	210	99	x	x	x	x	x	x
Amyl alcohol	180	82	200	93	210	99	180	82	200	93	210	99	150	66	200	93	180	82
Amyl chloride			x		x		x		190	88	210	99	x		x		x	
Aniline	150	66	140	60	140	60	x		230	110	250	121	x		140	60	60	60
Antimony trichloride	150	66	140	60	300	148	148	60	190	88	210	99			x		x	
Aqua regia 3:1					x		x		250	121	250	121			x		x	
Barium carbonate	190	88	200	93	300	148	180	82	190	88	210	99	180	82	160	71	180	82
Barium chloride	190	88	250	121	250	121	180	82	190	88	210	99	150	66	200	93	200	93
Barium hydroxide	190	88	250	121	250	121	180	82	190	88	210	99	150	66	200	93	200	93
Barium sulfate	190	88	200	93	300	148	180	82	190	88	210	99	180	82	160	71	180	82
Barium sulfide	190	88	200	93	140	60	140	60	190	88	210	99	150	66	200	93	200	93
Benzaldehyde	90	32	x	x	150	66	x	x	x	x	210	99	x	x	x	x	x	x
Benzene	x	x	x	x	x	x	x	x	190	88	210	99	x	x	x	x	x	x
Benzene sulfonic acid 10%	90	32	x	x	x	x	x	x	170	77	210	99	x	x	100	38	x	x
Benzoic acid	150	66	200	93	x	x	140	60	190	88	310	154	150	66	200	93	200	93
Benzyl alcohol	190	88	140	60	x	x	x	x	350	177	210	99	x	x	x	x	140	60
Benzyl chloride	x	x	x	x	x	x	x	x	110	43	210	99	x	x	x	x	x	x
Borax	190	88	200	93	300	148	210	99	190	88	210	99	150	66	200	93	180	82

(continued)

Table 1-1 *Continued*

Chemical	Butyl		Hypalon		EPDM		EPT		Viton A		Kalrez		Natural rubber		Neoprene		Buna N	
	°F	°C	°F	°C	°F	°C	°F	°C	°F	°C	°F	°C	°F	°C	°F	°C	°F	°C
Boric acid	190	88	290	143	190	88	140	60	190	88	210	99	150	66	200	93	180	82
Bromine gas, dry			60	16	x	x	x	x			210	99			x	x	x	x
Bromine gas, moist			60	16	x	x	x	x							x	x	x	x
Bromine liquid			60	16	x	x	x	x	350	177	140	60			x	x	x	x
Butadiene			x	x	x	x	x	x	190	88	210	99			140	60	200	93
Butyl acetate	x	x	60	16	140	60	x	x	x	x	210	99	x	x	60	16	x	x
Butyl alcohol	140	60	250	121	200	93	180	82	250	121	240	116	150	66	200	93	x	x
n-Butylamine							x	x	x	x	210	99					80	27
Butyl phthalate							x	x	80	27	210	99			x	x	x	x
Butyric acid	x	x	x	x	140	60	x	x	120	49	x	x	x	x	x	x	x	x
Calcium bisulfide	120	49	250	121			x	x	190	88	210	99			x	x	180	82
Calcium bisulfite	150	66	90	32	x	x	x	x	190	88	210	99	120	49	180	82	200	93
Calcium carbonate	190	88	90	32	210	99	140	60	190	88	200	93	180	82	60	16	180	82
Calcium chlorate	190	88	200	93	140	60	140	60	190	88	210	99	150	66	200	93	200	93
Calcium chloride	190	88	200	93	210	99	180	82	190	88	210	99	150	66	200	93	180	82
Calcium hydroxide 10%	190	88	250	121	210	99	180	82	190	88	210	99	200	93	220	104	180	82
Calcium hydroxide, sat.	190	88	250	121	220	104	180	82	190	88	210	99	200	93	220	104	180	82
Calcium hypochlorite	190	88	250	121	210	99	180	82	190	88	210	99	200	93	220	104	80	27
Calcium nitrate	190	88	100	38	300	148	180	82	190	88	210	99	150	66	200	93	200	93
Calcium oxide			200	93	210	99					210	99			160	71	180	82
Calcium sulfate	100	38	250	121	300	148	180	82	200	93	210	99	180	82	160	71	180	82
Caprylic acid			x	x	x	x	x	x	190	88	210	99	x	x				
Carbon bisulfide			200	93	250	121	180	82			210	99	150	66	x	x	x	x
Carbon dioxide, dry	190	88	200	93	250	121	180	82	x	x	210	99	150	66	200	93	200	93
Carbon dioxide, wet	190	88	200	93	250	121	180	82	x	x	210	99	150	66	200	93	200	93
Carbon disulfide	190	88	230	110	250	121	180	82	x	x	210	99	150	66	x	x	200	93

Chemical																
Carbon monoxide	x	x	x	x	x	x	88	190	x	x	x	x	x	x	x	x
Carbon tetrachloride	82	180	93	200	x	x	88	190	82	180	121	250	93	200	90	32
Carbonic acid	x	x	x	x	x	x	177	350	x	x	x	x	x	x	x	x
Cellosolve	x	x	x	x	x	x	x	x	x	x	148	300	x	x	150	66
Chloroacetic acid, 50% water	x	x	x	x	x	x	x	x	71	160	x	x	x	x	150	66
Chloroacetic acid	x	x	x	x	x	x	x	x	x	x	x	x	32	90	160	71
Chlorine gas, dry	x	x	x	x	x	x	x	x	x	x	x	x	x	x	x	x
Chlorine gas, wet	x	x	x				88	190	x	x	x	x	x	x	x	x
Chlorine liquid	x	x	x	x	x	x	88	190	x	x	x	x	x	x	90	x
Chlorobenzene	x	x	x	x	x	x	88	190	x	x	x	x	x	x	x	x
Chloroform	x	x	x	x	x	x	88	190	x	x	x	x	x	x	x	x
Chlorosulfonic acid	x	x	x	x	x	x	x	x	x	x	x	x	x	x	x	x
Chromic acid 10%	88	190	60	140	x	x	177	350	x	x	x	x	150	66	x	38
Chromic acid 50%	88	190	38	100	x	x	177	350	x	x	x	x	160	71	x	x
Chromyl chloride																
Citric acid 15%	82	180	99	210	43	110	88	190	99	210	121	250	93	200	200	82
Citric acid, conc.	82	180	99	210	66	150	x	x	99	210	121	250	93	200	200	82
Copper acetate	82	180	93	200	x	x	88	190	38	100	x	x	x	x	160	82
Copper carbonate	x	x	x	x	66	150	88	190	99	210	93	200	x	x	150	93
Copper chloride	93	200	121	250	71	160	88	190	99	210	121	250	71	160	160	82
Copper cyanide	82	180	121	250	66	150	88	190	99	210	104	220	121	250	150	93
Copper sulfate	93	200	121	250	x	x	88	190	x	x	93	200	x	x	200	x
Cresol	x	x	x	x			x	x	99	210	x	x	x	x	x	99
Cupric chloride 5%	99	210	93	200			82	180	99	210	99	210	99	210	210	99
Cupric chloride 50%	82	180	93	200			82	180	x	x	71	160	71	160	160	82
Cyclohexane	82	180	x	x	x	x	88	190	x	x	x	x	x	x	x	82
Cyclohexanol	x	x	x	x			88	190	x	x	x	x	x	x	x	x
Dichloroacetic acid			x	x	x	x			x	x	x	x	x	x	x	x
Dichloroethane	x	x	x	x	66	150	88	190	x	x	x	x	x	x	150	71
Ethylene glycol	93	200	93	200	66	150	177	350	93	200	121	250	93	200	200	93
Ferric chloride	93	200	121	250	66	150	88	190	104	220	121	250	71	160	200	93
Ferric chloride, 50% water	82	180	99	210	x	x	82	180	99	210	99	210	71	160	160	82

(continued)

Table 1-1 Continued

Chemical	Butyl °F	Butyl °C	Hypalon °F	Hypalon °C	EPDM °F	EPDM °C	EPT °F	EPT °C	Viton A °F	Viton A °C	Kalrez °F	Kalrez °C	Natural rubber °F	Natural rubber °C	Neoprene °F	Neoprene °C	Buna N °F	Buna N °C
Ferric nitrate 10–50%	190	88	250	121	210	99	180	82	190	88	210	99	150	66	200	93	200	93
Ferrous chloride	190	88	250	121	200	93	180	82	180	82	210	99	150	66	90	32	200	93
Ferrous nitrate	190	88			210	99	180	82	210	99			150	66	200	93	200	93
Fluorine gas, dry	x	x	140	60			x	x	x	x	x	x	x	x	x	x	x	x
Fluorine gas, moist					60	16	100	38	x	x	x	x			x	x	x	x
Hydrobromic acid, dil.	150	66	90	32	90	32	140	60	190	88	210	99	100	38	x	x	x	x
Hydrobromic acid 20%	160	71	100	38	140	60	140	60	190	88	210	99	110	43	x	x	x	x
Hydrobromic acid 50%	110	43	100	38	140	60	140	60	190	88	210	99	150	66	x	x	x	x
Hydrochloric acid 20%	x	x	160	71	100	38	x	x	350	177	210	99	150	66	90	32	x	x
Hydrochloric acid 38%	x	x	140	60	90	32	x	x	350	177	210	99	160	71	90	32	130	54
Hydrocyanic acid 10%	140	60	90	32	200	93	x	x	190	88	210	99	90	32	x	x	x	x
Hydrofluoric acid 30%	350	177	90	32	60	16	140	60	210	99	210	99	100	38	200	93	200	93
Hydrofluoric acid 70%	150	66	90	32	x	x	x	x	350	177	210	99	x	x	200	93	x	x
Hydrofluoric acid 100%	x	x	90	32	x	x	x	x	60	16	210	99	x	x	x	x	x	x
Hypochlorous acid	x	x	x		300	148	140	60	190	88	210	99	150	66	x	x	x	x
Iodine solution 10%					140	60	140	60	190	88	190	88			80	27	x	x
Ketones, general					x	x			x	x	210	99	x	x	x	x	80	27
Lactic acid 25%	120	49	x	x	140	60	210	99	190	88	210	99	x	x	140	60	x	x
Lactic acid, concentrated	120	49	140	60			210	99	150	66	210	99	80	27	90	32	x	x
Magnesium chloride	200	93	80	27	250	121	180	82	180	82	150	66	150	66	210	99	180	82
Malic acid	x	x	250	121	x	x	80	26	190	88	210	99	80	27			180	82
Manganese chloride			180	82			210	99	180	82	210	99			200	93	180	82
Methyl chloride	90	32	x	x	x	x	x	x	190	88	210	99	x	x	x	x	100	38
Methyl ethyl ketone	100	38	x	x	80	27	x	x	x	x	210	99	x	x	x	x	x	x
Methyl isobutyl ketone	80	27	x	x	60	16	x	x	x	x	210	99	x	x	x	x	x	x
Muriatic acid	x	x	140	60			x	x	350	177	210	99			x	x	x	x
Nitric acid 5%	160	71	100	38	60	16	x	x	190	88	210	99	x	x	x	x	x	x
Nitric acid 20%	160	71	100	38	60	16	x	x	190	88	210	99	x	x	x	x	x	x
Nitric acid 70%	90	32	x	x	x	x	x	x	190	88	160	71	x	x	x	x	x	x

Chemical																		
Nitric acid, anhydrous	x	x	x	x	x	x	x	x	x	x	x	x	x	x	x	x	x	x
Nitrous acid, concentrated	120	49	x	x	x	x	100	38	210	99	x	x	x	x	x	x	x	x
Oleum	x	x	x	x	x	x	190	88	210	99	x	x	x	x	x	x	x	x
Perchloric acid 10%	150	66	100	38	140	60	190	88	210	99	150	66	66	x	x	x	x	x
Perchloric acid 70%	90	32	90	32	x	x	140	60	210	99	x	x	43	x	x	x	x	x
Phenol	150	66	x	x	x	x	80	27	210	99	x	x	x	x	x	x	x	x
Phosphoric acid 50–80%	150	66	200	93	140	60	180	82	190	88	110	43	x	200	93	130	130	54
Picric acid	x	x	80	27	300	148	140	60	190	88	x	x	x	200	x	x	x	x
Potassium bromide 30%	250	121	250	121	210	99	180	99	190	88	160	71	71	160	71	180	180	82
Salicylic acid	80	27	x	x	x	x	x	x	80	x	x	x	x	x	x	x	x	x
Silver bromide 10%									210	99								
Sodium carbonate	180	82	250	121	300	148	180	82	190	88	180	82	99	200	82	200	200	93
Sodium chloride	180	82	240	116	140	60	180	82	190	88	130	54	99	200	93	180	180	82
Sodium hydroxide 10%	180	82	250	121	210	99	210	99	210	x	150	66	99	200	93	160	160	71
Sodium hydroxide 50%	190	88	250	121	180	82	200	93	210	x	150	66	99	200	93	150	150	66
Sodium hydroxide conc	180	82	250	121	180	82	80	27	210	x	150	66	99	200	93	150	150	66
Sodium hypochlorite 20%	130	54	250	121	300	148	x	x	190	88	90	32		x	x	x	x	x
Sodium hypochlorite	90	32			300	148	x	x	190	88	90	32		x	x	x	x	x
Sodium sulfide to 50%	150	66	250	121	300	148	210	99	190	88	150	66	99	200	66	180	180	82
Stannic chloride	150	66	90	32	300	148	210	99	180	82	150	66	99	210	66	180	180	82
Stannous chloride	150	66	200	93	280	138	210	99	190	88	150	66	99	160	66	180	180	82
Sulfuric acid 10%	150	66	250	121	150	66	210	99	350	171	150	66	116	200	66	150	150	66
Sulfuric acid 50%	150	66	250	121	150	66	210	99	350	177	150	38	99	200	93	200	200	93
Sulfuric acid 70%	100	38	160	71	140	60	210	99	350	177	100	x	66	200	93	x	x	x
Sulfuric acid 90%	x	x	x	x	x	x	80	27	350	177	x	x	66	x	x	x	x	x
Sulfuric acid 98%	x	x	110	43	x	x	x	x	350	177	x	x		x	x	x	x	x
Sulfuric acid 100%	x	x	x	x	x	x	x	x	190	88	x	x		x	x	x	x	x
Sulfuric acid, fuming	x	x	x	x	x	x	180	82	x	x	x	x	99	x	x	x	x	x
Sulfurous acid	150	66	200	93	x	x	180	82	190	88	x	x	99	180	x	x	x	x
Thionyl chloride	x	x	x	x	x	x	x	x	x	x	x	x	99	x	x	x	x	x
Toluene	x	x	x	x	80	27	x	x	190	88	x	x	27	x	x	x	150	66
Trichloroacetic acid	x	x	x	x	x	x	x	x	190	88	x	x	99	x	x	x	x	x
White liquor	x	x	x	x	300	148	180	82	190	88	x	x	99	140	66	140	140	60
Zinc chloride	190	88	250	121	300	82	210	82	210	99	150	66	99	160	71	190	190	88

Source: Schweitzer, Philip A. (1991). *Corrosion Resistance Tables*, Marcel Dekker, Inc., New York, Vols. 1 and 2.

2

Thermoplastic Piping Systems

2.1 DATA AND DESIGN CONSIDERATIONS

Combination of corrosion resistance and low cost has been the reason for the wide application of plastic piping systems. The U.S. pipe market grew from 4.9 billion feet in 1977 to 6.6 billion feet in 1988 and is expected to reach 9.8 billion feet by the year 2000.

Two general types of plastic resins are used in the manufacture of plastic piping systems: thermoplastic and thermosetting. Thermoplastic resins can be repeatedly reformed by the application of heat, similar to metallic materials. Thermosetting resins, once cured, cannot be changed again in shape. This permits pipe made of thermoplastic resins to be bent or shaped by the application of heat, whereas pipe made of thermosetting resins can not be so processed. Table 2-1 shows the abbreviations used for the various plastic materials.

Included within the thermoplastic resins is a group known as fluoroplasts, which is made up of such materials as PTFE, PVDF, ECTFE, CTFE, ETFE, PFA, and FEP. As a class these materials offer excellent corrosion resistance at temperatures from −328°F (−200°C) to 500°F (260°C).

Plastic piping systems do not have specific compositions. In order to produce the piping system, it is necessary to add various ingredients to the base resin to aid in the manufacturing operation and/or to alter mechanical or corrosion-resistant properties. In so doing, one property is improved at the expense of another. Each piping system carries the name of the base resin from which it is manufactured. Because of the compounding and various formulations, mechanical and corrosion-resistant properties can vary from one manufacturer to another. The tabulated values given in this book are averages and are meant to be used as guides. Specific recommendations for an application should be obtained from the manufacturer of the piping system.

2.1.1   Structure of Plastics

The periodic table provides a guide to the general differences in the corrosion resistance of the various plastic materials. The table arranges the basic elements of nature not only by atomic structure, but by chemical nature as well, with the elements placed into classes with similar properties, i.e., elements and compounds that exhibit similar chemical behavior. These classes are the alkali metals, alkaline earth metals, transition metals, rare earth series, actinide series, other metals, nonmetals, and noble (inert) gases.

Of particular interest and importance to thermoplastics is the category known as halogens. These elements include fluorine, chlorine, bromine, and iodine, and they are the most electronegative elements in the periodic table, making them the most likely to attract an electron from another element and become a stable structure. Fluorine is the most electronegative halogen. Because of this, fluorine bonds strongly with carbon and hydrogen atoms, but not well with itself. The carbon-fluorine bond is predominant in PVDF and is responsible for the important properties of these materials. These are among the strongest known organic compounds. The fluorine acts like a protective shield for other bonds of lesser strength within the main chain of the polymer. The carbon-hydrogen bond, of which plastics such as PP and PE are composed, is considerably weaker. The carbon-chlorine bond, a key bond in PVC, is weaker yet.

The arrangement of the elements in the molecule, the symmetry of the structure, and the degree of branching of the polymer chains are as important as the specific elements contained in the molecule. Plastics containing the carbon-hydrogen bonds, such as PP and PE, and carbon-chlorine bonds, such as PVC, ECTFE, and CTFE, are different in the important property of chemical resistance from a fully fluorinated plastic such as PTFE.

2.1.2   Corrosion of Plastics

The mechanism of corrosion of plastic materials differs somewhat from that of metallic materials. Metallic materials experience a specific corrosion rate resulting from an electrochemical reaction. Thus it is possible to predict the life of a metallic pipe in contact with a specific corrodent. In contrast, plastic materials do not exhibit a corrosion rate—they are usually either completely resistant to chemical attack or they deteriorate rapidly.

Plastics are attacked either by chemical reaction or by solvation. Solvation is the penetration of the plastic by a corrodent, which causes softening, swelling, and ultimate failure. Corrosion of plastics can be classified in the following ways as to the attack mechanism:

1. Disintegration or degradation of a physical nature, due to absorption, permeation, solvent action, or other factors
2. Oxidation, where chemical bonds are attacked
3. Hydrolysis, where ester linkages are attacked
4. Radiation
5. Thermal degradation, involving depolymerization and possibly repolymerization
6. Dehydration (rather uncommon)
7. Any combination of the above

Results of such attacks will appear in the form of softening, charring, crazing, delamination, blistering, embrittlement, discoloration, dissolving, or swelling.

Even though all plastics are attacked in the same manner, because of unique molecular structure, certain of the chemically resistant types suffer negligible attack. Even though knowledge of the resin structure may be a guide as to the resin's corrosion resistance, testing should always be conducted because of the complicated nature of the attack.

Two other factors affect the corrosion resistance of thermoset resins: cure and nature of the construction of the laminate. Improper or insufficient cure time will adversely affect the corrosion resistance, while proper cure procedures and proper cure time will greatly improve the corrosion resistance.

Selection of the laminate to be used is critical. It must also be resistant to the corrodent being encountered. For example, in the presence of hydrofluoric acid, glass reinforcing would not be used.

2.1.3   Deflection Temperatures

The heat deflection (distortion) temperature (HDT) test is one in which a bar of the plastic in question is heated uniformly in a closed chamber while a load of 66 psi or 264 psi is placed in the center of the horizontal bar. The HDT is the temperature at which a slight deflection of 0.25 mm at the center is noted. The HDT indicates how much mass (weight) the object must be constructed of to maintain the desired form stability and strength rating and provides a measure of the rigidity of the polymer under a load as well as temperature. Table 2-2 lists the HDT of the more common polymers.

2.1.4   Tensile Strength of Plastics

The tensile strength of a material is calculated by dividing the maximum load applied to the material before its breaking by the original cross-sectional area of the test piece. Tensile strength is a measure of the stress required to deform a material prior to breakage, in contrast to toughness, which is a measure of the energy required to break a material. Stress is defined as the force applied over an area on which it operates.

Tensile strength alone should not be used to determine the ability of a plastic to resist deformation and retain form. Other mechanical properties such as elasticity, ductility, creep resistance, hardness, and toughness must also be taken into account. Table 2-3 lists the tensile strength of plastics.

2.1.5   Working Pressures of Thermoplastic Piping Systems

The governing factor in the determination of the allowable working pressure of thermoplastic piping systems is the hoop or circumferential stress. The two most useful forms for expressing this stress are:

$$S = \frac{P[D_o - t]}{2t}$$

or

$$P = \frac{2St}{D_o - t}$$

where:

S = stress (psi)
P = internal pressure (psi)
D_o = outside pipe diameter (in.)
t = minimum wall thickness (in.)

Standard methods have been developed to determine the long-term hydrostatic strength of plastic pipe. From these tests, S values are calculated and a hydrostatic design basis (HDB) is established. This HDB is then multiplied by a service factor to obtain the hydrostatic design stress.

The service factor is based on two groups of conditions. The first is based on manufacturing and testing variables, such as normal variations in material, manufacture, dimensions, good handling techniques, and evaluation procedures. Normal variations within this group are usually within 1 to 10%. The second group takes into account the application or use, including such items as installation, environment (both inside and outside of the pipe), temperature, hazard involved, live expectancy desired, and degree of reliability selected.

A piping system handling water at 73°F (23°C) would have a service factor of 0.4. This means that the pressure rating of the pipe would be 1.25 times the operating pressure.

When a plastic piping system is to operate at elevated temperatures or is to handle a corrodent other than water, a service factor other than 0.4 must be used. Temperature effects have been studied and temperature correction factors developed by the manufacturers. These factors are used with the allowable operating pressures at 73°F (23°C) and are discussed under the heading for each individual system. The determination of allowable operating pressure for various corrodent systems is best left to the piping manufacturer.

After the service design factor has been determined, the maximum allowable operating pressure can be calculated from the following equation:

$$P_W = \frac{2S_W t}{D_o - t}$$

where:

P_W = pressure rating (psi)
S_W = hydrostatic design basis × service factor (psi)
  t = wall thickness (in.)
D_o = outside diameter of pipe (in.)

Thermoplastic piping systems are available in schedule 40 or schedule 80 designations, and have wall thicknesses corresponding to that of steel pipe with the same designations. Table 2-4 shows the dimensions of schedule 40 and schedule 80 thermoplastic pipe.

Because of the wide range of differences in the physical properties of the various thermoplastics, there is no correlation between the schedule number and allowable working pressure. For example, depending upon the specific thermoplastic, the operating pressure of a 1-inch schedule 80 pipe at 73°F (23°C) can vary from 130 to 630 psi. In addition, within the same plastic piping system the allowable operation pressure can vary from 850 psi for ½-inch-diameter schedule 80 pipe to 230 psi for 12-inch-diameter schedule 80 pipe.

Because of these problems, the industry is attempting to correct this condition by the use of Standard Dimension Ratio (SDR) ratings. These ratios are so arranged that a piping system designed to these dimensions will maintain a uniform pressure rating at a specified temperature regardless of pipe diameters. The allowable operating pressure varies between different piping materials for the same SDR rating.

Allowable working pressures and temperature correction factors for each particular system are discussed under the heading of that system.

2.1.6　Joining of Thermoplastic Pipe

A variety of methods are available for joining thermoplastic piping. The specific method selected will depend upon the function of the pipe, the corrodent being handled, and the material characteristics of the pipe. The available methods may be divided into two general categories, namely, permanent joint techniques and nonpermanent joint techniques. Included in the permanent joint techniques are:

Solvent cementing
Butt fusion
Socket fusion

Nonpermanent methods include:

Threading
Flanging
Bell-ring-gasket joint
Compression insert joint
Grooved end mechanical joint

As would be expected, there are advantages and disadvantages to the use of each system as well as specific types of installations for which each method is best suited.

Solvent Cementing

This is the most popular method used for the joining of PVC, CPVC, ABS, and other styrene materials. No special tools are required, and the solvent-cemented joint is as strong as the pipe itself. Sufficient drying time must be allowed before the joint can be tested or pressurized.

One word of caution: Specific solvent cements are recommended for specific piping systems and specific piping sizes. Manufacturers recommendations for specific primers and cements should always be followed. So-called universal cements should be avoided.

The advantages of this method of joining are:

1. Pipe is less expensive
2. Optimal connection for type of material
3. Pull-out resistant connection
4. Large selection of fittings and valves available
5. Pressure resistant up to burst pressure
6. Good chemical resistance (joint does not become a "weak" link in the system)
7. No thread cutting
8. Easy installation

The disadvantages of this technique are:

1. Instructions must be followed carefully
2. Connection cannot be disassembled
3. Leaky joints difficult to repair
4. Revision to the piping system more difficult since all joints are permanent

Butt Fusion

This method results in a joint having excellent strength and permits the piping system to be put into service as soon as the last joint has cooled. Materials joined by this technique are those made from the polyolefins, such as polypropylene, polyethylene, polybutylenes, and polyvinylidene fluoride. In this method the pipe ends to be butted together are heated and partially melted, butted together, and fused.

The advantages of this technque are:

1. Optimal connection method for the polyolefins
2. Pull-out resistance
3. Pressure resistance beyond the burst pressure of the pipe
4. Excellent chemical resistance (a "weak" link is not formed at the joint)

The primary disadvantages are:

1. Cost of the fusion equipment
2. Bulkiness of the equipment required
3. Connection cannot be disassembled
4. Instructions for fusing must be followed carefully

Socket Fusion

This technique is also used for the polyolefins such as polypropylene, polyethylene, polybutylene, and polyvinylidene fluoride. Two different methods are available to produce these thermally fused joints: electric-resistance fusion and socket heat fusion.

The first of these methods utilizes heat from an electrified copper coil to soften the outside surface of the pipe end and the inside surface of the fitting socket. Some manufacturers furnish the fittings with the coils imbedded. Where this is not the case, the coils are attached manually. This technique is primarily used for polypropylene piping systems in acid waste drainage installations.

The primary advantage of this method is that a piping system can be dry fitted and assembled before permanent joints are made.

The disadvantages to this system are:

1. Imperfect heat distribution possible, resulting in low joint strength and possible corrosion resistance problems
2. System cannot be disassembled
3. Joining instructions must be followed closely
4. Cumbersome equipment required

Socket heat fusion is the preferred technique to use for systems handling corrodents. This method involves the use of an electrically heated tool, which softens the outside surface of the pipe and the inside surface of the fitting. The joint produced is as strong as the pipe or fitting. This technique is used on all of the polyolefins.

The advantages of this method are:

1. Flexibility of installation (heat tool may be held by hand or in a bench vise while making a joint)
2. Joint is as strong as the pipe or fitting
3. Equipment needed is least expensive and least cumbersome of any of the fusion techniques
4. Optimal connection for polyolefin materials
5. Pull-out–resistant connection
6. Excellent chemical resistance

The primary disadvantage stems from one of the advantages. The high level of flexibility afforded by this technique requires that the installer have a high degree of skill and dexterity. Also, the connection cannot be disassembled.

Threaded Joints

Threading of thermoplastic piping can be utilized provided schedule 80 pipe or pipe of thicker wall is used and the pipe diameter does not exceed 4 inches. With a diameter of less than 4 inches, the pipe out of roundness can be more easily controlled.

Even with smaller diameters, leakage at the joints poses a problem. Consequently, threaded joints are only recommended for use in applications where leakage will not pose a hazard or in low-pressure systems.

The only advantage to threaded joints is the fact that the system can easily be dismantled for periodic cleaning and/or modification.

The disadvantages to a threaded system are numerous:

1. Allowable operating pressure of the pipe reduced by 50%
2. Threaded joints in the polyolefins pose high threat of leakage if operating pressure exceeds 20 psig due to the low modulus of elasticity of these materials
3. Heavier, more expensive pipe must be used (schedule 80 vs. schedule 40, which is used for solvent-cementing or thermal fusion)
4. Allowing for expansion and contraction due to thermal changes difficult

Flanging

Flanges are available for most thermoplastic piping systems. They are affixed to the pipe by any of the previous jointing techniques discussed. A flanged joint is primarily used for one or more of the following conditions:

1. Connection to pumps, equipment, or metallic piping systems
2. Temporary piping system
3. Process lines that require periodic dismantling
4. To reduce field labor time and/or expertise, since these joints can be made up in the shop and sent to the field for installation
5. When weather conditions or lack of utilities prevent other methods from being used

When making a flanged joint, it is important to use the correct bolting torque. Table 2-5 shows the recommended bolting torques based on well-lubricated bolts, which will give a bolt stress of 10,000–15,000 psi. It is also necessary to provide a gasket between mating flanges of appropriate corrosion resistant properties and compressibility.

The disadvantages of flanged joints are:

1. High material and labor costs
2. Bulkiness

Until recently, flanging was the only method whereby thermoplastic pipe could be connected to metallic pipe. Now a transition union is available. The union consists of a metallic male threaded section, which joins a female threaded connection on the metallic pipe. The other end of the union consists of a plastic tailpiece with either a socket or female threaded end. A plastic nut screws onto the tailpiece, which secures the metal end connector tightly against an O-ring face seal on the tailpiece.

Bell-Ring-Gasket Joints

The most common use of this joint is on underground PVC pressure-rated piping systems handling water. This joint can also be used to connect PVC pipe to metallic pipe.

An elastomeric ring is retained in a groove in the female joint section. The ring becomes compressed as the pipe is inserted into the joint. It is common practice to anchor, in concrete, these joints at every directional change of the piping to prevent the pipe from backing out of the joint or heaving. This joint would not be used on underground pipe used to convey corrosive materials since it is very difficult to achieve a bubble-tight joint seal.

The advantages of this joint are:

1. Simplicity (quick to make)
2. Ease of making with reduced labor cost

The disadvantages are:

1. Difficult to make leak free
2. Danger of pull-out

Grooved-End Mechanical Joint

A groove is cut or rolled around the perimeter near the ends of each pipe to be joined together. A metal coupling consisting of two identical sleeves, each of which has a ridge corresponding to the groove in the pipe ends, is fitted over the pipe. The two halves are connected by means of a bolt and hinge. Retained between the pipe and sleeves is an elastomeric ring, which forms the seal.

This type of joint is especially useful with PVC or CPVC pipe, since the modulus of elasticity of these materials and the material strength is sufficient to retain the integrity of the grooves.

The grooved-end mechanical joint is used primarily for field assembly where frequent dismantling of the lines is required because of the ease with which the joint may be disassembled and reassembled. It is also used for temporary piping installations.

The primary disadvantage is the difficulty in finding an elastomer that will be compatible with the corrodent being handled.

Compression Insert Joints

This type of joint is not generally used when handling corrodents. Its primary application is in water service such as for irrigation systems, swimming pools, and agricultural service, and it is used with flexible PVC or PE tubing in sizes under 2 inches. The joint consists of a barbed fitting inserted into the tubing and held in place by means of a stainless steel clamp attached around the tubing to maintain pressure on the fitting barb.

2.1.7   Effect of Thermal Change on Thermoplastic Pipe

As the temperature changes above and below the installation temperature, plastic piping undergoes dimensional changes. The amount of expansion or contraction that occurs is dependent upon the piping material and its coefficient of linear expansion, the temperature differential, and the length of pipe run between direction changes. Thermoplastic piping materials have a considerably larger coefficient of thermal expansion than do metallic piping materials, consequently temperature changes that would not affect metallic materials can be detrimental to plastic materials if not compensated for. The change in length resulting from a thermal change can be calculated using the following formula:

$$\Delta L = \frac{Y[T_1 - T_2]}{10} \times \frac{L}{100}$$

where:

ΔL = dimensional change due to thermal expansion or contraction (in.)
Y = expansion coefficient (in./10°F/100 ft)

$T_1 - T_2$ = difference between installation temperature and the maximum or minimum system temperature, whichever provides the greatest differential (°F)

L = length of pipe run between changes in direction (ft)

If the movement that results from thermal changes is restricted by equipment or supports, the resulting stresses and forces can cause damage to the equipment or to the pipe itself. The following formulas provide an estimate of the magnitude of the resultant forces:

$$S = EC \ (T_1 - T_2)$$

where:

S = stress (psi)
E = modulus of elasticity (psi)
C = coefficient of thermal expansion (in./in./°F $\times$ 10^{-5})
$(T_1 - T_2)$ = difference between the installation temperature and the maximum or minimum system temperature

and

$$F = S \times A$$

where:

F = force (lb)
S = stress (psi)
A = cross sectional area (in.^{2}) $= \left[\left(\dfrac{\text{Pipe OD}}{2} \right)^2 - \left(\dfrac{\text{Pipe ID}}{2} \right)^2 \right] 3.14$

The stresses and forces resulting from thermal expansion or contraction can be reduced or eliminated by providing flexibility in the piping system. This can be accomplished by incorporating sufficient changes in direction to provide the flexibility. Most piping installations are automatically installed with sufficient changes in direction to compensate for these stresses.

The following formula can be used to determine the required flexibility in leg R (see Fig. 2-1). Since both pipe legs will expand and contract, the shortest leg must be selected for the flexibility test when analyzing inherent flexibility in naturally occurring offsets.

$$R = 2.041 \ \sqrt{D \ \Delta L}$$

where:

R = length of opposite leg to be flexed (ft)
D = actual outside diameter of pipe (in.)
ΔL = dimensional change in adjacent leg due to thermal expansion or contraction (in.)

Flexural offsets must be incorporated into systems where straight runs of pipe are long or where the ends are restricted from movement. An expansion loop, such as shown in Fig. 2-2, can be calculated using the following formula:

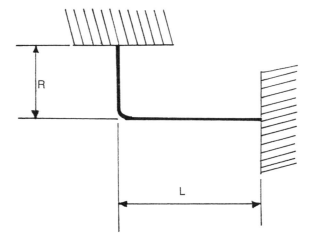

Figure 2-1 Rigidly installed pipe with a change of direction.

$$R^1 = 2.041 \sqrt{\dfrac{D \, \Delta L}{2}}$$

or

$$R^1 = 1.443 \sqrt{D \, \Delta L}$$

Keep in mind that rigid supports or restraints should not be placed within the length of an expansion loop, offset, or bend.

Thermal expansion stresses and forces can also be compensated for by the installation of expansion joints. Initially these joints were produced in natural rubber, but at the present time they are available in a wide range of elastomeric materials such as neoprene, Buna-N, Butyl, EPDM, and TFE. When using these joints care must be taken that the material of construction of the tube section is compatible with the fluid being handled. Figure 2-3 shows the typical construction of an expansion joint. These joints provide axial compression, axial elongation, angular movement, torsional movement, and lateral movement. Operating pressures and temperatures are dependent upon the specific elastomer.

2.1.8   Burial of Thermoplastic Pipe

The primary criterion for the safe design of a buried thermoplastic pipe system is that of maximum allowable deflection. On rare occasions buckling of the pipe and/or excessive loading may control the design. In the latter case the safe stress under continuous compression may conservatively be assumed to equal the hydrodynamic design stress for the end use conditions. Control of the deflection of the pipe, along with proper installation procedures, then becomes the key factor in an underground pipe installation. The percent of deflection can be calculated as follows:

$$\% \text{ deflection} = \dfrac{\text{Calculated deflection} \times 100}{\text{Outside diameter of pipe}}$$

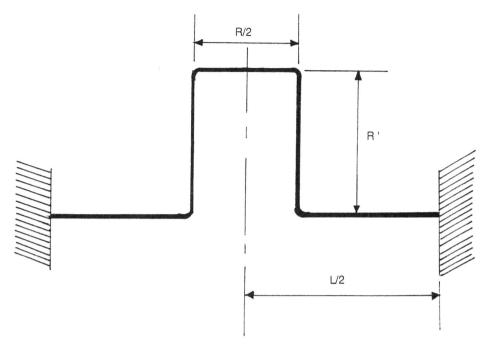

Figure 2-2 Rigidly installed pipe with an expansion loop.

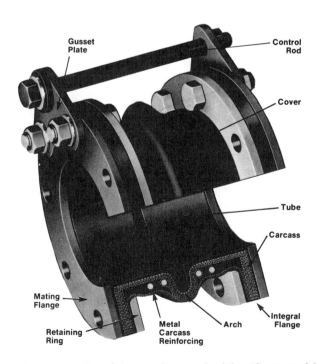

Figure 2-3 Typical construction of elastomeric expansion joint. (Courtesy of the Metraflex Co., Chicago, IL.)

The calculated deflection is found from the following equation:

$$x = \frac{K_x DW}{[0.149 \times PS + 0.061 \times E^1]}$$

where:

OD = outside diameter of pipe (in.)
K_x = bedding factor (see Table 2-9)
D = deflection lag factor (see Table 2-8)
W = weight per lineal inch (lb/in.)
PS = pipe stiffness (psi)
E^1 = soil modulus (see Table 2-6 and 2-7)
E = flexural modulus (psi)
H = height of cover (ft)
SD = soil density (lb/ft^3) (If specific weight of soil is unknown, assume a soil density of 120 lb/ft^3.)

$$W = \frac{H \times OD \times SD}{144} + [\text{soil pressure} \times OD]$$

$$PS = \frac{E \times I}{0.149 \, R^3}$$

where:

t = average wall thickness (in.)
I = moment of inertia = $t^3/12$
R = mean radii of pipe (in.) $= \frac{(OD - t)}{2}$

Pipe that is to be used for potable water service should be installed a minimum of 12 inches below the expected frost line.

The width of the trench must be sufficient to allow convenient installation of the pipe, but at the same time be as narrow as possible. Table 2-10 lists trench widths for various pipe sizes found adequate in most cases. However, sizes may have to be varied because of local terrain and/or specific applications. These minimum trench widths may be utilized by joining the pipe outside of the trench and lowering it into place after adequate joint strength has been attained.

The trench bottom should be relatively smooth and free of rocks. If hardpan, boulders, or rocks are encountered, it is advisable to provide a minimum of 4 inches of sand or tamped earth beneath the pipe to act as a cushion and to protect the pipe from damage.

Before installing any section of pipe, the pipe should be carefully inspected for cuts, scratches, gouges, nicks, buckling, or any other imperfection. Any section of pipe or fitting that has an imperfection should not be used.

Thermal expansion and contraction must be taken into account, particularly in hot weather. On a hot summer day the sun's rays could cause the wall of the pipe to reach a temperature of 150°F (66°C). If this pipe were to be installed, contraction would occur overnight, which would cause weakening or failure of the joints. To avoid this the pipe is "snaked" along in the trench. The loops formed will compensate for the contraction of the pipe, thus protecting the joints. For example, 100 feet of PVC pipe will contract or expand ¾ inch for every 20°F temperature change. Should the pipe wall reach 150°F (66°C) during the day and cool off to 70°F (21°C) during the night, the pipe would contract 3 inches. $\Delta T = 80°F$ (44°C). Snaking of the pipe will allow this to happen without imposing any stress on the piping. Table 2-11 gives the required offset and loop length when snaking the pipe for each 10°F (5°C) temperature change.

It is preferable to backfill the trench with the installed pipe during the early morning hours before the pipe has had an opportunity to expand as a result of the sun. The pipe should be continuously and uniformly supported over its entire length on firm stable material. Do not use blocks to support the pipe or to change grade. Backfill materials should be free of rocks with a particle size of ½ inch or less. After 6–8 inches of cover over the pipe has been placed, it should be compacted by hand or with a mechanical tamper. The remainder of the backfill can be placed in uniform layers until the trench is completely filled, leaving no voids under or around, rocks, or clumps of earth. Heavy tampers should only be used on the final backfill.

Additional details for installation of underground piping can be found in (1) ASTM D2774, *Underground Installation of Thermoplastic Pressure Piping*, and (2) ASTM D2321, *Underground Installation of Flexible Thermoplastic Sewer Pipe*.

2.1.9 Supporting Thermoplastic Pipe

The general principles of pipe support used for metallic pipe can be applied to plastic pipe with a few special considerations. Spacing of supports for thermoplastic pipe is closer than for metallic pipe and varies for each specific system. The support spacing for each plastic material is given under the chapter for that specific piping system. It is critical that the support spacings given be followed to provide a piping system that will give long uninterrupted service. Care should be taken not to install plastic pipes near or on steam lines or other hot surfaces.

Supports and hangers can be by means of clamps, saddle, angle, spring, or other standard types. The most commonly used are clevis hangers. It is advisable to select those with broad, smooth bearing surfaces, which will minimize the danger of stress concentrations. Relatively narrow or sharp support surfaces can damage the pipe or impose stresses on the pipe. If desired, protective sleeves may be installed between the hanger and pipe (see Fig. 2-4). When the recommended support spacing is relatively close (2–3 feet or less), it will be more economical to provide a continuous support. For this purpose channel iron or inverted angle iron can be used (see Fig. 2-5). This procedure should also be followed when the line is operating at elevated temperatures.

The support hangers should not rigidly grip the pipe, but should permit axial movement to allow for thermal expansion. All valves should be individually supported and braced against operating torques. All fittings except couplings should also be individually supported. See Fig. 2-6 for a typical valve support. Rigid clamping or anchor points should be located at valves and fittings close to a pipeline change in direction (see Figs. 2-7, 2-8, and 2-9). Figure 2-10 shows a means of supporting a horizontal line from the bottom.

Vertical runs should be supported by means of a saddle at the bottom and the use of long U-bolts as guides on the vertical. These guides should not grip the pipe but be loose enough to permit axial movement but not traverse movement.

It is important that vibration from pumps or other equipment not be transmitted to the

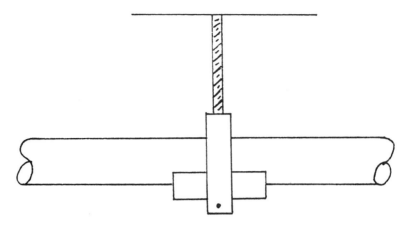

Figure 2-4 Typical clevis hanger support with protective sleeve.

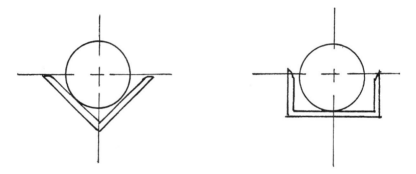

Figure 2-5 Continuous support of pipe using angle or channel iron.

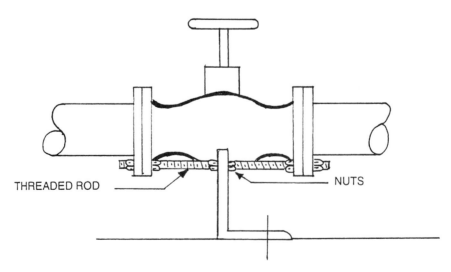

THREADED ROD

NUTS

Figure 2-6 Typical valve support.

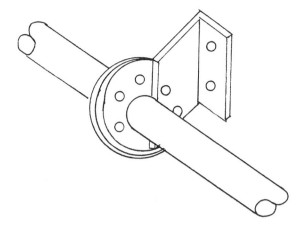

Figure 2-7 Anchoring making use of a flanged joint.

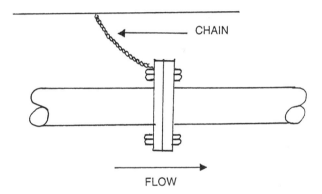

Figure 2-8 Anchoring making use of a flanged joint.

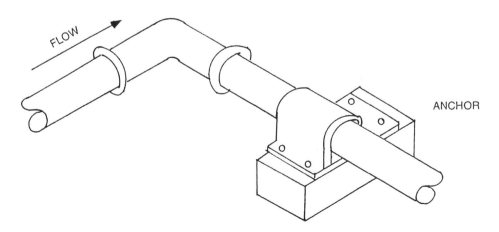

Figure 2-9 Anchoring at an elbow.

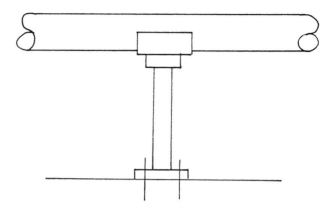

Figure 2-10   Typical support of a pipe from the bottom.

pipe. These vibrations can be isolated by the use of flexible couplings installed at the juncture of the pipe and pump or equipment.

2.1.10   Fire Hazards

Potential fire hazards should also be considered when selecting the specific plastic material to be used. Although the fluid being conveyed may not be combustible, there is always the danger of a fire from other sources. In the physical and mechanical properties table located under the heading of each specific thermoplastic material are two entries relating to this topic. The first is the limiting oxygen index percent. This is a measure of the minimum oxygen level required to support combustion of the thermoplast. The second is the flame spread classification. These ratings are based on common tests as outlined by the Underwriters Laboratories and are defined as follows:

Flame Spread Rating	Classification
0–25	Noncombustible
25–50	Fire retardant
50–75	Slow burning
75–200	Combustible
over 200	Highly combustible

2.2   POLYVINYL CHLORIDE (PVC) PIPE

Polyvinyl chloride is the most widely used of any of the thermoplastic materials. It has been used for over 30 years in such areas as chemical processing, industrial plating, water supply systems, chemical drainage, and irrigation systems.

PVC is polymerized vinyl chloride, which is produced from acetylene and anhydrous hydrochloric acid. The structure is as follows:

$$
\begin{array}{cccc}
H & CL & H & Cl \\
| & | & | & | \\
-C- & C- & C- & C- \\
| & | & | & | \\
H & H & H & H
\end{array}
$$

PVC is stronger and more rigid than other thermoplastic materials. It has a high tensile strength and modulus of elasticity. With a design stress of 2000 psi, PVC has the highest long-term hydrostatic strength at 73°F (23°C) of any of the major thermoplastics used for piping systems. Table 2-12 lists the physical and mechanical properties of PVC.

Two types of PVC are used to produce pipe: normal impact (type 1) and high impact (type 2). Type 1 is a rigid unplasticized PVC having normal impact with optimum chemical resistance. Type 2 has optimum impact resistance and normal chemical resistance. To obtain improved impact properties, modification of the basic polymer is required, thus compromising chemical resistance. This is characteristic of many thermoplastics. The high-impact pipe is designed to take greater shock loads than normal-impact pipe. Type 1 pipe has a maximum allowable temperature rating of 150°F (66°C), while type 2 pipe has a maximum allowable temperature rating of 140°F (60°C). Most PVC pipe is of the high-impact variety.

2.2.1 Pipe Data

PVC piping is available in ¼ inch to 16 inches nominal diameter in schedule 40 and schedule 80 wall thicknesses. Dimensions are shown in Table 2-4. The maximum allowable working pressures are given in Table 2-13 for schedule 40 and schedule 80. Note from the table that the operating pressure decreases as the nominal diameter increases. To overcome this, PVC piping is also available in Standard Dimension Ratings (SDR). For any one SDR specification the allowable operating pressure is constant at a given temperature, regardless of pipe diameter. The dimension ratio is derived by dividing the outside diameter by the minimum wall thickness of the pipe.

There are six SDR standards for PVC pipe:

SDR 13.5
SDR 18
SDR 21
SDR 25
SDR 26
SDR 32.5

Dimensions of these SDR-rated pipe systems are given in Tables 2-14A and 2-14B. The range of pipe sizes for each SDR system varies, as seen in the tables. This is because of the purpose for which each rating was designed. Table 2-15 supplies the maximum allowable operating pressure for each rating.

A complete range of pipe sizes is not available in all SDR ratings. SDR 13.5, which has the highest operating pressure, is only available in sizes of ½ inch through 4 inches. SDR 14 and SDR 21 both have the same pressure rating, however, wall thicknesses vary for the same size pipe. SDR 14 is available in sizes of 4–12 inches and has the heavier wall. This system is recommended for underground work. SDR 21 is available in sizes of ¾ inch through 12 inches and is usually recommended for aboveground installation. For the same reasons, SDR 18 and SDR 25 pipe is recommended for burial, with the latter being available from 4 through 24 inches. Most of the larger-diameter buried pipe is used to handle potable water and meets either AWWA C900 or AWWA C905 specifications.

A transparent schedule 40 PVC pipe system is also available in sizes of ¼ inch through 6 inches. It is joined by solvent cementing. Maximum operating pressures are the same as for regular PVC pipe. This transparent material is manufactured by Thermoplastic Processes Inc. and is sold under the tradename of Excelon R-4000.

Flanged PVC piping systems have a somewhat different pressure rating than do other schedule 40 and schedule 80 pipes. The rating is constant through 8-inch diameters. However, the pressure rating decreases with increasing temperatures (see Table 2-16).

As with all thermoplastic piping systems, the maximum allowable pressure rating decreases with increasing temperature. Table 2-17 supplies the temperature correction factors to be used with PVC pipe in schedule 40, schedule 80, and the SDR ratings. They are applied in the following manner:

Assume a 2-inch schedule 80 solvent welded, type 1 line is to operate at 110°F (43°C); at 73.4°F (23°C) the maximum allowable operating pressure from Table 2-13 is 400 psi. The temperature correction factor from Table 2-17 is 0.80, therefore the maximum allowable operating pressure at 110°F (43°C) is 400(0.80) = 320 psi.

On occasion PVC pipe is used under such conditions that it is operating under a negative internal pressure, such as in pump suction lines and vacuum lines. For this type of service, solvent-cemented joints should be used. Under continuous pumping PVC lines can be evacuated to 5 μm, but when the system is shut off, the pressure will rise to and stabilize at approximately 10,000 μm or 10 mmHg at 73°F (23°C). For this type of service the collapse pressure rating of the pipe must be taken into account (see Table 2-18). These values determine the allowable pressure differential between internal and external pressure. If a temperature other than 74.3°F (23°C) is to be used, the correction factors in Table 2-17 must be applied.

2.2.2 Joining, Installation, and Support

The preferred method of joining PVC pipe is by means of solvent-cementing. Other methods that can be used are:

Threaded joints on schedule 80 pipe
Flanged joints
Bell and ring gasket joints for underground piping
Grooved-end mechanical joints
Compression insert joints

For an explanation of these joints, see pages 23–24. Installation of PVC pipe should be in accordance with the principles on pages 24–32.

Support spacing at various operating temperatures for the different types of PVC piping can be found in Tables 2-19 through 2-24. These support spacings are for uninsulated pipe only, with fluids having a specific gravity of 1.35 or less. All valves, meters, or other accessory items installed in the line must be independently supported. If the support spacing indicated is 3 feet or less, it will be more economical to provide continuous support for the line.

2.2.3 Corrosion Resistance

Unplasticized PVC resists attack by most acids and strong alkalies as well as gasoline, kerosene, aliphatic alcohols, and hydrocarbons. It is particularly useful in the handling of inorganic materials such as hydrochloric acid. It has been approved by the National Sanitation Foundation for the handling of potable water. When modified, PVC's resistance to oxidizing and highly alkaline mediums is compromised.

PVC may be attacked by aromatics, chlorinated organic compounds, and lacquer solvents. Table 2-25 shows the compatibility of PVC with selected corrodents. This chart relates only to the chemical compatibility of PVC with corrodents. Mechanical and physical properties and allowable operating pressures of PVC pipe at the temperatures shown must also be taken into account. PVC has found wide application as process piping, laboratory drainage piping, potable water supply piping, and vent line piping.

2.3 FIBERGLASS-ARMORED PVC PIPE

The dual construction of a fiberglass outer covering and a PVC inner liner has definite advantages, particularly for buried piping systems. PVC has relatively low physical properties as compared to a fiberglass pipe. Fiberglass-armored PVC pipe is essentially a fiberglass pipe with a PVC liner. In mechanical design the PVC liner is ignored for structural purposes. All mechanical strength comes from the fiberglass pipe. The PVC liner provides the corrosion resistance, which is superior to that of the fiberglass. Other liners can also be used. Physical and mechanical properties are given in Table 2-26.

2.3.1 Pipe Data

This piping is produced for underground as well as above-ground applications with the former being the predominate usage for water distribution systems, fire lines, forced sewer lines, and industrial pipe lines.

Underground piping is available in diameters of 4–12 inches in two standard pressure ratings of 250 and 350 psi. These pressure ratings are sustained working pressures including surge pressure at 73°F (23°C). As operating temperatures increase, the allowable operating pressure decreases. Temperature correction factors can be found in Table 2-27. To use these factors, multiply the SDR rating at 70°F (21°C) by the appropriate factor for the actual operating temperature. For example, if the operating temperature is 100°F (38°C), the correction factor is 0.80. Assuming that pressure class 250 pipe is to be used, then the allowable operating pressure at 100°F (38°C) will be 250 (0.80) = 200 psi.

The above-ground piping is manufactured in sizes of 2–6 inches. Maximum operating pressures and temperatures are given in Table 2-28.

2.3.2 Joining, Installation, and Support

Underground piping is manufactured with an integral bell and spigot end. An elastomeric ring is "locked" in the ring groove at the time of manufacture. The purpose of the ring is twofold: it aids in the centering of the pipe and it is used to make a tight seal. Unfortunately, fittings are not available, but the pipe can be used directly with cast iron dimensional fittings, both slip joint and mechanical joint. When using mechanical joint fittings, caution should be exercised when tightening the bolt. Overtorquing can damage the pipe. A maximum torque of 50 ft-lb should be used.

Installation is in accordance with the details of buried fiberglass pipe found on pages 24–32.

Above-ground piping is joined in a two-step procedure. The inner PVC liner is solvent welded, after which the entire joint is wrapped with fiberglass tape saturated with the proper epoxy adhesive, which is supplied by the pipe manufacturer. Support spacing for this pipe is given in Table 2-29.

2.3.3 Corrosion Resistance

The internal corrosion resistance is that of PVC as shown in Table 2-15, while the external resistance is shown in Table 3-15.

2.4   CHLORINATED POLYVINYL CHLORIDE (CPVC) PIPE

PVC is chlorinated to increase the chlorine content to approximately 67% from 56.8% to produce CPVC. This additional chlorine increases the heat deflection temperature from 160°F (71°C) to 212°F (100°C). This also permits a higher allowable operating temperature. The chemical structure is as follows:

$$
\begin{array}{cccc}
\text{H} & \text{H} & \text{Cl} & \text{H} \\
| & | & | & | \\
-\text{C} - & \text{C} - & \text{C} - & \text{C} - \\
| & | & | & | \\
\text{H} & \text{Cl} & \text{H} & \text{Cl}
\end{array}
$$

CPVC has been used for over 25 years for handling corrosive materials in the chemical process industry. It is also widely used as condensate return lines to convey hot water, particularly in areas where external corrosion is present. The physical and mechanical properties are shown in Table 2-30.

2.4.1   Pipe Data

CPVC pipe is available in diameters of ¼ inch through 12 inches in both schedule 40 and schedule 80. Dimensions are shown in Table 2-4. The maximum allowable operating pressures at 73°F (23°C) are given in Tables 2-31 and 32. As with all thermoplastic piping systems, the allowable operating pressure is decreased as the temperature increases. Table 2-33 provides the factors by which the operating pressures must be reduced at elevated temperatures. To use the temperature correction factors in Table 2-33, multiply the allowable working pressure at 70°F (21°C) by the factor for the actual operating temperature. For example, if the operating temperature were 140°F (60°C) for a 2-inch schedule 80 solvent-welded line, the maximum allowable operating pressure at 140°F (60°C) would be (400)(0.5) = 200 psi.

Table 2-34 provides the collapse pressure rating of schedule 80 pipe. This is useful when the piping will be operating under negative pressure such as in a pump suction line. The pressure shown is the differential between the external and internal pressures.

2.4.2   Joining, Installation, and Support

The preferred method of joining CPVC pipe is by means of solvent cementing. However, schedule 80 pipe may be threaded, and flanges are also available. Threaded joints should not be used at temperatures above 150°F (66°C). Flanges should be limited to making connections to pumps, equipment, or accessories installed in the line or where frequent disassembly of the line may be necessary.

Installation of CPVC pipe should be in accordance with the principles on pages 24–32.

Recommended maximum support spacing for schedule 40 pipe is given in Table 2-35 and for schedule 80 pipe in Table 2-36.

2.4.3   Corrosion Resistance

The corrosion resistance of CPVC is somewhat similar to that of PVC, but there are differences, therefore CPVC cannot be used in all applications where PVC is suitable. In general CPVC can not be used to handle most polar organic materials including chlorinated or aromatic hydrocarbons, esters and ketones.

CPVC can be used to handle most acids, alkalis, salts, halogens, and many corrosive wastes. Table 2-37 lists the compatibility of CPVC with selected corrodents. It is important to recognize that Table 2-37 only provides the compatibility of CPVC with selected corrodents. At elevated temperatures, although CPVC may be resistant to a specific corrodent, the operating conditions may be such that a CPVC piping system may not be suitable. The physical and mechanical properties and allowable operating pressures under the operating conditions must be checked.

2.5   POLYPROPYLENE (PP) PIPE

Polypropylene is one of the most common and versatile thermoplastics used. Polypropylene is closely related to polyethylene. They are both members of a group known as "polyolefins," composed only of carbon and hydrogen. Other common thermoplasts, such as PVC and PTFE, also contain chlorine and fluorine, respectively.

If unmodified, polypropylene is the lightest of the common thermoplasts, having a specific gravity of approximately 0.91. In addition to the light weight, its major advantages are high heat resistance, stiffness, and chemical resistance.

Within the structural nature of PP a distinction is made between isotactic PP and atactic PP. Isotactic PP accounts for 97% of PP, wherein the polymer units are highly ordered:

$$
\begin{array}{cccc}
\text{H} & & \text{H} & \\
| & & | & \\
-\ \text{C} & \!\!\!\!-\!\!\!\!-\!\!\!\!- & \text{C} & \\
| & & | & \\
\text{H} & \text{H} -\!\!\!\!- & \text{C} & -\ \text{H} \\
& & | & \\
& & \text{H} &
\end{array}
$$

Atactic PP is a viscous liquid–type PP having a PP polymer matrix. In some cases PP is actually a combination of PE and PP formed during a second stage of the polymerization. This form is designated copolymer PP and produces a product that is less brittle than the homopolymer. The copolymer is able to withstand impact forces down to –20°F (–29°C), while the homopolymer is extremely brittle below 40°F (4°C).

The homopolymers are generally long-chain, high molecular weight molecules with a minimum of random molecular orientation, thus optimizing their chemical, thermal, and physical properties. Homopolymers are preferred over copolymer materials for demanding chemical, thermal, and physical conditions. Increased impact resistance is characteristic of most copolymers. However, tensile strength and stiffness are substantially reduced, increasing the potential for distortion and cold flow, particularly at elevated temperatures.

Piping systems produced from the homopolymer have an upper temperature limit of 180°F (82°C), while those produced from the copolymer have an upper temperature limit of 200°F (93°F). The minimum operating temperatures are also affected, with the homopolymer limited to 50°F (10°C) and the copolymer to 15°F (4°C). Table 2-38 shows the physical and mechanical properties of homopolymer PP, while Table 2-39 provides the physical and mechanical properties of the copolymer.

2.5.1   Pipe Data

Polypropylene pipe is produced from both the copolymer and the homopolymer in schedule 40, schedule 80, and pressure-rated systems.

The maximum allowable operating pressures of schedule 40 and schedule 80 fusion welded homopolymer PP pipe at 74°F (23°C) are shown in Table 2-40, while the maximum allowable operating pressure for threaded schedule 80 homopolymer pipe at 70°F (21°C) is given in Table 2-41. These pressure ratings must be reduced if the operating temperature exceeds 70°F (21°C). Table 2-42 provides the temperature correction factors to be used at elevated temperatures. Threaded PP pipe is not to be used at elevated temperatures. The temperature correction factor for homopolymer PP pipe at 120°F (49°C) is 0.75. To determine the allowable operating pressure of 2-inch schedule 40 pipe at 120°F (49°C), which has an allowable operating pressure of 140 psi at 70°F (21°C), multiply 140 by 0.75 to give an allowable operating pressure of 105 psi at 120°F (49°C).

Table 2-43 provides the maximum allowable operating pressure of flanged homopolymer PP systems based on operating temperature.

When a piping system is used under negative pressure, such as a pump suction line, care should be taken not to exceed the collapse pressure rating of the pipe. These values are given in Table 2-44 for schedule 80 homopolymer PP pipe.

Homopolymer polypropylene pipe is also available in standard pressure ratings of 45 psi, 90 psi, and 150 psi. Pipe sizes available at each of these ratings are as follows:

Rating	Size (in.)
45 psi	4–32
90 psi	2–32
150 psi	½–20

Copolymer PP pipe is available in SDRs of 32, with an operating pressure of 45 psi at 73.4°F (23°C) and SDR 11, which is rated at 150 psi at 73.4°F (23°C). Temperature correction factors for the SDR-rated pipe may be found in Table 2-45. These factors are used in the same way as those for the schedule 40 and schedule 80 pipe.

2.5.2   Joining, Installation, and Support

Polypropylene pipe can be joined by thermal fusion, threading, and flanging. The preferred method is thermal fusion. Threading is restricted to schedule 80 pipe. Flanged connections are used primarily for connecting to pumps, instruments, valves, or other flanged accessories. See pages 21–23 for information on joining methods.

Installation of PP pipe should follow the general rules established for thermoplastic piping as given on pages 24–32.

Support spacing for homopolymer PP pipe is given in Table 2-46, while Table 2-47 gives the recommended support spacing for copolymer PP pipe. All accessory items in the pipe line, such as valves, instruments, etc., must be independently supported.

2.5.3   Corrosion Resistance

PP piping is particularly useful in salt water disposal lines, crude oil piping, low pressure gas gathering systems, sulfur bearing compounds, caustics, solvents, acids, and other organic chemicals. It is not recommended for use with oxidizing-type acids, detergents, low-boiling hydrocarbons, alcohols, aromatics, and some chlorinated organic materials. Table 2-48 provides a listing of the compatibility of PP with selected corrodents.

Unpigmented PP is degraded by ultraviolet light. When using Table 2-48, it should be kept in mind that this table is only providing the compatibility of polypropylene with selected corrodents. Operating conditions of pressure, temperature, or other considerations may prohibit the use of a polypropylene pipe system. The mechanical requirements of the system must also be checked. To protect PP it must be pigmented, shielded, or otherwise stabilized.

2.6 POLYETHYLENE (PE) PIPE

Polyethylene is a thermoplastic material which varies from type to type depending upon the molecular structure, its crystallinity, molecular weight, and molecular weight distribution. These variations are made possible through the changes in the polymerization conditions used in the manufacturing process. PE is formed by polymerizing ethylene gas obtained from petroleum hydrocarbons.

The terms low, high, and medium density refer to the ASTM designations based on the unmodified polyethylene. Low-density PE has a specific gravity of 0.91–0.925; medium-density PE has a specific gravity of 0.926–0.940; high-density PE has a specific gravity of 0.941–0.959. The densities, being related to the molecular structure, are indicators of the properties of the final product.

Ultra high molecular weight (UHMW) polyethylene has an extremely long molecular chain. According to ASTM specifications, UHMW PE must have a molecular weight of at least 3.1 million. The molecular weight of pipe-grade PE is in the range of 500,000.

Regardless of the type, the basic molecular unit of PE is:

$$\begin{array}{ccc} H & & H \\ | & & | \\ -\,C & -\, C & - \\ | & & | \\ H & & H \end{array}$$

When PE is exposed to ultraviolet radiation, usually from the sun, photo or light oxidation will occur. Therefore, it is necessary to incorporate carbon black into the resin to stabilize it. Other types of stabilizers will not provide complete protection.

Table 2-49 lists the physical and mechanical properties of extra high molecular weight (EHMW) polyethylene.

2.6.1 Pipe Data

Polyethylene pipe is available in several categories. An industrial grade such as Plexico 3408 EHMW pipe is stabilized with both UV stabilizers and a minimum of 2% carbon black to protect against degradation due to ultraviolet light. This stabilized system may be used above ground without fear of loss of pressure rating or other important properties due to ultraviolet light degradation. Pipe is available in sizes of ½ inch through 36 inches in standard pressure ratings from 255 psi to 50 psi at 73°F (23°C) (see Table 2-50). In pressure operations it has a maximum allowable temperature rating of 140°F (60°C). For nonpressure applications it may be used up to 180°F (82°C). As with other thermoplastic piping systems, as the temperature increases above 73°F (23°C), the pressure rating must be reduced. Table 2-51 provides the temperature correction factors to be used at elevated temperatures. These factors can be used in the following manner. Assume SDR 9 pipe rated for 200 psi at 73°F (23°C) is to be used at a temperature of 120°F (49°C). From Table

2-51 we find the correction factor of 0.63. Therefore the maximum allowable operating pressure at 120°F/49°C is (200)(0.63) = 126 psi.

A high-temperature pipe, Plexico HTPE, is also available. Its pressure/temperature ratings are given in Table 2-52.

Pipe for burial underground is furnished color coded by means of four longitudinal stripes in blue, red, or yellow. These are designated by Plexico as Bluestripe for potable water, Redstripe for underground fire main systems, and Yellowstripe for gas distribution systems. This pipe is manufactured from the same 3408 EHMW resin. The colored stripes form an integral part of the pipe and have the same properties as the pipe itself. The Redstripe piping system has Factory Mutual approval for use as underground fire mains. Pipe is designed to meet class 150 (150 psi at 73°F [23°C]) and class 200 (200 psi at 73°F [23°C]) with a temperature rating of from −50°F (−46°C) to 140°F (60°C). Pressure/temperature ratings are given in Table 2-53.

Ryerson Plastics also produces a line of polyethylene in two grades under the trade name of Driscopipe. Grade 8600 is recommended for use wherever environmental stress cracking or abrasion resistance are major concerns. It has superior toughness and fatigue strength and high impact strength. It is also protected from ultraviolet degradation and is available in the following ratings.

SDR	Allowable pressure (psi)	Size range available (in.)
32.5	65	5–42
25.3	65	10–48
15.5	110	2–24
11.0	160	3/4–27
9.33	190	3/4–6
8.3	220	8
7.3	254	14, 18

For less demanding applications, grade 1000 is recommended and is available in the following ratings:

SDR	Allowable pressure (psi)	Size range available (in.)
32.5	51	3–54
26.0	64	3–54
21.0	80	3–36
19.0	89	3–36
17.0	100	2–36
15.5	110	2–32
13.5	128	2–30
11.0	160	1–24
9.0	200	2–18
7.0	267	2–14

2.6.2 Joining, Installation, and Support

Polyethylene pipe is joined by thermal fusion techniques, as described in Sec. 2.1.6. Specific instructions should be requested from the manufacturer.

Installation should follow the general principles outlined in Sec. 2.1.9 for above-ground pipe and Sec. 2.1.8 for the burial of thermoplastic pipe Specific details are supplied by the pipe manufacturer.

Support spacings of above-ground pipe may be found in Table 2-54. It is important that these spacings not be exceeded.

2.6.3 Corrosion Resistance

Polyethylene pipe has a wide range of corrosion resistance ranging from the handling of potable water to corrosive wastes. Although used for above-ground installations, its primary area of application is in underground piping to handle corrosives. It is particularly useful in this application because of its ability to resist external corrosion. In addition, it does not support biological growth. Table 2-55 provides the compatibility of EHMW PE with selected corrodents. This table provides only the compatibility of EHMW polyethylene with selected corrodents. Just because an application is shown to be satisfactory does not mean that polyethylene piping would be satisfactory for the application. Mechanical and physical properties, temperature, and pressure ratings of the pipe must be checked to verify that the pipe is satisfactory for the application.

2.7 ETHYLENE CHLOROTRIFLUORETHYLENE (ECTFE) PIPE

Ethylene chlorotrifluorethylene is a 1:1 alternating copolymer of ethylene and chlorotrifluorethylene. This chemical structure gives the polymer a unique combination of properties. It possesses excellent chemical resistance, a broad-use temperature range from cryogenic to 340°F (171°C) with continuous service to 300°F (149°C) and has excellent abrasion resistance. The chemical structure is:

$$
\begin{array}{cccc}
H & H & F & F \\
| & | & | & | \\
-C & -C & -C & -C \\
| & | & | & | \\
H & H & F & Cl
\end{array}
$$

ECTFE is a tough material with excellent impact strength over its entire operating temperature range. Outstanding in this respect are properties related to impact at low temperature. In addition to these excellent impact properties, ECTFE also possesses good tensile, flexural, and wear-related properties. It is also one of the most radiation-resistant polymers. Other important properties include a low coefficient of friction and the ability to be pigmented. Table 2-56 gives the physical and mechanical properties. ECTFE is sold under the trade name of Halar by Ausimont.

2.7.1 Pipe Data

ECTFE piping is available in sizes of 1 inch through 3 inches in an SDR pressure-rated system of 160 psi at 68°F (20°C). Table 2-57 shows the correction factors to be used to determine the maximum allowable operating pressure at elevated temperatures.

From Table 2-57, the temperature correction factor for 140°F (60°C) is found to be 0.65. Applying this factor to the allowable 150 psi at 73°F (23°C) gives 150(0.65) = 97.5 psi at 140°F (60°C). Note that above 256°F (125°C) the piping should only be used for gravity flow without any positive pressure.

2.7.2 Joining Installation and Support

All pipe and fittings are joined by the butt fusion method as described on page 21 with installation as shown on pages 24–32. Table 2-58 provides the recommended support spacing based on operating temperatures.

2.7.3 Corrosion Resistance

The chemical resistance of ECTFE is outstanding. It is resistant to most of the common corrosive chemicals encountered in industry. Included in this list of chemicals are strong mineral and oxidizing acids, alkalies, metal etchants, liquid oxygen, and practically all organic solvents except hot amines (aniline, dimethylamine, etc.). Severe stress tests have shown that ECTFE is not subject to chemically induced stress cracking from strong acids, bases, or solvents. Some halogenated solvents can cause ECTFE to become slightly plasticized when it comes into contact with them. Under normal circumstances this does not impair the usefulness of the polymer. Upon removal of the solvent from contact and upon drying, its mechanical properties return to their original values, indicating that no chemical attack has taken place. ECTFE, like other fluoropolymers, will be attacked by metallic sodium and potassium.

Table 2-59 provides the compatibility of ECTFE with selected corrodents. Although the table may indicate that ECTFE can be used with a specific corrodent, allowable operating pressures and temperatures of the piping system must also be checked, since the table only refers to the corrosion resistance of ECTFE.

2.8 VINYLIDENE FLUORIDE (PVDF) PIPE

PVDF is a crystalline, high molecular weight polymer of vinylidene fluoride containing 59% fluorine. It has a very linear chemical structure and is similar to PTFE, with the exception that it is not fully fluorinated. The chemical structure is:

$$
\begin{array}{ccc}
\text{F} & & \text{H} \\
| & & | \\
-\;\text{C} & -\;\text{C} & - \\
| & & | \\
\text{F} & & \text{H}
\end{array}
$$

It has a high tensile strength and high heat deflection temperature and is resistant to the permeation of gases. Approval has been granted by the Food and Drug Administration for repeated use in contact with food, as in food-handling and processing equipment.

The physical and mechanical properties of PVDF may be found in Table 2-60. Much of the strength and chemical resistance of PVDF is retained within a temperature range of –40 to 320°F (–40 to 160°C).

PVDF is manufactured under the trade name of Kynar by Elf Atochem, Solef by Solvay, Hylar by Ausimont USA, and Super Pro 230 & ISO by Asahi/America.

2.8.1 Pipe Data

PVDF pipe is available in schedule 40, schedule 80, and two pressure-rated systems: 150 and 230 psi at 73.4°F (23°C). These piping systems can be operated continuously at 280°F (138°C). Table 2-61 supplies the maximum allowable operating pressure for schedule 40 and schedule 80 pipe. This pipe is available in sizes of ½ inch through 6 inches. Table

2-62 provides the temperature correction factors when the pipe is used at elevated temperatures. These factors are to be applied to the maximum allowable operating pressure at 70°F (21°C). The correction factor is used in the following manner. A 2-inch schedule 80 solvent-welded pipe at 70°F (21°C) has an allowable operating pressure of 275 psi. The temperature correction factor from Table 2-62 at 140°F (60°C) is 0.670. Therefore, the allowable operating pressure of 2-inch solvent-welded pipe at 140°F (60°C) is (275) (0.670) = 184 psi. Table 2-63 supplies the maximum allowable operating pressure of flanged PVDF systems based on operating temperatures.

If pipe is to be used in negative pressure applications, care should be exercised that the collapse pressure rating not be exceeded. Table 2-64 gives the collapse pressure rating for schedule 80 pipe.

The 230 psi pressure-rated pipe system as manufactured by Asahi/America is available in sizes of ⅜ inch through 4 inches as standard with sizes through 12 inches available on special order. The 150 psi rated system is available in sizes 2½ through 12 inches. These systems are marketed under the trade names of Super Pro 230 and Super Pro 150. Table 2-65 provides the pressure rating correction factors when the piping is to be operated at elevated temperatures.

2.8.2　Joining, Installation, and Support

PVDF pipe can be joined by threading, fusion welding, and flanging. Of these methods, fusion welding is the preferred technique (see pages 21–23). Threading can only be done using schedule 80 pipe. It is recommended that flanging be limited to connections to equipment, pumps, instruments, etc.

Installation and piping support is the same as for any thermoplastic piping system. See pages 24–32 for details of installation and support. Table 2-66 provides the recommended support spacing for schedule 40 and schedule 80 pipe, while Table 2-67 provides the same information for Super Pro 230 and Super Pro 150.

2.8.3　Corrosion Resistance

PVDF is chemically resistant to most acids, bases, and organic solvents. It also has the ability to handle wet or dry chlorine, bromine, and other halogens.

PVDF should not be used with strong alkalies, fuming acids, polar solvents, amines, ketones, and esters. When used with strong alkalies, it stress cracks. Table 2-68 lists the compatibility of PVDF with selected corrodents.

Most thermoplastic piping materials require the addition of stabilizers to reduce the effects of sunlight and oxygen. PVDF is an exception, since it does not suffer any significant degradation when used outdoors. Consequently it is not necessary to add UV stabilizers and antioxidants to the formulation.

2.9　POLYVINYLIDENE CHLORIDE (SARAN) PIPE

Polyvinylidene chloride pipe is manufactured by Dow Chemical and sold under the trade name of Saran. Because of the poor physical properties of Saran, applications are somewhat limited. The pipe system has a relatively low allowable operating pressure, which decreases rapidly as operating temperatures increase above ambient. Consequently, Saran-lined steel pipe is more commonly used than solid Saran pipe. Table 2-69 provides the physical and mechanical properties of Saran.

Saran pipe meets FDA regulations for use in food processing and for potable water as well as with those regulations prescribed by the meat inspection division of the U.S. Department of Agriculture for transporting fluids used in meat production.

2.9.1 Pipe Data

Saran pipe is available only in schedule 80 in diameters of ½ inch through 6 inches. The dimensions of Saran pipe differ from those of other thermoplastic piping systems (see Table 2-70). The maximum allowable operating pressures at various operating temperatures are given in Table 2-71.

2.9.2 Joining, Installing, and Support

Saran pipe is joined only by threading. No other method has proven successful. Installation should follow the details given on pages 24–32. Table 2-72 provides the recommended support spacing for the piping. With a support spacing of 2 feet or less it would be more economical to use a continuous support as shown in Figure 2-5.

2.9.3 Corrosion Resistance

Saran has a fairly good range of corrosion resistance. It has found wide application in the handling of plating chemicals and acid wastes.

It is not suitable for use with strong sulfuric or nitric acid solutions, strong bases, or ketones. Table 2-73 gives the compatibility of Saran with selected corrodents.

2.10   CHEM-AIRE* ABS PIPE

Chem-Aire is a new thermoplastic piping system specifically designed for compressed air. It is manufactured from a specially engineered formulation of acrylonitrile butadiene styrene (ABS) that has been modified extensively. The end result is a homogeneous shatter-resistant piping system with outstanding strength, ductility, and impact resistance. (See Table 2-74 for the physical and mechanical properties of Chem-Aire ABS pipe.)

ABS is a vast family of compounds whose properties can be varied extensively. The use of a specially modified ABS resin for this application does not imply that all ABS resins are suitable. Other ABS piping systems designed for liquid handling must not be used for compressed air. On the other hand, this piping system is suitable for liquid handling.

As industrial automation increases, the need for clean compressed air systems has increased. Black iron or galvanized steel piping systems commonly used for compressed air generate rust, corrosion products, and other debris, which creates operational problems. If stainless steel or copper piping systems are used, these problems are minimized or eliminated. However, the installed cost of stainless steel or copper becomes prohibitive. Chem-Aire ABS piping systems eliminate the problems of contamination in the line at a reasonable cost.

2.10.1   Pipe Data

Chem-Aire pipe is designed to iron pipe size outside dimensions with wall dimensions conforming to a SDR of 9.0. This results in a constant 185 psi pressure rating for all sizes

---

*Trademark of Chemtrol Division of Nibco Inc.

in the temperature range of −20 to 110°F (−29 to 43°C). As with all thermoplastic piping systems, the allowable operating pressure decreases as the operating temperature exceeds 110°F (43°C). The maximum allowable operating temperature is 140°F (60°C). Table 2-75 lists the allowable operating pressures at various temperatures.

When used on compressed air lines, it is important that the heat of compression be fully dissipated so that the maximum temperature of 140°F (60°C) is not exceeded.

Pipe and fittings are available in nominal pipe sizes of ½ inch through 2 inches.

2.10.2   Joining, Installation, and Support

Chem-Aire pipe and fittings are joined by means of solvent cementing. Socket-type solvent-cemented flanges are available for connection to flanged equipment or to join sections of pipe that may have to be dismantled. Chem-Aire pipe cannot be threaded.

Installation should follow good piping practices. Pipe should never be "sprung" into place. Support should be as discussed in Sec. 2.1.9. Support spacing is given in Table 2-76.

2.10.3   Corrosion Resistance

The compatibility of ABS with selected corrodents is shown in Table 2-77. If an oil-free compressor is not used, the lubricant must be compatible with the pipe. Most lubricating compressor oils are compatible. However, most synthetic oils can damage the ABS pipe and should therefore be avoided.

Table 2-1   Abbreviations Used for Plastics

ABS	Acrylonitrile-butadiene-styrene
CPVC	Chlorinated polyvinyl chloride
CR	Chloroprene rubber (Neoprene)
CSM	Chlorine sulfonyl polyethylene (Hypalon)
EP	Epoxide epoxy
ECTFE	Ethylene-chlorotrifluorethylene
EPDM	Ethylene propylene rubber
ETFE	Ethylene tetrafluoroethylene
FEP	Perfluorethylenepropylene
FPM	Fluorine rubber (Viton[a])
HDPE	High density polyethylene
HP	Isobutene isoprene (butyl) rubber
NBR	Nitrile (butadiene) rubber
NR	Natural rubber
PFA	Perfluoralkoxy resin
PA	Polyamide
PB	Polybutylene
PC	Polycarbonate
PCTFE	Polychlorotrifluoroethylene
PF	Phenol-formaldehyde
PP	Polypropylene
PTFE	Polytetrafluoroethylene (Teflon[a])
PVC	Polyvinyl chloride
PVDC	Polyvinylidene chloride
PVDF	Polyvinylidene fluoride

[a]Registered trademark of E. I. DuPont.

Table 2-2   Heat Distortion Temperature of the Common Polymers

	Pressure		
Polymer	66 psi	264 psi	Melt point
PTFE	250°F/121°C	132°F/56°C	620°F/327°C
PVC	135°F/57°C	140°F/60°C	285°F/141°C
LDPE	—	104°F/40°C	221°F/105°C
UHMWPE	155°F/68°C	110°F/43°C	265°F/129°C
PP	225°F/107°C	120°F/49°C	330°F/166°C
PFA	164°F/73°C	118°F/48°C	590°F/310°C
FEP	158°F/70°C	124°F/51°C	554°F/290°C
PVDF	298°F/148°C	235°F/113°C	352°F/178°C
ECTFE	240°F/116°C	170°F/77°C	464°F/240°C
CTFE	258°F/126°C	167°F/75°C	424°F/218°C
ETFE	220°F/104°C	165°F/74°C	518°F/270°C

Table 2-3 Tensile Strength of Plastics at 73°F (25°C) at Break

Plastic	Strength (psi)
PVDF	8000
ETFE	6500
CTFE	4500–6000
PFA	4000–4300
ECTFE	7000
PTFE	2500–6000
FEP	2700–3100
PVC	6000–7500
PE	1200–4550
PP	4500–6000
UHMW PE	5600

Table 2-4 Thermoplastic Pipe Dimensions

Nominal pipe size (in.)	Schedule 40 (in.)			Schedule 80 (in.)		
	Outside diameter	Inside diameter	Wall thickness	Outside diameter	Inside diameter	Wall thickness
¼	.540	.364	.088	.540	.302	.119
⅜	.675	.493	.091	.675	.423	.126
½	.840	.622	.109	.840	.546	.147
¾	1.050	.824	.113	1.050	.742	.154
1	1.315	1.049	.133	1.315	.957	.179
1¼	1.660	1.380	.140	1.660	1.278	.191
1½	1.900	1.610	.145	1.900	1.500	.200
2	2.375	2.067	.154	2.375	1.939	.218
2½	2.875	2.469	.203	2.875	2.323	.276
3	3.500	3.068	.216	3.500	2.900	.300
4	4.500	4.026	.237	4.500	3.826	.337
6	6.625	6.065	.280	6.625	5.761	.432
8	8.625	7.981	.322	8.625	7.625	.500
10	10.750	10.020	.365	10.750	9.564	.593
12	12.750	11.938	.406	12.750	11.376	.687
14	14.000	13.126	.437			
16	16.000	15.000	.500			

Table 2-5 Recommended Bolt Torques for Plastic Flanges

Flange size (in.)	Bolt diameter (in.)	Torque (ft-lb)
½	½	10–15
¾	½	10–15
1	½	10–15
1¼	½	10–15
1½	½	10–15
2	⅝	20–30
2½	⅝	20–30
3	⅝	20–30
4	⅝	20–30
6	¾	35–40
8	¾	35–40
10	⅞	55–75
12	1	80–110

Table 2-6 Values of E_1 Soil Modulus for Various Soils

Soil type and pipe bedding material	E_1 for degree of compaction of bedding (lb/in.^{2})			
	Dumped	Slight <85% Proctor <40% rel. den.	Moderate 85–95% Proctor 40–70% rel. den.	High >95% Proctor >70% rel. den.
Fine-grained soils (LL > 50) with medium to high plasticity. CH,MH,CH-MH	No data available—consult a soil engineer or use $E_1 = 0$			
Fine-grained soils (LL < 50) with medium to no plasticity. CL,ML,ML-CL with <25% coarse-grained particles	50	200	400	1000
Fine-grained soils (LL < 50) with no plasticity. CL,ML, ML–CL with >25% coarse grained particles	100	400	1000	2000
Coarse-grained soils with fines. GM,GC,SM,SC contains >12% fines	100	400	1000	2000
Coarse-grained soils with little or no fines. GW,SW,GP,SP contains <12% fines (or any borderline soil beginning with one of these symbols GM-GC,GC-SC)	200	1000	2000	3000
Crushed rock	1000		3000	

See Table 2-7 for Unified Soil Classification Group Symbols.

Table 2-7 Unified Soil Classification Group Symbols[a]

GW	Well-graded gravels, gravel-sand mixtures, little or no fines
GP	Poorly graded gravels, gravel-sand mixtures, little or no fines
GM	Silty gravels, poorly graded gravel-sand-silt mixtures
GO	Clayey gravels, poorly graded gravel-sand-clay mixtures
SW	Well-graded sands, gravelly sands, little or no fines
SP	Poorly graded sands, gravelly sands, little or no fines
SM	Silty sands, poorly graded sand-silt mixtures
SC	Clayey sands, poorly graded sand-clay mixtures
ML	Inorganic silts and very fine sand, silty or clayey fine sands
CL	Inorganic clays of low to medium plasticity
MH	Inorganic silts, micaeous or diatomaceous fine sandy or silty soils, elastic silts
CH	Inorganic clays of high plasticity, fat clays
OL	Organic silts and organic silt clays of low plasticity
OH	Organic clays of medium to high plasticity
Pt	Peat and other highly organic soils

[a]ASTM D-2487-83 Classification of Soils for Engineering Purposes.

Table 2-8 Values of Deflection Lag Factor D

Installation condition	D
Burial depth <5 ft with moderate to high degree of compaction (85% or greater Proctor, ASTM D-698 or 50% or greater relative density ASTM 2049)	2.0
Burial depth <5 ft with dumped or slight degrees of compaction (Proctor, ASTM D-698, less than 85% relative density; ASTM D-2049, less than 40%)	1.5
Burial depth >5 ft with moderate to high degree of compaction	1.5
Burial depth >5 ft with dumped or slight degree of compaction	1.25

Table 2-9 Bedding Factor K_x

Type of installation	K_x
Shaped bottom with tamped backfill material placed at the sides of the pipe, 95% Proctor density or greater	0.083
Compacted coarse-grained bedding and backfill material placed at the side of the pipe, 70–100% relative density	0.083
Shaped bottom, moderately compacted backfill material placed at the sides of the pipe, 85–95% Proctor density	0.103
Coarse-grained bedding, lightly compacted backfill material placed at the sides of the pipe, 40–70% relative density	0.103
Flat bottom, loose material placed at the sides of the pipe (not recommended); <35% Proctor density, <40% relative density	0.110

Table 2-10 Minimum Trench Widths

Pipe size (in.)	Trench width (in.)
≤3	8
4 and 6	12
8	16

Table 2-11 Offset and Loop Length Requirements

Loop length (ft)	Maximum temperature variation, °F (°C)									
	10 5.6	20 11.2	30 16.8	40 22.4	50 28	60 39.6	70 39.2	80 44.8	90 50.4	100 56
20	3	4	5	5	6	6	7	7	8	8
50	7	9	11	13	14	16	17	18	19	20
100	13	18	22	26	29	32	35	37	40	42

Table 2-12 Physical and Mechanical Properties of PVC

Property	Type 1	Type 2
Specific gravity	1.45	1.38
Water absorption (24 hr at 73°F (23°C), %	0.04	0.05
Tensile strength at 73°F (23°C), psi	6,800	5,500
Modulus of elasticity in tension at 73°F (23°C) × 10^5	5.0	4.2
Compressive strength, psi	10,000	7,900
Flexural strength, psi	14,000	11,000
Izod impact strength, notched at 73°F (23°C)	0.88	12.15
Coefficient of thermal expansion		
in./in.-°F × 10^{-5}	4.0	6.0
in./10°F/100 ft	0.40	0.60
Thermal conductivity Btu/hr/sq ft/°F/in	1.33	1.62
Heat distortion temperature, °F (°C)		
at 66 psi	130/54	135/57
at 264 psi	155/68	160/71
Resistance to heat, °F (°C) at continuous drainage	150/66	140/60
Limiting oxygen index, %		43
Flame spread		15-20
Underwriters lab rating (Sub 94)		94V-0

Table 2-13 Pressure (psi) of Schedule 40 and Schedule 80 PVC Pipe at 73.4°F (23°C)

Nominal pipe size (in.)	Schedule 40, solvent welded		Schedule 80			
			Solvent Welded		Threaded	
	Type 1	Type 2	Type 1	Type 2	Type 1	Type 2
¼	780	390	1130	565	575	280
⅜	620	310	920	460	460	230
½	600	300	850	425	420	210
¾	480	240	690	345	340	170
1	450	225	630	315	320	160
1¼	370	185	520	260	260	130
1½	330	165	470	235	240	120
2	280	140	400	200	235	120
2½	300	150	420	210	200	100
3	260	130	370	185	190	95
3½	240	120	345	170	210	105
4	220	110	320	160	185	90
5	195	100	290	145	175	85
6	180	90	280	140	NR	NR
8	160	80	250	120	NR	NR
10	140	70	230	115	NR	
12	130	65	230	115	NR	
14	130	65				
16	130	65				

NR = Not recommended.

Table 2-14A Dimensions of PVC Pipe Made to SDR Standards

Nominal pipe size (in.)	Outside diameter (in.)	SDR 13.5		SDR 21		SDR 26		SDR 32.5	
		Inside diameter (in.)	Wall thickness (in.)	Inside diameter (in.)	Wall thickness (in.)	Inside diameter (in.)	Wall thickness (in.)	Inside diameter (in.)	Wall thickness (in.)
½	0.840	0.716	0.062						
¾	1.050	0.894	0.078	0.930	0.060				
1	1.315	1.121	0.007	1.189	0.063				
1¼	1.660	1.414	0.123	1.502	0.079	1.532	0.064		
1½	1.900	1.618	0.141	1.720	0.090	1.754	0.073	1.78	0.060
2	2.375	2.023	0.176	2.149	0.113	2.193	0.091	2.229	0.073
2½	2.875	2.449	0.213	2.601	0.137	2.655	0.110	2.699	0.088
3	3.500	2.982	0.259	3.166	0.167	3.230	0.135	3.284	0.108
4	4.500	3.834	0.333	4.072	0.214	4.154	0.173	4.224	0.138
6	6.625			5.993	0.316	6.115	0.255	6.217	0.204
8	8.625			7.803	0.411	7.961	0.332	8.095	0.265
10	10.750			9.728	0.511	9.924	0.413	10.088	0.331
12	12.750			11.538	0.606	11.770	0.490	11.966	0.392

Table 2-14B Dimensions of PVC Pipe Made to SDR Standards

Nominal pipe size (in.)	Outside diameter (in.)	SDR 14		SDR 18		SDR 25	
		Inside diameter (in.)	Wall thickness (in.)	Inside diameter (in.)	Wall thickness (in.)	Inside diameter (in.)	Wall thickness (in.)
4	4.80	4.07	0.343	4.23	0.267	4.39	0.192
6	6.90	5.86	0.493	6.09	0.383	6.30	0.276
8	9.05	7.68	0.646	7.98	0.503	8.28	0.362
10	11.10	9.42	0.793	9.79	0.617	10.16	0.444
12	13.20	11.20	0.943	11.65	0.733	12.08	0.528
14	15.3					14.08	0.612
16	17.4					16.01	0.696
18	19.5					17.94	0.780
20	21.6					19.87	0.864
24	25.8					23.74	1.032

Table 2-15 Maximum Operating Pressure (psi) of SDR-Rated Pipe at 73.4°F (23°C)

SDR rating	Maximum pressure
13.5	315
14	200
18	150
21	200
25	165
26	160
32.5	125

Table 2-16 Maximum Allowable Operating Pressure of Flanged PVC Systems Based on Operating Temperatures

Operating temperature (°F/°C)	Maximum allowable operating pressure (psi)
100/38	150
110/43	135
120/49	110
130/54	75
140/60	50

Threaded flanges >2½ inches must be backwelded in order to operate at ratings shown.

Table 2-17 Temperature Correction Factors for PVC Pipe Referred to 73.4°F (23°C)

Operating temperature (°F/°C)	Factor	
	Type 1	Type 2
50/10	1.20	1.20
60/16	1.10	1.10
70/21	1.05	1.05
73.4/23	1.00	1.00
80/27	0.95	0.95
90/32	0.90	0.90
100/38	0.85	0.85
110/43	0.80	0.70
115/46	0.75	0.55
120/49	0.70	0.45
125/52	0.70	0.35
130/54	0.65	0.25
140/60	0.60	NR
150/66	0.55	NR

NR = Not recommended.

Table 2-18 Collapse Ratings of Type 1 PVC
Pipe at 73.4°F (23°C)

Pipe size (in.)	Schedule 40	Schedule 80
½	450	575
¾	285	499
1	245	469
1¼	160	340
1½	120	270
2	75	190
2½	100	220
3	70	155
4	45	115
6	25	80
8	16	50
10	12	43
12	9	39

Table 2-19 Support Spacing for Schedule 40 PVC Pipe[a]

Nominal pipe size (in.)	Support spacing (ft) at °F (°C)					
	20 (−7)	60 (16)	80 (27)	100 (38)	120 (49)	140 (60)
¼	4.5	4.0	3.5	3.5	2.0	2.0
⅜	4.5	4.0	4.0	3.5	2.5	2.0
½	5.0	4.5	4.5	4.0	2.5	2.5
¾	5.5	5.0	4.5	4.0	2.5	2.5
1	6.0	5.5	5.0	4.5	3.0	2.5
1¼	6.0	5.5	5.5	5.0	3.0	3.0
1½	6.5	6.0	5.5	5.0	3.5	3.0
2	6.5	6.0	5.5	5.0	3.5	3.0
2½	7.5	7.0	6.5	6.0	4.0	3.5
3	8.0	7.0	7.0	6.0	4.0	3.5
4	8.5	7.5	7.0	6.5	4.5	4.0
6	9.5	8.5	8.0	7.5	5.0	4.5
8	10.0	9.0	8.5	8.0	5.0	4.5
10	11.0	10.0	9.0	8.5	5.5	5.0
12	12.0	11.5	10.5	9.5	6.5	5.5
14	13.0	12.0	11.0	10.0	7.0	5.0
16	14.0	13.0	12.0	11.0	8.0	6.0

[a]Based on fluid having a specific gravity of 1.35 and the lines being uninsulated.

Table 2-20 Support Spacing for Schedule 80 PVC Pipe[a]

Nominal pipe size (in.)	Support spacing (ft) at °F (°C)					
	20 (−7)	60 (16)	80 (27)	100 (38)	120 (49)	140 (60)
1/4	4.5	4.0	4.0	3.5	2.5	2.0
3/8	5.0	4.5	4.0	4.0	2.5	2.5
1/2	5.5	5.0	4.5	4.5	3.0	2.5
3/4	6.0	5.5	5.0	4.5	3.0	2.5
1	6.5	6.0	5.5	5.0	3.5	3.0
1 1/4	7.0	6.0	6.0	5.5	3.5	3.0
1 1/2	7.0	6.5	6.0	5.5	3.5	3.5
2	7.5	7.0	6.5	6.0	4.0	3.5
2 1/2	8.5	7.5	7.5	6.5	4.5	4.0
3	9.0	8.0	7.5	7.0	4.5	4.0
4	10.0	9.0	8.5	7.5	5.0	4.5
6	11.0	10.0	9.5	9.0	6.0	5.0
8	12.5	11.0	10.5	9.5	6.5	5.5
10	13.5	12.0	11.5	10.0	7.0	6.0
12	14.5	13.0	12.0	10.5	7.5	6.5

[a]Based on fluid having a specific gravity of 1.35 and the lines being uninsulated.

Table 2-21 Support Spacing for SDR 13.5 PVC Pipe[a]

Nominal pipe size (in.)	Support spacing (ft) at °F (°C)					
	20 (−7)	60 (16)	80 (27)	100 (38)	120 (49)	140 (60)
1/4	4.5	4.0	4.0	3.5	C	C
3/4	5.0	4.5	4.0	3.5	C	C
1	5.5	5.0	4.5	4.0	2.5	C
1 1/4	5.5	5.0	5.0	4.5	2.5	2.5
1 1/2	6.5	6.0	5.5	5.0	3.5	3.0
2	7.0	6.5	6.0	5.5	3.5	3.0
2 1/2	8.0	7.0	7.0	6.0	4.0	3.5
3	8.5	7.5	7.0	6.5	4.0	3.5
4	9.5	8.5	8.0	7.0	5.0	4.5

[a]Based on fluid having a specific gravity of 1.35 and the pipelines being uninsulated.
C = Continuous support.

Table 2-22 Support Spacing for SDR-21 PVC Pipe[a]

Nominal pipe size (in.)	Support spacing (ft) at °F (°C)					
	20 (−7)	60 (16)	80 (27)	100 (38)	120 (49)	140 (60)
¾	2.5	2.0	2.0	C	C	C
1	3.0	2.5	2.5	2.0	C	C
1¼	3.5	3.0	3.0	2.5	C	C
1½	4.0	3.5	3.0	2.5	2.0	C
2	4.0	3.5	3.0	3.0	2.0	2.0
2½	4.5	4.0	4.0	3.5	2.0	2.0
3	5.0	4.0	3.5	3.5	2.5	2.0
4	8.0	7.0	6.5	6.0	4.0	3.5
5	7.0	6.5	4.0	4.0	2.0	1.5
6	6.0	5.5	5.0	5.0	3.0	2.5
8	6.5	6.0	5.5	5.0	3.5	3.0
10	7.0	6.5	6.0	5.5	3.5	3.0
12	7.5	7.0	6.5	5.5	4.0	3.5

[a]Based on fluid having a specific gravity of 1.35 and the lines being uninsulated.
C = Continuous support.

Table 2-23 Support Spacing for SDR-26 PVC Pipe[a]

Nominal pipe size (in.)	Support spacing (ft) at °F (°C)					
	20 (−7)	60 (16)	80 (27)	100 (38)	120 (49)	140 (60)
1	3.0	2.5	2.5	2.5	C	C
1¼	3.0	2.5	2.5	2.5	C	C
1½	3.0	3.0	2.5	2.5	C	C
2	3.0	3.0	2.5	2.5	C	C
2½	3.5	3.5	3.0	2.5	2.0	C
3	5.0	4.0	4.0	3.5	2.5	2.0
4	6.0	5.5	4.5	4.0	3.0	2.0
5	6.0	5.5	5.0	4.5	3.0	2.5
6	6.5	6.0	5.5	5.0	3.5	3.0
8	7.0	6.0	6.0	5.5	3.5	3.0
10	7.5	7.0	6.0	6.0	4.0	3.5
12	8.5	8.0	7.5	6.5	4.5	4.0
14	9.0	8.5	7.5	7.0	5.0	2.5
16	10.0	9.0	8.5	7.5	5.5	4.0
18	11.0	9.5	9.0	8.0	6.0	4.5
20	12.0	10.0	9.5	8.5	6.5	5.0
24	13.0	11.0	10.0	9.0	7.0	5.5

[a]Based on fluid having a specific gravity of 1.35 and the lines being uninsulated.
C = Continuous support.

Table 2-24 Support Spacing for SDR-41 PVC Pipe[a]

Nominal pipe size (in.)	Support spacing (ft) at °F (°C)					
	20 (−7)	60 (16)	80 (27)	100 (38)	120 (49)	140 (60)
4	3.5	3.0	2.5	2.5	C	C
6	4.0	3.5	3.0	3.0	2.0	C
8	4.0	3.5	3.5	3.0	2.0	C
10	4.5	4.0	3.5	3.5	2.5	2.0
12	5.0	5.0	4.5	4.0	2.5	2.5
14	5.5	5.0	4.5	4.0	3.0	2.0
16	6.0	5.0	5.0	4.5	3.0	2.5
18	6.5	6.0	5.5	5.0	3.5	2.5
20	7.5	6.0	6.0	5.0	4.0	3.0
24	8.0	6.5	6.5	5.5	4.0	3.0

[a]Based on fluid having a specific gravity of 1.35 and the lines being uninsulated.
C = Continuous support.

Table 2-25 Compatibility of Type 2 PVC with Selected Corrodents.
The chemicals listed are in the pure state or in a saturated solution unless otherwise indicated. Compatibility is shown to the maximum allowable temperature for which data are available. Incompatibility is shown by an x. A blank space indicates that data are unavailable.

Chemical	Maximum temp.		Chemical	Maximum temp.	
	°F	°C		°F	°C
Acetaldehyde	x	x	Ammonium carbonate	140	60
Acetamide	x	x	Ammonium chloride 10%	140	60
Acetic acid 10%	100	38	Ammonium chloride 50%	140	60
Acetic acid 50%	90	32	Ammoinum chloride, sat.	140	60
Acetic acid 80%	x	x	Ammonium fluoride 10%	90	32
Acetic acid, glacial	x	x	Ammonium fluoride 25%	90	32
Acetic anhydride	x	x	Ammonium hydroxide 25%	140	60
Acetone	x	x	Ammonium hydroxide, sat.	140	60
Acetyl chloride	x	x	Ammonium nitrate	140	60
Acrylic acid	x	x	Ammonium persulfate	140	60
Acrylonitrile	x	x	Ammonium phosphate	140	60
Adipic acid	140	60	Ammonium sulfate 10–40%	140	60
Allyl alcohol	90	32	Ammonium sulfide	140	60
Allyl chloride	x	x	Ammonium sulfite		
Alum	140	60	Amyl acetate	x	x
Aluminum acetate	100	38	Amyl alcohol	x	x
Aluminum chloride, aqueous	140	60	Amyl chloride	x	x
Aluminum chloride, dry			Aniline	x	x
Aluminum fluoride	140	60	Antimony trichloride	140	60
Aluminum hydroxide	140	60	Aqua regia 3:1	x	x
Aluminum nitrate	140	60	Barium carbonate	140	60
Aluminum oxychloride	140	60	Barium chloride	140	60
Aluminum sulfate	140	60	Barium hydroxide	140	60
Ammonia gas	140	60	Barium sulfate	140	60
Ammonium bifluoride	90	32	Barium sulfide	140	60

Table 2-25 *Continued*

Chemical	Maximum temp. °F	°C	Chemical	Maximum temp. °F	°C
Benzaldehyde	x	x	Chromyl chloride		
Benzene	x	x	Citric acid 15%	140	60
Benzene sulfonic acid 10%	140	60	Citric acid, conc.	140	60
Benzoic acid	140	60	Copper acetate		
Benzyl alcohol	x	x	Copper carbonate	140	60
Benzyl chloride			Copper chloride	140	60
Borax	140	60	Copper cyanide	140	60
Boric acid	140	60	Copper sulfate	140	60
Bromine gas, dry	x	x	Cresol	x	x
Bromine gas, moist	x	x	Cupric chloride 5%		
Bromine liquid	x	x	Cupric chloride 50%		
Butadiene	60	16	Cyclohexane		
Butyl acetate	x	x	Cyclohexanol	x	x
Butyl alcohol	x	x	Dichloroacetic acid	120	49
n-Butylamine	x	x	Dichloroethane (ethylene di-chloride)	x	x
Butyl phthalate					
Butyric acid	x	x	Ethylene glycol	140	60
Calcium bisulfide	140	60	Ferric chloride	140	60
Calcium bisulfite	140	60	Ferric chloride 50% in water		
Calcium carbonate	140	60	Ferric nitrate 10–50%	140	60
Calcium chlorate	140	60	Ferrous chloride	140	60
Calcium chloride	140	60	Ferrous nitrate	140	60
Calcium hydroxide 10%	140	60	Fluorine gas, dry	x	x
Calcium hydroxide, sat.	140	60	Fluorine gas, moist	x	x
Calcium hypochlorite	140	60	Hydrobromic acid, dilute	140	60
Calcium nitrate	140	60	Hydrobromic acid 20%	140	60
Calcium oxide	140	60	Hydrobromic acid 50%	140	60
Calcium sulfate	140	60	Hydrochloric acid 20%	140	60
Caprylic acid			Hydrochloric acid 38%	140	60
Carbon bisulfide	x	x	Hydrocyanic acid 10%	140	60
Carbon dioxide, dry	140	60	Hydrofluoric acid 30%	120	49
Carbon dioxide, wet	140	60	Hydrofluoric acid 70%	68	20
Carbon disulfide	x	x	Hydrofluoric acid 100%		
Carbon monoxide	140	60	Hypochlorous acid	140	60
Carbon tetrachloride	x	x	Iodine solution 10%		
Carbonic acid	140	60	Ketones, general	x	x
Cellosolve	x	x	Lactic acid 25%	140	60
Chloracetic acid, 50% water			Lactic acid, concentrated	80	27
Chloracetic acid	105	40	Magnesium chloride	140	60
Chlorine gas, dry	140	60	Malic acid	140	60
Chlorine gas, wet	x	x	Manganese chloride		
Chlorine, liquid	x	x	Methyl chloride	x	x
Chlorobenzene	x	x	Methyl ethyl ketone	x	x
Chloroform	x	x	Methyl isobutyl ketone	x	x
Chlorosulfonic acid	60	16	Muriatic acid	140	60
Chromic acid 10%	140	60	Nitric acid 5%	100	38
Chromic acid 50%	x	x	Nitric acid 20%	140	60

Table 2-25 *Continued*

Chemical	Maximum temp.		Chemical	Maximum temp.	
	°F	°C		°F	°C
Nitric acid 70%	70	140	Sodium hypochlorite, con-centrated	140	60
Nitric acid, anhydrous	x	x	Sodium sulfide to 50%	140	60
Nitrous acid, concentrated	60	16	Stannic chloride	140	60
Oleum	x	x	Stannous chloride	140	60
Perchloric acid 10%	60	16	Sulfuric acid 10%	140	60
Perchloric acid 70%	60	16	Sulfuric acid 50%	140	60
Phenol	x	x	Sulfuric acid 70%	140	60
Phosphoric acid 50–80%	140	60	Sulfuric acid 90%	x	x
Picric acid	x	x	Sulfuric acid 98%	x	x
Potassium bromide 30%	140	60	Sulfuric acid 100%	x	x
Salicylic acid	x	x	Sulfuric acid, fuming	x	x
Silver bromide 10%	105	40	Sulfurous acid	140	60
Sodium carbonate	140	60	Thionyl chloride	x	x
Sodium chloride	140	60	Toluene	x	x
Sodium hydroxide 10%	140	60	Trichloroacetic acid	x	x
Sodium hydroxide 50%	140	60	White liquor	140	60
Sodium hydroxide, con-centrated	140	60	Zinc chloride	140	60
Sodium hypochlorite 20%	140	60			

Source: Schweitzer, Philip A. (1991). *Corrosion Resistance Tables,* Marcel Dekker, Inc., New York, Vols. 1 and 2.

Table 2-26 Physical and Mechanical Properties of Fiberglass-Armored PVC Pipe

	Glass fiber (outer)	PVC (inner)
Specific gravity	1.75	1.38
Tensile strength at 80°F (27°C), psi	9910	7000
Compressive strength at 80°F (27°C), psi	13,800	9600
Coefficient of expansion, in./in./°F	1.39×10^{-5}	[a]
Thermal conductivity, Btu/hr/sq ft/°F/in.	1.8	1.0

[a]Thermal expansion of the PVC is held in restraint to approximately that of the glass fiber.

Table 2-27 Temperature Correction Factors for Underground Fiberglass-Armored PVC Pipe

Operating temperature (°F/°C)	Pressure class, psi	
	250	350
73/23	1.0	1.0
80/27	0.92	0.93
90/32	0.85	0.85
100/38	0.80	0.80
110/43	0.70	0.70
120/49	0.60	0.60
130/54	0.55	0.55
140/60	0.50	0.50

Table 2-28 Maximum Operating Pressure and Temperature of Above-Ground Fiberglass-Armored PVC Pipe

Nominal pipe size (in.)	Operating temperature (°F/°C)	Operating pressure (psi)
2	200/93	150
2½	200/93	150
3	200/93	150
4	200/93	100
6	200/93	100

Table 2-29 Support Spacing for Above-Ground Fiberglass-Armored PVC Pipe[a]

Nominal pipe size (in.)	Support spacing (ft)
2	9.0
2½	9.5
3	10.0
4	10.5
6	11.5

[a]Based on fluid having a specific gravity of 1.35 and the lines being uninsulated.

Table 2-30 Physical and Mechanical Properties of CPVC

Specific gravity	1.55
Water absorption 24 hr at 73°F (23°C), %	0.03
Tensile strength at 73°F (23°C), psi	8000
Modulus of elasticity in tension at 73°F (23°C) $\times 10^5$	4.15
Compressive strength at 73°F (23°C), psi	9000
Flexural strength, psi	15,100
Izod impact strength at 73°F (23°C)	1.5
Coefficient of thermal expansion	
in./in.-°F $\times 10^{-5}$	3.4
in./10°F/100 ft	0.034
Thermal conductivity Btu/hr/sq ft/°F/in.	0.95
Heat distortion temperature, °F/°C	
at 66 psi	238/114
at 264 psi	217/100
Resistance to heat at continuous drainage, °F/°C	200/93
Limiting oxygen index, %	60
Flame spread	15
Underwriters lab rating (U.L. 94)	VO;5VA;5VB

Source: Courtesy of B. F. Goodrich, Speciality Polymers and Chemical Division.

Table 2-31 Maximum Allowable Operating Pressure of CPVC Pipe at 73°F (23°C)

Nominal pipe size (in.)	Schedule 40 Solvent welded	Schedule 80 Solvent welded	Threaded
¼	640	900	450
⅜	620	800	365
½	600	850	340
¾	480	690	275
1	450	630	250
1¼	370	520	205
1½	330	470	185
2	280	400	160
2½	300	420	170
3	260	370	150
4	220	320	130
5	180	290	115
6	180	280	115
8	160	250	NR
10	140	230	NR
12	130	230	NR

NR = Not recommended.

Table 2-32 Maximum Allowable Operating Pressure of Flanged CPVC Systems Based on Operating Temperature[a]

Operating temperature (°F/°C)	Maximum allowable operating pressure (psi)
100/38	150
110/43	145
120/49	135
130/54	125
140/60	110
150/66	100
160/71	90
170/77	80
180/82	70
190/88	60
200/93	50
210/99	40

[a]Threaded flange sizes 2½ through 6 inches must be backwelded in order to operate at the ratings shown above.

Table 2-33 Temperature Correction Factors for CPVC Pipe

Operating temperature		Factor
°F	°C	
70	21	1.00
80	27	1.00
90	32	0.91
100	38	0.82
120	49	0.65
140	60	0.50
160	71	0.40
180	82	0.25
200	93	0.20

Source: Courtesy of B. F. Goodrich Specialty Polymers Chemical Division.

Table 2-34 Collapse Pressure Rating of Schedule 80 CPVC Pipe at 73°F (25°C)

Nominal pipe size (in.)	Pressure rating (psi)
½	575
¾	499
1	469
1¼	340
1½	270
2	190
2½	220
3	155
4	115
6	80
8	50

Table 2-35 Support Spacing for Schedule 40 CPVC Pipe[a]

Nominal pipe size (in.)	Support spacing (ft) at °F(°C)						
	60 (16)	100 (38)	120 (49)	140 (60)	160 (71)	180 (82)	200 (93)
¼	4.5	4.0	3.5	3.5	2.0	2.0	C
⅜	4.5	4.0	4.0	3.5	2.5	2.0	C
½	5.0	4.5	4.5	4.0	2.5	2.5	C
¾	5.5	5.0	4.5	4.0	2.5	2.5	C
1	6.0	5.5	5.0	4.5	3.0	2.5	C
1¼	6.0	5.5	5.5	5.0	3.0	3.0	1.5
1½	6.5	6.0	5.5	5.0	3.5	3.0	1.5
2	6.5	6.0	5.5	5.0	3.5	3.0	1.5
2½	7.5	7.0	6.5	6.0	4.0	3.5	1.5
3	8.0	7.0	7.0	6.0	4.0	3.5	2.0
4	8.5	7.5	7.0	6.5	4.5	4.0	2.0
5	9.5	8.5	8.0	7.5	5.0	4.5	2.0
6	9.5	8.5	8.0	7.5	5.0	4.5	2.0
8	10.0	9.0	8.5	8.0	5.0	4.5	2.0
10	10.5	9.5	9.0	8.5	5.5	5.0	2.0
12	11.0	10.0	9.5	9.0	6.0	5.5	2.5

[a]Based on fluid having a specific gravity of 1.35 and the lines being uninsulated.

C = Continuous support.

Table 2-36 Support Spacing for Schedule 80 CPVC Pipe[a]

Nominal pipe size (in.)	Support spacing (ft) at °F(°C)						
	60 (16)	100 (38)	120 (49)	140 (60)	160 (71)	180 (82)	200 (93)
¼	4.5	4.0	4.0	3.5	2.5	2.0	C
⅜	5.0	4.5	4.5	4.0	2.5	2.5	C
½	5.5	5.0	4.5	4.5	3.0	2.5	C
¾	5.5	5.5	5.0	4.5	3.0	2.5	C
1	6.0	6.0	5.5	5.0	3.5	3.0	1.5
1¼	6.5	6.0	6.0	5.5	3.5	3.0	1.5
1½	7.0	6.5	6.0	5.5	3.5	3.5	2.0
2	7.0	7.0	6.5	6.0	4.0	3.5	2.0
2½	8.0	7.5	7.5	6.5	4.5	4.0	2.5
3	8.0	8.0	7.5	7.0	4.5	4.0	2.5
4	8.5	9.0	8.5	7.5	5.0	4.5	2.5
5	10.0	9.5	9.0	8.0	5.5	5.0	3.0
6	10.0	9.5	9.0	8.0	5.5	5.0	3.0
8	11.0	10.5	10.0	9.0	6.0	5.5	3.5
10	11.5	11.0	10.5	9.5	6.5	6.0	4.0
12	12.5	12.0	11.5	10.5	7.5	6.5	4.5

[a]Based on fluid having a specific gravity of 1.00 and the lines being uninsulated.
C = Continuous support.

Table 2-37 Compatibility of CPVC with Selected Corrodents.
The chemicals listed are in the pure state or in a saturated solution unless otherwise indicated. Compatibility is shown to the maximum allowable temperature for which data are available. Incompatibility is shown by an x. A blank space indicates that data are unavailable.

Chemical	Maximum temp.		Chemical	Maximum temp.	
	°F	°C		°F	°C
Acetaldehyde	x	x	Aluminum hydroxide	200	93
Acetamide			Aluminum nitrate	200	93
Acetic acid 10%	90	32	Aluminum oxychloride	200	93
Acetic acid 50%	x	x	Aluminum sulfate	200	93
Acetic acid 80%	x	x	Ammonia gas, dry	200	93
Acetic acid, glacial	x	x	Ammonium bifluoride	140	60
Acetic anhydride	x	x	Ammonium carbonate	200	93
Acetone	x	x	Ammonium chloride 10%	180	82
Acetyl chloride	x	x	Ammonium chloride 50%	180	82
Acrylic acid	x	x	Ammonium chloride, sat.	200	93
Acrylonitrile	x	x	Ammonium fluoride 10%	200	93
Adipic acid	200	93	Ammonium fluoride 25%	200	93
Allyl alcohol 96%	200	93	Ammonium hydroxide 25%	x	x
Allyl chloride	x	x	Ammonium hydroxide, sat.	x	x
Alum	200	93	Ammonium nitrate	200	93
Aluminum acetate	100	38	Ammonium persulfate	200	93
Aluminum chloride, aqueous	200	93	Ammonium phosphate	200	93
Aluminum chloride, dry	180	82	Ammonium sulfate 10–40%	200	93
Aluminum fluoride	200	93	Ammonium sulfide	200	93

Table 2-37 *Continued*

Chemical	Maximum temp. °F	Maximum temp. °C	Chemical	Maximum temp. °F	Maximum temp. °C
Ammonium sulfite	160	71	Cellosolve	180	82
Amyl acetate	x	x	Chloracetic acid, 50% water	100	38
Amyl alcohol	130	54	Chloracetic acid	x	x
Amyl chloride	x	x	Chlorine gas, dry	140	60
Aniline	x	x	Chlorine gas, wet	x	x
Antimony trichloride	200	93	Chlorine, liquid	x	x
Aqua regia 3:1	80	27	Chlorobenzene	x	x
Barium carbonate	200	93	Chloroform	x	x
Barium chloride	180	82	Chlorosulfonic acid	x	x
Barium hydroxide	180	82	Chromic acid 10%	210	99
Barium sulfate	180	82	Chromic acid 50%	210	99
Barium sulfide	180	82	Chromyl chloride	180	82
Benzaldehyde	x	x	Citric acid 15%	180	82
Benzene	x	x	Citric acid, conc.	180	82
Benzene sulfonic acid 10%	180	82	Copper acetate	80	27
Benzoic acid	200	93	Copper carbonate	180	82
Benzyl alcohol	x	x	Copper chloride	210	99
Benzyl chloride	x	x	Copper cyanide	180	82
Borax	200	93	Copper sulfate	210	99
Boric acid	210	99	Cresol	x	x
Bromine gas, dry	x	x	Cupric chloride 5%	180	82
Bromine gas, moist	x	x	Cupric chloride 50%	180	82
Bromine liquid	x	x	Cyclohexane	x	x
Butadiene	150	66	Cyclohexanol	x	x
Butyl acetate	x	x	Dichloroacetic acid, 20%	100	38
Butyl alcohol	140	60	Dichloroethane (ehtylene di-chloride)	x	x
n-Butylamine	x	x			
Butyl phthalate			Ethylene glycol	210	99
Butyric acid	140	60	Ferric chloride	210	99
Calcium bisulfide	180	82	Ferric chloride 50% in water	180	82
Calcium bisulfite	210	99			
Calcium carbonate	210	99	Ferric nirate 10–50%	180	82
Calcium chlorate	180	82	Ferrous chloride	210	99
Calcium chloride	180	82	Ferrous nitrate	180	82
Calcium hydroxide 10%	170	77	Fluorine gas, dry	x	x
Calcium hydroxide, sat.	210	99	Fluorine gas, moist	80	27
Calcium hypochlorite	200	93	Hydrobromic acid, dilute	130	54
Calcium nitrate	180	82	Hydrobromic acid 20%	180	82
Calcium oxide	180	82	Hydrobromic acid 50%	190	88
Calcium sulfate	180	82	Hydrochloric acid 20%	180	82
Caprylic acid	180	82	Hydrochloric acid 38%	170	77
Carbon bisulfide	x	x	Hydrocyanic acid 10%	80	27
Carbon dioxide, dry	210	99	Hydrofluoric acid 30%	x	x
Carbon dioxide, wet	160	71	Hydrofluoric acid 70%	90	32
Carbon disulfide	x	x	Hydrofluoric acid 100%	x	x
Carbon monoxide	210	99	Hypochlorous acid	180	82
Carbon tetrachloride	x	x	Iodine solution 10%		
Carbonic acid	180	82	Ketones, general	x	x

Table 2-37 *Continued*

Chemical	Maximum temp. °F	Maximum temp. °C	Chemical	Maximum temp. °F	Maximum temp. °C
Lactic acid 25%	180	82	Sodium chloride	210	99
Lactic acid, concentrated	100	38	Sodium hydroxide 10%	190	88
Magnesium chloride	230	110	Sodium hydroxide 50%	180	82
Malic acid	180	82	Sodium hydroxide, concentrated	190	88
Manganese chloride	180	82			
Methyl chloride	x	x	Sodium hypochlorite 20%	190	88
Methyl ethyl ketone	x	x	Sodium hypochlorite, concentrated	180	82
Methyl isobutyl ketone	x	x			
Muriatic acid	170	77	Sodium sulfide to 50%	180	82
Nitric acid 5%	180	82	Stannic chloride	180	82
Nitric acid 20%	160	71	Stannous chloride	180	82
Nitric acid 70%	180	82	Sulfuric acid 10%	180	82
Nitric acid, anhydrous	x	x	Sulfuric acid 50%	180	82
Nitrous acid, concentrated	80	27	Sulfuric acid 70%	200	93
Oleum	x	x	Sulfuric acid 90%	x	x
Perchloric acid 10%	180	82	Sulfuric acid 98%	x	x
Perchloric acid 70%	180	82	Sulfuric acid 100%	x	x
Phenol	140	60	Sulfuric acid, fuming	x	x
Phosphoric acid 50–80%	180	82	Sulfurous acid	180	82
Picric acid	x	x	Thionyl chloride	x	x
Potassium bromide 30%	180	82	Toluene	x	x
Salicylic acid	x	x	Trichloroacetic acid, 20%	140	60
Silver bromide 10%	170	77	White liquor	180	82
Sodium carbonate	210	99	Zinc chloride	180	82

Source: Schweitzer, Philip A. (1991). *Corrosion Resistance Tables,* Marcel Dekker, Inc., New York, Vols. 1 and 2.

Table 2-38 Physical and Mechanical Properties of Homopolymer PP

Specific gravity	0.905
Water absorption 24 hr at 73°F (23°C), %	0.02
Tensile strength at 73°F (23°C), psi	5,000
Modulus of elasticity in tension at 73°F (23°C) $\times 10^5$	1.7
Compressive strength, psi	9.243
Flexural strength, psi	7,000
Izod impact strength, notched at 73°F (23°C)	1.3
Coefficient of thermal expansion	
in./in.-°F $\times 10^{-5}$	5.0
in./10°F/100 ft	0.05
Thermal conductivity, Btu/hr/sq ft/°F/in.	1.2
Heat distortion temperature, °F/°C	
at 66 psi	220/104
at 264 psi	140/60
Resistance to heat at continuous drainage, °F/°C	180/82
Limiting oxygen index, %	17
Flame spread	slow burning
Underwriters lab rating (Sub. 94)	94HB

Table 2-39 Physical and Mechanical Properties of Copolymer PP

Specific gravity	0.91
Water absorption 24 hr at 73°F (23°C), %	0.03
Tensile strength at 73°F (23°C), psi	4000
Modulus of elasticity in tension at 73°F (23°C) $\times$ 10^5	1.5
Compressive strength, psi	8500
Izod impact strength, notched at 73°F (23°C)	8
Coefficient of thermal expansion	
in./in.-°F $\times$ 10^{-5}	6.1
in./10°F/100 ft	0.061
Thermal conductivity, Btu/hr/sq ft/°F/in.	1.3
Heat distortion temperature, °F/°C	
at 66 psi	225/107
at 264 psi	120/49
Resistance to heat at continuous drainage, °F/°C	200/93
Flame spread	slow burning

Table 2-40 Operating Pressure of Homopolymer PP Fusion Welded Pipe at 70°F (23°C)

Nominal pipe size (in.)	Maximum allowable working pressure, psi	
	Schedule 40	Schedule 80
½	290	410
¾	240	330
1	225	300
1¼	185	250
1½	165	230
2	140	200
2½	130	185
3	130	190
4	115	165
6	90	140
8	80	130
10	75	115
12	60	110

Table 2-41 Maximum Operating Pressure for Threaded
Schedule 80 Homopolymer PP Pipe at 70°F (23°C)

Nominal pipe size (in.)	Maximum operating pressure (psi)
½	20
¾	20
1	20
1¼	20
1½	20
2	20
3	20
4	20

Table 2-42 Temperature Correction Factors
for Schedule 40 and Schedule 80 Fusion-Welded
Homopolymer PP Pipe Referred to 70°F (23°C)

Operating temperature (°F/°C)	Pipe size	
	½–8 in.	10 and 12 in.
40/4	1.220	1.230
50/10	1.150	1.150
60/16	1.080	1.090
70/23	1.000	1.000
80/27	0.925	0.945
90/32	0.875	0.870
100/38	0.850	0.800
110/43	0.800	0.727
120/49	0.750	0.654
130/54	0.700	0.579
140/60	0.650	0.505
150/66	0.575	0.430
160/71	0.500	0.356

Table 2-43 Maximum Allowable Operating Pressure of Flanged
Homopolymer PP Systems Based on Operating Temperatures

Operating temperature (°F/°C)	Maximum allowable operating pressure (psi)
100/38	150
110/93	140
120/49	130
130/54	118
140/60	105
150/66	93
160/71	80
170/77	70
180/82	50

Table 2-44 Collapse Pressure Rating of Schedule 80 Homopolymer PP Pipe at 73°F (23°C)

Nominal pipe size (in.)	psi
½	230
¾	200
1	188
1¼	136
1½	108
2	76
3	62
4	46
6	32

Table 2-45 Temperature Correction Factors for SDR-Rated Copolymer PP Pipe Referred to 73.4°F (23°C)

Temperature (°F/°C)	Correction factor
73/23	1.00
100/38	0.64
140/60	0.40
180/82	0.28
200/93	0.10

Table 2-46 Support Spacing for Homopolymer PP Pipe Uninsulated Conveying Specific Gravity 1.35 Material

Nominal pipe size (in.)	Support spacing (ft) at °F(°C):					
	Schedule 40			Schedule 80		
	70 (21)	120 (49)	150 (66)	70 (21)	120 (49)	150 (66)
½	4.0	3.0	C	5.0	3.5	C
¾	4.0	3.0	C	5.0	3.5	C
1	4.5	3.0	C	5.5	4.0	C
1¼	4.5	3.5	C	5.5	4.0	2.5
1½	5.0	3.5	C	5.5	4.0	2.5
2	5.0	3.5	2.0	6.0	4.5	2.5
2½	5.5	4.0	2.5	6.5	4.5	3.0
3	6.0	4.0	2.5	7.0	5.0	3.0
4	6.0	4.5	3.0	7.5	5.0	3.5
6	6.5	5.0	3.0	8.5	6.0	4.0
8	7.0	5.0	3.5	9.0	6.0	4.0
10	C	C	C	C	C	C
12	C	C	C	C	C	C

C = Continuous support.
All valves, instruments, and accessory equipment must be independently supported.

Table 2-47 Support Spacing for Copolymer PP Pipe (Uninsulated) Conveying Specific Gravity 1.35 Material

| Nominal pipe size (in.) | Support spacing (ft) at °F(°C): | | | |
| | SDR-32 | | SDR-11 | |
	70(21)	120(49)	70(21)	120(49)
⅜	1.7	1.3	3.0	2.3
½	1.7	1.3	3.0	2.3
¾	1.7	1.3	3.0	2.3
1	1.9	1.4	3.5	2.6
1¼	2.2	1.6	4.0	3.0
1½	2.5	1.9	4.5	3.3
2	2.8	2.1	5.0	3.7
3	3.0	2.3	5.5	4.0
4	3.3	2.5	6.0	4.5
6	3.9	3.0	7.0	5.2
8	4.1	3.0	7.5	5.6
10	4.7	3.5	8.5	6.3
12	5.2	3.9	9.5	7.1
14	5.5	4.0	10.0	7.5
16	5.8	4.3	10.8	7.8
18	6.3	4.7	11.5	8.6
20	6.6	4.9	12.0	9.0
24	7.4	5.5	13.5	10.0

Continuous support should be used above 120°F (49°C). All valves, instruments, or other accessory equipment installed in the pipe line to be independently supported.

Table 2-48 Compatibility of PP with Selected Corrodents
The chemicals listed are in the pure state or in a saturated solution unless otherwise indicated. Compatibility is shown to the maximum allowable temperature for which data are available. Incompatibility is shown by an x. A blank space indicates that data are unavailable.

| Chemical | Maximum temp. | | Chemical | Maximum temp. | |
	°F	°C		°F	°C
Acetaldehyde	120	49	Allyl chloride	140	60
Acetamide	110	43	Alum	220	104
Acetic acid 10%	220	104	Aluminum acetate	100	38
Acetic acid 50%	200	93	Aluminum chloride, aqueous	200	93
Acetic acid 80%	200	93	Aluminum chloride, dry	220	104
Acetic acid, glacial	190	88	Aluminum fluoride	200	93
Acetic anhydride	100	38	Aluminum hydroxide	200	93
Acetone	220	104	Aluminum nitrate	200	93
Acetyl chloride	x	x	Aluminum oxychloride	220	104
Acrylic acid	x	x	Aluminum sulfate		
Acrylonitrile	90	32	Ammonia gas	150	66
Adipic acid	100	38	Ammonium bifluoride	200	93
Allyl alcohol	140	60	Ammonium carbonate	220	104

Table 2-48 *Continued*

Chemical	Maximum temp. °F	°C	Chemical	Maximum temp. °F	°C
Ammonium chloride 10%	180	82	Calcium hydroxide, sat.	220	104
Ammonium chloride 50%	180	82	Calcium hypochlorite	210	99
Ammonium chloride, sat.	200	93	Calcium nitrate	210	99
Ammonium fluoride 10%	210	99	Calcium oxide	220	104
Ammonium fluoride 25%	200	93	Calcium sulfate	220	104
Ammonium hydroxide 25%	200	93	Caprylic acid	140	60
Ammonium hydroxide, sat.	200	93	Carbon bisulfide	x	x
Ammonium nitrate	200	93	Carbon dioxide, dry	220	104
Ammonium persulfate	220	104	Carbon dioxide, wet	140	60
Ammonium phosphate	200	93	Carbon disulfide	x	x
Ammonium sulfate 10–40%	200	93	Carbon monoxide	220	104
Ammonium sulfide	220	104	Carbon tetrachloride	x	x
Ammonium sulfite	220	104	Carbonic acid	220	104
Amyl acetate	x	x	Cellosolve	200	93
Amyl alcohol	200	93	Chloracetic acid, 50% water	80	27
Amyl chloride	x	x	Chloracetic acid	180	82
Aniline	180	82	Chlorine gas, dry	x	x
Antimony trichloride	180	82	Chlorine gas, wet	x	x
Aqua regia 3:1	x	x	Chlorine, liquid	x	x
Barium carbonate	200	93	Chlorobenzene	x	x
Barium chloride	220	104	Chloroform	x	x
Barium hydroxide	200	93	Chlorosulfonic acid	x	x
Barium sulfate	200	93	Chromic acid 10%	140	60
Barium sulfide	200	93	Chromic acid 50%	150	66
Benzaldehyde	80	27	Chromyl chloride	140	60
Benzene	140	60	Citric acid 15%	220	104
Benzene sulfonic acid 10%	180	82	Citric acid, conc.	220	104
Benzoic acid	190	88	Copper acetate	80	27
Benzyl alcohol	140	60	Copper carbonate	200	93
Benzyl chloride	80	27	Copper chloride	200	93
Borax	210	99	Copper cyanide	200	93
Boric acid	220	104	Copper sulfate	200	93
Bromine gas, dry	x	x	Cresol	x	x
Bromine gas, moist	x	x	Cupric chloride 5%	140	60
Bromine liquid	x	x	Cupric chloride 50%	140	60
Butadiene	x	x	Cyclohexane	x	x
Butyl acetate	x	x	Cyclohexanol	150	66
Butyl alcohol	200	93	Dichloroacetic acid	100	38
n-Butylamine	90	32	Dichloroethane (ethylene di-chloride)	80	27
Butyl phthalate	180	82			
Butyric acid	180	82	Ethylene glycol	210	99
Calcium bisulfide	210	99	Ferric chloride	210	99
Calcium bisulfite	210	99	Ferric chloride 50% in water	210	99
Calcium carbonate	210	99	Ferric nitrate 10–50%	210	99
Calcium chlorate	220	104	Ferrous chloride	210	99
Calcium chloride	220	104	Ferrous nitrate	210	99
Calcium hydroxide 10%	200	93	Fluorine gas, dry	x	x

Table 2-48 *Continued*

Chemical	Maximum temp. °F	°C		Maximum temp. °F	°C
Fluorine gas, moist	x	x	Phosphoric acid 50–80%	210	99
Hydrobromic acid, dilute	230	110	Picric acid	140	60
Hydrobromic acid 20%	200	93	Potassium bromide 30%	210	99
Hydrobromic acid 50%	190	88	Salicylic acid	130	54
Hydrochloric acid 20%	220	104	Silver bromide 10%	170	77
Hydrochloric acid 38%	200	93	Sodium carbonate	220	104
Hydrocyanic acid 10%	150	66	Sodium chloride	200	93
Hydrofluoric acid 30%	180	82	Sodium hydroxide 10%	220	104
Hydrofluoric acid 70%	200	93	Sodium hydroxide 50%	220	104
Hydrofluoric acid 100%	200	93	Sodium hydroxide, concentrated	140	60
Hypochlorous acid	140	60			
Iodine solution 10%	x	x	Sodium hypochlorite 20%	120	49
Ketones, general	110	43	Sodium hypochlorite, concentrated	110	43
Lactic acid 25%	150	66			
Lactic acid, concentrated	150	66	Sodium sulfide to 50%	190	88
Magnesium chloride	210	99	Stannic chloride	150	66
Malic acid	130	54	Stannous chloride	200	93
Manganese chloride	120	49	Sulfuric acid 10%	200	93
Methyl chloride	x	x	Sulfuric acid 50%	200	93
Methyl ethyl ketone	x	x	Sulfuric acid 70%	180	82
Methyl isobutyl ketone	80	27	Sulfuric acid 90%	180	82
Muriatic acid	200	93	Sulfuric acid 98%	120	49
Nitric acid 5%	140	60	Sulfuric acid 100%	x	x
Nitric acid 20%	140	60	Sulfuric acid, fuming	x	x
Nitric acid 70%	x	x	Sulfurous acid	180	82
Nitric acid, anhydrous	x	x	Thionyl chloride	100	38
Nitrous acid, concentrated	x	x	Toluene	x	x
Oleum	x	x	Trichloroacetic acid	150	66
Perchloric acid 10%	140	60	White liquor	220	104
Perchloric acid 70%	x	x	Zinc chloride	200	93
Phenol	180	82			

Source: Schweitzer, Philip A. (1991). *Corrosion Resistance Tables*, Marcel Dekker, Inc., New York, Vols. 1 and 2.

Table 2-49 Physical and Mechanical Properties of PE EHMW

Specific gravity	0.94–0.96
Water absorption 24 hrs at 73°F (23°C), %	<0.01
Tensile strength at 73°F (23°C), psi	3100–3500
Modulus of elasticity in tension at 73°F (23°C) $\times 10^5$	1.18
Flexural modulus, psi $\times 10^5$	1.33
Izod impact strength, notched at 73°F (23°C)	0.4–6.0
Coefficient of thermal expansion	
in./in.-°F $\times 10^{-5}$	11.1
in./10°F/100 ft	0.111
Thermal conductivity, Btu/hr/sq ft/°F/in.	0.269
Heat distortion temperature, °F/°C	
at 66 psi	150/66
at 264 psi	250/121
Resistance to heat at continuous drainage, °F/°C	180/82
Flame spread	slow burning

Table 2-50 Pressure Rating of Plexico 3408 EHMW PE Pipe at 73°F (23°C)

SDR	Pressure rating (psi)
7.3	255
9.0	200
11.0	160
13.5	130
15.5	110
17	100
21	80
26	65
32.5	50

Courtesy of Plexico Inc., subsidiary of Chevron Chemical Company, Franklin Park, IL.

Table 2-51 Temperature Correction Factors for 73°F (23°C) Rating of Plexico 3408 EHMW Pipe

Operating temperature (°F/°C)	Correction factors
40/4	1.2
60/16	1.80
73/23	1.00
100/38	0.78
120/49	0.63
140/60	0.50

Courtesy of Plexico Inc., subsidiary of Chevron Chemical Company, Franklin Park, IL.

Table 2-52 Pressure Rating of Plexico High Temperature PE Pipe

	Rating (psi) at °F (°C)			
SDR	120 (49)	140 (60)	160 (71)	176 (80)
7.3	175	160	125	100
9.0	140	125	100	80
11.0	110	100	80	65
13.5	90	80	65	50
15.5	75	70	55	45
17.0	70	60	50	40
21.0	55	50	40	30
26.0	45	40	30	25
32.5	35	30	25	20

Note: Pressure ratings based on water service for an estimated 10-year service life for temperatures above 140°F (60°C). Courtesy of Plexico Inc., subsidiary of Chevron Chemical Company, Franklin Park, IL.

Table 2-53 Pressure Temperature Ratings of Redstripe EHMW PE Pipe

	Service Pressure Ratings (psi)			
Class	73°F (23°C)	100°F (38°C)	120°F (49°C)	140°F (60°C)
150	150	125	100	80
200	200	170	135	110

Courtesy of Plexico Inc., subsidiary of Chevron Chemical Company, Franklin Park, IL.

Table 2-54 Support Spacing for Plexico 3408 UHMW-PE Pipe at 70°F (21°C)

Nominal pipe size (in.)	Support spacing (ft) at SDR:								
	7.3	9	11	13.5	15.5	17	21	26	32.5
2	5.4	5.2	5.0	4.8					
3	6.5	6.3	6.1	5.9	5.7	5.6			
4	7.4	7.1	6.9	6.6	6.5	6.4	6.1		
6	9.0	8.7	8.4	8.1	7.8	7.7	7.4	7.0	6.7
8	10.2	9.9	9.6	9.2	9.0	8.8	8.4	8.0	7.6
10	11.4	11.1	10.7	10.3	10.0	9.8	9.4	9.0	8.5
12	12.4	12.0	11.6	11.2	10.9	10.7	10.2	9.8	9.3
14	13.0	12.6	12.2	11.7	11.4	11.2	10.7	10.2	9.7
16	13.9	13.5	13.0	12.5	12.2	12.0	11.5	10.9	10.4
18	14.8	14.3	13.8	13.3	12.9	12.7	12.2	11.6	11.0
20	15.6	15.1	14.6	14.0	13.6	13.4	12.8	12.2	11.6
22	16.3	15.8	15.3	14.7	14.3	14.0	13.4	12.8	12.2
24	17.1	16.5	16.0	15.4	14.9	14.7	14.0	13.4	12.7
26		17.2	16.6	16.0	15.6	15.3	14.6	13.9	13.3
28			17.2	16.6	16.1	15.8	15.2	14.5	13.8
30			17.8	17.2	16.7	16.4	15.7	15.0	14.2
32			18.4	17.7	17.3	16.9	16.2	15.5	14.7
34				18.3	17.8	17.5	16.7	15.9	15.2
36				18.8	18.3	18.0	17.2	16.4	15.6

Based on fluid having a specific gravity of 1 and the lines being uninsulated.
Courtesy of Plexico Inc., subsidiary of Chevron Chemical Company, Franklin Park, IL.

Table 2-55 Compatibility of EHMW PE with Selected Corrodents
   The chemicals listed are in the pure state or in a saturated solution unless otherwise indicated. Compatibility is shown to the maximum allowable temperature for which data are available. Incompatibility is shown by an x. A blank space indicates that data are unavailable.

Chemical	Maximum temp.		Chemical	Maximum temp.	
	°F	°C		°F	°C
Acetaldehyde, 40%	90	32	Allyl chloride	80	27
Acetamide			Alum	140	60
Acetic acid 10%	140	60	Aluminum acetate		
Acetic acid 50%	140	60	Aluminum chloride, aqueous	140	60
Acetic acid 80%	80	27	Aluminum chloride, dry	140	60
Acetic acid, glacial			Aluminum fluoride	140	60
Acetic anhydride	x	x	Aluminum hydroxide	140	60
Acetone	120	49	Aluminum nitrate		
Acetyl chloride			Aluminum oxychloride		
Acrylic acid			Aluminum sulfate	140	60
Acrylonitrile	150	66	Ammonia gas	140	60
Adipic acid	140	60	Ammonium bifluoride		
Allyl alcohol	140	60	Ammonium carbonate	140	60

Table 2-55 *Continued*

Chemical	°F	°C	Chemical	°F	°C
Ammonium chloride 10%	140	60	Calcium hydroxide, sat.	140	60
Ammonium chloride 50%	140	60	Calcium hypochlorite	140	60
Ammonium chloride, sat.	140	60	Calcium nitrate	140	60
Ammonium fluoride 10%	140	60	Calcium oxide	140	60
Ammonium fluoride 25%	140	60	Calcium sulfate	140	60
Ammonium hydroxide 25%	140	60	Caprylic acid		
Ammonium hydroxide, sat.	140	60	Carbon bisulfide	x	x
Ammonium nitrate	140	60	Carbon dioxide, dry	140	60
Ammonium persulfate	140	60	Carbon dioxide, wet	140	60
Ammonium phosphate	80	27	Carbon disulfide	x	x
Ammonium sulfate 10–40%	140	60	Carbon monoxide	140	60
Ammonium sulfide	140	60	Carbon tetrachloride	x	x
Ammonium sulfite			Carbonic acid	140	60
Amyl acetate	140	60	Cellosolve		
Amyl alcohol	140	60	Chloracetic acid, 50% water	x	x
Amyl chloride	x	x	Chloracetic acid	x	x
Aniline	130	54	Chlorine gas, dry	80	27
Antimony trichloride	140	60	Chlorine gas, wet, 10%	120	49
Aqua regia 3:1	130	54	Chlorine, liquid	x	x
Barium carbonate	140	60	Chlorobenzene	x	x
Barium chloride	140	60	Chloroform	80	27
Barium hydroxide	140	60	Chlorosulfonic acid	x	x
Barium sulfate	140	60	Chromic acid 10%	140	60
Barium sulfide	140	60	Chromic acid 50%	90	32
Benzaldehyde	x	x	Chromyl chloride		
Benzene	x	x	Citric acid 15%	140	60
Benzene sulfonic acid 10%	140	60	Citric acid, conc.	140	60
Benzoic acid	140	60	Copper acetate		
Benzyl alcohol	170	77	Copper carbonate		
Benzyl chloride			Copper chloride	140	60
Borax	140	60	Copper cyanide	140	60
Boric acid	140	60	Copper sulfate	140	60
Bromine gas, dry	x	x	Cresol	80	27
Bromine gas, moist	x	x	Cupric chloride 5%	80	27
Bromine liquid	x	x	Cupric chloride 50%		
Butadiene	x	x	Cyclohexane	130	54
Butyl acetate	90	32	Cyclohexanol	170	77
Butyl alcohol	140	60	Dichloroacetic acid	73	23
n-Butylamine	x	x	Dichloroethane (ethylene di-chloride)	x	x
Butyl phthalate	80	27			
Butyric acid	130	54	Ethylene glycol	140	60
Calcium bisulfide	140	60	Ferric chloride	140	60
Calcium bisulfite	80	27	Ferric chloride 50% in water	140	60
Calcium carbonate	140	60	Ferric nitrate 10–50%	140	60
Calcium chlorate	140	60	Ferrous chloride	140	60
Calcium chloride	140	60	Ferrous nitrate	140	60
Calcium hydroxide 10%	140	60	Fluorine gas, dry	x	x

Table 2-55 *Continued*

Chemical	Maximum temp.		Chemical	Maximum temp.	
	°F	°C		°F	°C
Fluorine gas, moist	x	x	Phosphoric acid 50–80%	100	38
Hydrobromic acid, dilute	140	60	Picric acid	100	38
Hydrobromic acid 20%	140	60	Potassium bromide 30%	140	60
Hydrobromic acid 50%	140	60	Salicylic acid		
Hydrochloric acid 20%	140	60	Silver bromide 10%		
Hydrochloric acid 38%	140	60	Sodium carbonate	140	60
Hydrocyanic acid 10%	140	60	Sodium chloride	140	60
Hydrofluoric acid 30%	80	27	Sodium hydroxide 10%	170	77
Hydrofluoric acid 70%	x	x	Sodium hydroxide 50%	170	77
Hydrofluoric acid 100%	x	x	Sodium hydroxide, con-		
Hypochlorous acid			centrated		
Iodine solution 10%	80	27	Sodium hypochlorite 20%	140	60
Ketones, general	x	x	Sodium hypochlorite, con-	140	60
Lactic acid 25%	140	60	centrated		
Lactic acid, concentrated	140	60	Sodium sulfide to 50%	140	60
Magnesium chloride	140	60	Stannic chloride	140	60
Malic acid	100	38	Stannous chloride	140	60
Manganese chloride	80	27	Sulfuric acid 10%	140	60
Methyl chloride	x	x	Sulfuric acid 50%	140	60
Methyl ethyl ketone	x	x	Sulfuric acid 70%	80	27
Methyl isobutyl ketone	80	27	Sulfuric acid 90%	x	x
Muriatic acid	140	60	Sulfuric acid 98%	x	x
Nitric acid 5%	140	60	Sulfuric acid 100%	x	x
Nitric acid 20%	140	60	Sulfuric acid, fuming	x	x
Nitric acid 70%	x	x	Sulfurous acid	140	60
Nitric acid, anhydrous	x	x	Thionyl chloride	x	x
Nitrous acid, concentrated			Toluene	x	x
Oleum			Trichloroacetic acid	140	60
Perchloric acid 10%	140	60	White liquor		
Perchloric acid 70%	x	x	Zinc chloride	140	60
Phenol	100	38			

Source: Schweitzer, Philip A. (1991). *Corrosion Resistance Tables*, Marcel Dekker, Inc., New York, Vols. 1 and 2.

Table 2-56 Physical and Mechanical Properties of E-CTFE

Specific gravity	1.68
Water absorption 24 hr at 73°F (23°C), %	<0.01
Tensile strength at 73°F (23°C), psi	4500
Modulus of elasticity in tension at 73°F (23°C) $\times 10^5$	2.4
Flexural strength, psi	7000
Izod impact strength notched at 73°F (23°C)	no break
Linear coefficient of thermal expansion, in./in.-°F at:	
−22 to 122°F (−30 to 50°C)	4.4×10^{-5}
122 to 185°F (50 to 80°C)	5.6×10^{-5}
185 to 257°F (85 to 125°C)	7.5×10^{-5}
257 to 356°F (125 to 180°C)	9.2×10^{-5}
Thermal conductivity, Btu/hr/sq ft/°F/in.	1.07
Heat distortion temperature, °F/°C	
at 66 psi	195/91
at 264 psi	151/66
Resistance to heat at continuous drainage, °F/°C	300–340/150–171
Limiting oxygen index	60
Underwriters lab rating (Sub 94)	V-O

Table 2-57 Temperature Correction Factors for E-CTFE Pipe

Temperature		Correction
°F	°C	factor
68	20	1.0
83	30	0.90
104	40	0.82
121	50	0.73
140	60	0.65
158	70	0.54
176	80	0.39
194	90	0.27
212	100	0.20
256	125	0.10
292	150	drainage
340	170	pressure only

Table 2-58 Maximum Recommended Support Spacing at Service Temperature of ECTFE Pipe

Nominal pipe size (in.)	Spacing at	
	68°F (20°C)	248°F (140°C)
1	3 ft 8 in.	2 ft 6 in.
2	4 ft 11 in.	3 ft 1 in.
3	5 ft 9 in.	3 ft 9 in.

Table 2-59 Compatibility of E-CTFE with Selected Corrodents.
The chemicals listed are in the pure state or in a saturated solution unless otherwise indicated.
Compatibility is shown to the maximum allowable temperature for which data are available.
Incompatibility is shown by an x. A blank space indicates that data are unavailable.

Chemical	Maximum temp. °F	Maximum temp. °C	Chemical	Maximum temp. °F	Maximum temp. °C
Acetaldehyde			Antimony trichloride	100	38
Acetamide			Aqua regia 3:1	250	121
Acetic acid 10%	250	121	Barium carbonate	300	149
Acetic acid 50%	250	121	Barium chloride	300	149
Acetic acid 80%	150	66	Barium hydroxide	300	149
Acetic acid, glacial	200	93	Barium sulfate	300	149
Acetic anhydride	100	38	Barium sulfide	300	149
Acetone	150	66	Benzaldehyde	150	66
Acetyl chloride	150	66	Benzene	150	66
Acrylic acid			Benzene sulfonic acid 10%	150	66
Acrylonitrile	150	66	Benzoic acid	250	121
Adipic acid	150	66	Benzyl alcohol	300	149
Allyl alcohol			Benzyl chloride	300	149
Allyl chloride	300	149	Borax	300	149
Alum	300	149	Boric acid	300	149
Aluminum acetate			Bromine gas, dry	x	x
Aluminum chloride, aqueous	300	149	Bromine gas, moist		
			Bromine liquid	150	66
Aluminum chloride, dry			Butadiene	250	121
Aluminum fluoride	300	149	Butyl acetate	150	66
Aluminum hydroxide	300	149	Butyl alcohol	300	149
Aluminum nitrate	300	149	*n*-Butylamine		
Aluminum oxychloride	150	66	Butyl phthalate		
Aluminum sulfate	300	149	Butyric acid	250	121
Ammonia gas	300	149	Calcium bisulfide	300	149
Ammonium bifluoride	300	149	Calcium bisulfite	300	149
Ammonium carbonate	300	149	Calcium carbonate	300	149
Ammonium chloride 10%	290	143	Calcium chlorate	300	149
Ammonium chloride 50%	300	149	Calcium chloride	300	149
Ammonium chloride, sat.	300	149	Calcium hydroxide 10%	300	149
Ammonium fluoride 10%	300	149	Calcium hydroxide, sat.	300	149
Ammonium fluoride 25%	300	149	Calcium hypochlorite	300	149
Ammonium hydroxide 25%	300	149	Calcium nitrate	300	149
Ammonium hydroxide, sat.	300	149	Calcium oxide	300	149
Ammonium nitrate	300	149	Calcium sulfate	300	149
Ammonium persulfate	150	66	Caprylic acid	220	104
Ammonium phosphate	300	149	Carbon bisulfide	80	27
Ammonium sulfate 10–40%	300	149	Carbon dioxide, dry	300	149
Ammonium sulfide	300	149	Carbon dioxide, wet	300	149
Ammonium sulfite			Carbon disulfide	80	27
Amyl acetate	160	71	Carbon monoxide	150	66
Amyl alcohol	300	149	Carbon tetrachloride	300	149
Amyl chloride	300	149	Carbonic acid	300	149
Aniline	90	32	Cellosolve	300	149

Table 2-59 *Continued*

Chemical	Maximum temp. °F	Maximum temp. °C	Chemical	Maximum temp. °F	Maximum temp. °C
Chloracetic acid, 50% water	250	121			
Chloracetic acid	250	121	Lactic acid, concentrated	150	66
Chlorine gas, dry	150	66	Magnesium chloride	300	149
Chlorine gas, wet	250	121	Malic acid	250	121
Chlorine, liquid	250	121	Manganese chloride		
Chlorobenzene	150	66	Methyl chloride	300	149
Chloroform	250	121	Methyl ethyl ketone	150	66
Chlorosulfonic acid	80	27	Methyl isobutyl ketone	150	66
Chromic acid 10%	250	121	Muriatic acid	300	149
Chromic acid 50%	250	121	Nitric acid 5%	300	149
Chromyl chloride			Nitric acid 20%	250	121
Citric acid 15%	300	149	Nitric acid 70%	150	66
Citric acid, conc.	300	149	Nitric acid, anhydrous	150	66
Copper acetate			Nitrous acid, concentrated	250	121
Copper carbonate	150	66	Oleum	x	x
Copper chloride	300	149	Perchloric acid 10%	150	66
Copper cyanide	300	149	Perchloric acid 70%	150	66
Copper sulfate	300	149	Phenol	150	66
Cresol	300	149	Phosphoric acid 50–80%	250	121
Cupric chloride 5%	300	149	Picric acid	80	27
Cupric chloride 50%	300	149	Potassium bromide 30%	300	149
Cyclohexane	300	149	Salicylic acid	250	121
Cyclohexanol	300	149	Silver bromide 10%		
Dichloroacetic acid			Sodium carbonate	300	149
Dichloroethane (ethylene di-chloride)			Sodium chloride	300	149
			Sodium hydroxide 10%	300	149
Ethylene glycol	300	149	Sodium hydroxide 50%	250	121
Ferric chloride	300	149	Sodium hydroxide, con-centrated	150	66
Ferric chloride 50% in water	300	149			
			Sodium hypochlorite 20%	300	149
Ferric nitrate 10–50%	300	149	Sodium hypochlorite, con-centrated	300	149
Ferrous chloride	300	149			
Ferrous nitrate	300	149	Sodium sulfide to 50%	300	149
Fluorine gas, dry	x	x	Stannic chloride	300	149
Fluorine gas, moist	80	27	Stannous chloride	300	149
Hydrobromic acid, dilute	300	149	Sulfuric acid 10%	250	121
Hydrobromic acid 20%	300	149	Sulfuric acid 50%	250	121
Hydrobromic acid 50%	300	149	Sulfuric acid 70%	250	121
Hydrochloric acid 20%	300	149	Sulfuric acid 90%	150	66
Hydrochloric acid 38%	300	149	Sulfuric acid 98%	150	66
Hydrocyanic acid 10%	300	149	Sulfuric acid 100%	80	27
Hydrofluoric acid 30%	250	121	Sulfuric acid, fuming	300	149
Hydrofluoric acid 70%	240	116	Sulfurous acid	250	121
Hydrofluoric acid 100%	240	116	Thionyl chloride	150	66
Hypochlorous acid	300	149	Toluene	150	66
Iodine solution 10%	250	121	Trichloroacetic acid	150	66
Ketones, general			White liquor	250	121
Lactic acid 25%	150	66	Zinc chloride	300	149

Source: Schweitzer, Philip A. (1991). *Corrosion Resistance Tables*, Marcel Dekker, Inc., New York, Vols. 1 and 2.

Table 2-60 Physical and Mechanical Properties of PVDF

Specific gravity	1.76
Water absorption 24 hr at 73°F (23°C), %	<0.04
Tensile strength at 73°F (23°C), psi	6000
Modulus of elasticity in tension at 73°F (23°C) $\times$ 10^5	2.1
Compressive strength, psi	11600
Flexural strength, psi	10750
Izod impact strength, notched at 73°F (23°C)	3.8
Coefficient of thermal expansion	
in./in.-°F $\times$ 10^{-5}	7.9
in./10°F/100 ft	0.079
Thermal conductivity, Btu/hr/sq ft/°F/in.	0.79
Heat distortion temperature, °F/°C	
at 66 psi	284/140
at 264 psi	194/90
Resistance to heat at continuous drainage, °F/°C	280/138
Limiting oxygen index, %	44
Flame spread	0
Underwriters lab rating (Sub94)	94V-0

Table 2-61 Maximum Allowable Operating Pressure of Schedule 40 and Schedule 80 PVDF Pipe at 70°F (21°C)

Nominal pipe size (in.)	Maximum Working Pressure (psi)		
	Schedule 40, fusion welded	Schedule 80, threaded	Schedule 80, fusion welded
½	360	290	580
¾	300	235	470
1	280	215	430
1 ¼	—	215	320
1 ½	220	160	320
2	180	135	275
3	NR	NR	250
4	NR	NR	220
6	NR	NR	190

NR = Not recommended.

Table 2-62 Temperature Correction Factors for Schedule 40 and Schedule 80 PVDF Pipe

Operating temperature (°F/°C)	Factor	Operating temperature (°F/°C)	Factor
30/-1	1.210	160/71	0.585
40/4	1.180	170/77	0.543
50/10	1.110	180/82	0.500
60/16	1.060	190/88	0.458
70/21	1.000	200/93	0.423
80/27	0.950	210/99	0.388
90/32	0.900	220/104	0.352
100/38	0.850	230/110	0.317
110/43	0.804	240/116	0.282
120/49	0.760	250/121	0.246
130/54	0.712	260/127	0.212
140/60	0.670	270/132	0.183
150/66	0.627	280/138	0.155

Table 2-63 Maximum Allowable Operating Pressure of Flanged PVDF Systems Based on Operating Temperatures

Operating temperature (°F/°C)	Maximum allowable operating pressure (psi)
100/38	150
110/43	150
120/49	150
130/54	150
140/60	150
150/66	140
160/71	133
170/77	125
180/82	115
190/88	106
200/93	97
210/99	90
240/116	60
280/138	25

Table 2-64 Collapse Pressure Rating of Schedule 80 PVDF Pipe at 73°F (23°C)

Nominal pipe size (in.)	Collapse pressure rating (psi)
½	391
¾	339
1	319
1 ½	183
2	129

Table 2-65 Pressure Rating Correction Factors for Elevated Operating Temperature for Super Pro[a] 230 and Super Pro[a] 150 Pipe

Temperature		Correction factor
°F	°C	
70	21	1.00
80	27	0.95
90	32	0.87
100	38	0.80
120	49	0.68
140	60	0.58
160	71	0.49
180	82	0.42
200	93	0.36
240	115	0.25
260	138	0.18

[a]Super Pro 230 and Super Pro 150 are trademarks of Asahi/America.
Source: Courtesy of Asahi/America.

Table 2-66 Support Spacing for Schedule 40 and Schedule 80 PVDF Pipe Based on Uninsulated Pipe Conveying a Fluid Having a Specific Gravity of 1.35

Nominal pipe size (in.)	Support spacing (ft) at °F (°C):									
	80 (27)	100 (38)	120 (49)	140 (60)	160 (71)	80 (27)	100 (38)	120 (49)	140 (60)	160 (71)
	Schedule 40					Schedule 80				
½	4.0	3.75	3.0	2.5	C	4.5	4.2	3.5	2.5	C
¾	4.0	3.75	3.1	2.5	C	4.5	4.5	3.7	3.0	C
1	4.5	4.25	3.5	2.5	C	5.0	4.7	4.0	3.0	C
1¼	4.5	4.25	3.5	2.5	C	5.0	4.7	4.0	3.0	C
1½	4.5	4.25	3.5	3.0	C	5.5	5.0	4.2	3.0	C
2	5.0	4.5	3.75	3.0	C	5.5	5.2	4.5	3.0	C
3	5.5	5.0	4.0	2.5	C	6.0	5.5	4.5	3.0	C
4	5.5	5.0	4.5	3.0	C	6.5	6.0	5.0	3.5	C
6	6.5	6.0	5.0	3.5	C	7.5	7.0	5.5	4.0	C

C = Continuous Support.

Table 2-67 Support Spacing for Super Pro 230 and Super Pro 150 Pipe

Nominal pipe size (in.)	Spacing (ft) at:	
	68°F (19°C)	120°F (49°C)
⅜	3.0	2.25
½	3.0	2.25
¾	3.0	2.25
1	3.5	2.6
1 ¼	4.0	3.0
1 ½	4.5	3.4
2	5.0	3.7
3	5.5	4.0
4	6.0	4.5
6	7.0	5.25
8	7.5	5.6
10	8.5	6.4
12	9.5	7.0

Super Pro is the trademark of Asahi/America. All valves, instruments, or accessory equipment installed in the pipe line must be supported independently.

Operating temperatures above 120°F (49°C) require continuous support.

Source: Courtesy of Asahi/America.

Table 2-68 Compatibility of PVDF with Selected Corrodents.
The chemicals listed are in the pure state or in a saturated solution unless otherwise indicated.
Compatibility is shown to the maximum allowable temperature for which data are available.
Incompatibility is shown by an x. A blank space indicates that data are unavailable.

Chemical	Maximum temp. °F	Maximum temp. °C	Chemical	Maximum temp. °F	Maximum temp. °C
Acetaldehyde	150	66	Antimony trichloride	150	66
Acetamide	90	32	Aqua regia 3:1	130	54
Acetic acid 10%	300	149	Barium carbonate	280	138
Acetic acid 50%	300	149	Barium chloride	280	138
Acetic acid 80%	190	88	Barium hydroxide	280	138
Acetic acid, glacial	190	88	Barium sulfate	280	138
Acetic anhydride	100	38	Barium sulfide	280	138
Acetone	x	x	Benzaldehyde	120	49
Acetyl chloride	120	49	Benzene	150	66
Acrylic acid	150	66	Benzene sulfonic acid 10%	100	38
Acrylonitrile	130	54	Benzoic acid	250	121
Adipic acid	280	138	Benzyl alcohol	280	138
Allyl alcohol	200	93	Benzyl chloride	280	138
Allyl chloride	200	93	Borax	280	138
Alum	180	82	Boric acid	280	138
Aluminum acetate	250	121	Bromine gas, dry	210	99
Aluminum chloride, aqueous	300	149	Bromine gas, moist	210	99
Aluminum chloride, dry	270	132	Bromine liquid	140	60
Aluminum fluoride	300	149	Butadiene	280	138
Aluminum hydroxide	260	127	Butyl acetate	140	60
Aluminum nitrate	300	149	Butyl alcohol	280	138
Aluminum oxychloride	290	143	*n*-Butylamine	x	x
Aluminum sulfate	300	149	Butyl phthalate	80	27
Ammonia gas	270	132	Butyric acid	230	110
Ammonium bifluoride	250	121	Calcium bisulfide	280	138
Ammonium carbonate	280	138	Calcium bisulfite	280	138
Ammonium chloride 10%	280	138	Calcium carbonate	280	138
Ammonium chloride 50%	280	138	Calcium chlorate	280	138
Ammonium chloride, sat.	280	138	Calcium chloride	280	138
Ammonium fluoride 10%	280	138	Calcium hydroxide 10%	270	132
Ammonium fluoride 25%	280	138	Calcium hydroxide, sat.	280	138
Ammonium hydroxide 25%	280	138	Calcium hypochlorite	280	138
Ammonium hydroxide, sat.	280	138	Calcium nitrate	280	138
Ammonium nitrate	280	138	Calcium oxide	250	121
Ammonium persulfate	280	138	Calcium sulfate	280	138
Ammonium phosphate	280	138	Caprylic acid	220	104
Ammonium sulfate 10–40%	280	138	Carbon bisulfide	80	27
Ammonium sulfide	280	138	Carbon dioxide, dry	280	138
Ammonium sulfite	280	138	Carbon dioxide, wet	280	138
Amyl acetate	190	88	Carbon disulfide	80	27
Amyl alcohol	280	138	Carbon monoxide	280	138
Amyl chloride	280	138	Carbon tetrachloride	280	138
Aniline	200	93	Carbonic acid	280	138

Table 2-68 *Continued*

Chemical	Maximum temp. °F	Maximum temp. °C	Chemical	Maximum temp. °F	Maximum temp. °C
Cellosolve	280	138			
Chloracetic acid, 50% water	210	99	Lactic acid, concentrated	110	43
Chloracetic acid	200	93	Magnesium chloride	280	138
Chlorine gas, dry	210	99	Malic acid	250	121
Chlorine gas, wet, 10%	210	99	Manganese chloride	280	138
Chlorine, liquid	210	99	Methyl chloride	x	x
Chlorobenzene	220	104	Methyl ethyl ketone	x	x
Chloroform	250	121	Methyl isobutyl ketone	110	43
Chlorosulfonic acid	110	43	Muriatic acid	280	138
Chromic acid 10%	220	104	Nitric acid 5%	200	93
Chromic acid 50%	250	121	Nitric acid 20%	180	82
Chromyl chloride	110	43	Nitric acid 70%	120	49
Citric acid 15%	250	121	Nitric acid, anhydrous	150	66
Citric acid, concentrated	250	121	Nitrous acid, concentrated	210	99
Copper acetate	250	121	Oleum	x	x
Copper carbonate	250	121	Perchloric acid 10%	210	99
Copper chloride	280	138	Perchloric acid 70%	120	49
Copper cyanide	280	138	Phenol	200	93
Copper sulfate	280	138	Phosphoric acid 50–80%	220	104
Cresol	210	99	Picric acid	80	27
Cupric chloride 5%	270	132	Potassium bromide 30%	280	138
Cupric chloride 50%	270	132	Salicylic acid	220	104
Cyclohexane	250	121	Silver bromide 10%	250	121
Cyclohexanol	210	99	Sodium carbonate	280	138
Dichloroacetic acid	120	49	Sodium chloride	280	138
Dichloroethane (ethylene dichloride)	280	138	Sodium hydroxide 10%	230	110
			Sodium hydroxide 50%	220	104
Ethylene glycol	280	138	Sodium hydroxide, concentrated[a]	150	66
Ferric chloride	280	138			
Ferric chloride 50% in water	280	138	Sodium hypochlorite 20%	280	138
			Sodium hypochlorite, concentrated	280	138
Ferric nitrate 10–50%	280	138	Sodium sulfide to 50%	280	138
Ferrous chloride	280	138	Stannic chloride	280	138
Ferrous nitrate	280	138	Stannous chloride	280	138
Fluorine gas, dry	80	27	Sulfuric acid 10%	250	121
Fluorine gas, moist	80	27	Sulfuric acid 50%	220	104
Hydrobromic acid, dilute	260	127	Sulfuric acid 70%	220	104
Hydrobromic acid 20%	280	138	Sulfuric acid 90%	210	99
Hydrobromic acid 50%	280	138	Sulfuric acid 98%	140	60
Hydrochloric acid 20%	280	138	Sulfuric acid 100%	x	x
Hydrochloric acid 38%	280	138	Sulfuric acid, fuming	x	x
Hydrocyanic acid 10%	280	138	Sulfurous acid	220	104
Hydrofluoric acid 30%	260	127	Thionyl chloride	x	x
Hydrofluoric acid 70%	200	93	Toluene	x	x
Hydrofluoric acid 100%	200	93	Trichloroacetic acid	130	54
Hypochlorous acid	280	138	White liquor	80	27
Iodine solution 10%	250	121	Zinc chloride	260	127
Ketones, general	110	43			
Lactic acid 25%	130	54			

[a]Material subject to cracking.

Source: Schweitzer, Philip A. (1991). *Corrosion Resistance Tables*, Marcel Dekker, Inc., New York, Vols. 1 and 2.

Table 2-69 Physical and Mechanical Properties of Polyvinylidene Chloride

Specific gravity	1.75–1.85
Water absorption 24 hr at 73°F (23°C), %	nil
Tensile strength at 73°F (23°C), psi	2700–3700
Coefficient of thermal expansion	
in./in.-°F (°C) $\times$ 10^{-5}	3.9 to 5
in./10°F/100 ft	0.039 to 0.05
Thermal conductivity, Btu/hr/sq ft/°F/in.	1.28
Flame spread	self-extinguishing

Table 2-70 Dimensions of Schedule 80 Saran Pipe

Nominal pipe size (in.)	Outside diameter (in.)	Inside diameter (in.)	Wall thickness (in.)
½	0.840	0.546	0.147
¾	1.050	0.742	0.154
1	1.315	0.957	0.179
1 ¼	1.660	1.278	0.191
1 ½	1.900	1.500	0.200
2	2.375	1.939	0.218
2 ½	2.875	2.277	0.299
3	3.500	2.842	0.329
4	4.500	3.749	0.376
6	6.625	5.875	0.375

Table 2-71 Maximum Operating Pressures of Saran Pipe at Various Temperatures

Nominal pipe size (in.)	Maximum operating pressure (psi) at °F (°C):						
	50(10)	68(20)	77(25)	86(30)	104(40)	140(60)	176(80)
½	270	235	220	200	170	120	100
¾	230	200	190	180	150	110	80
1	210	180	170	155	145	100	70
1 ¼	180	155	140	135	110	80	60
1 ½	160	140	130	120	100	70	50
2	140	120	110	105	90	65	45
2 ½	125	110	105	100	80	60	40
3	115	100	90	85	70	50	35
4	100	85	80	70	60	40	30
6	80	70	65	60	50	30	20

Table 2-72 Support Spacing for
Saran Pipe at 77°F (25°C)

Nominal pipe size (in.)	Spacing (ft)
½	C
¾	C
1	2.0
1 ¼	2.0
1 ½	2.0
2	2.0
2 ½	2.0
3	2.0
4	2.5
6	3.0

C = Continuous support.
Based on uninsulated lines carrying fluid
having a specific gravity of 1.35.

Table 2-73 Compatibility of Polyvinylidene Chloride (Saran) with Selected Corrodents
The chemicals listed are in the pure state or in a saturated solution unless otherwise indicated.
Compatibility is shown to the maximum allowable temperature for which data are available.
Incompatibility is shown by an x. A blank space indicates that data are unavailable.

Chemical	Maximum temp. °F	°C	Chemical	Maximum temp. °F	°C
Acetaldehyde	150	66	Aluminum sulfate	180	82
Acetamide			Ammonia gas	x	x
Acetic acid 10%	150	66	Ammonium bifluoride	140	60
Acetic acid 50%	130	54	Ammonium carbonate	180	82
Acetic acid 80%	130	54	Ammonium chloride 10%		
Acetic acid, glacial	140	60	Ammonium chloride 50%		
Acetic anhydride	90	32	Ammonium chloride, sat.	160	71
Acetone	90	32	Ammonium fluoride 10%	90	32
Acetyl chloride	130	54	Ammonium fluoride 25%	90	32
Acrylic acid			Ammonium hydroxide 25%	x	x
Acrylonitrile	90	32	Ammonium hydroxide, sat.	x	x
Adipic acid	150	66	Ammonium nitrate	120	49
Allyl alcohol	80	27	Ammonium persulfate	90	32
Allyl chloride			Ammonium phosphate	150	66
Alum	180	82	Ammonium sulfate 10–40%	120	49
Aluminum acetate			Ammonium sulfide	80	27
Aluminum chloride, aqueous	150	66	Ammonium sulfite		
Aluminum chloride, dry			Amyl acetate	120	49
Aluminum fluoride	150	66	Amyl alcohol	150	66
Aluminum hydroxide	170	77	Amyl chloride	80	27
Aluminum nitrate	180	82	Aniline	x	x
Aluminum oxychloride	140	60	Antimony trichloride	150	66

Table 2-73 *Continued*

Chemical	Maximum temp.		Chemical	Maximum temp.	
	°F	°C		°F	°C
Aqua regia 3 : 1	120	49	Chlorine gas, wet	80	27
Barium carbonate	180	82	Chlorine, liquid	x	x
Barium chloride	180	82	Chlorobenzene	80	27
Barium hydroxide	180	82	Chloroform	x	x
Barium sulfate	180	82	Chlorosulfonic acid	x	x
Barium sulfide	150	66	Chromic acid 10%	180	82
Benzaldehyde	x	x	Chromic acid 50%	180	82
Benzene	x	x	Chromyl chloride		
Benzene sulfonic acid 10%	120	49	Citric acid 15%	180	82
Benzoic acid	120	49	Citric acid, concentrated	180	82
Benzyl alcohol			Copper acetate		
Benzyl chloride	80	27	Copper carbonate	180	82
Borax			Copper chloride	180	82
Boric acid	170	77	Copper cyanide	130	54
Bromine gas, dry			Copper sulfate	180	82
Bromine gas, moist			Cresol	150	66
Bromine liquid	x	x	Cupric chloride 5%	160	71
Butadiene	x	x	Cupric chloride 50%	170	77
Butyl acetate	120	49	Cyclohexane	120	49
Butyl alcohol	150	66	Cyclohexanol	90	32
n-Butylamine			Dichloroacetic acid	120	49
Butyl phthalate	180	82	Dichloroethane (ethylene di-	80	27
Butyric acid	80	27	chloride)		
Calcium bisulfide			Ethylene glycol	180	82
Calcium bisulfite	80	27	Ferric chloride	140	60
Calcium carbonate	180	82	Ferric chloride 50% in water	140	60
Calcium chlorate	160	71	Ferric nitrate 10–50%	130	54
Calcium chloride	180	82	Ferrous chloride	130	54
Calcium hydroxide 10%	160	71	Ferrous nitrate	80	27
Calcium hydroxide, sat.	180	82	Fluorine gas, dry	x	x
Calcium hypochlorite	120	49	Fluorine gas, moist	x	x
Calcium nitrate	150	66	Hydrobromic acid, dilute	120	49
Calcium oxide	180	82	Hydrobromic acid 20%	120	49
Calcium sulfate	180	82	Hydrobromic acid 50%	130	54
Caprylic acid	90	32	Hydrochloric acid 20%	180	82
Carbon bisulfide	90	32	Hydrochloric acid 38%	180	82
Carbon dioxide, dry	180	82	Hydrocyanic acid 10%	120	49
Carbon dioxide, wet	80	27	Hydrofluoric acid 30%	160	71
Carbon disulfide	80	27	Hydrofluoric acid 70%		
Carbon monoxide	180	82	Hydrofluoric acid 100%	x	x
Carbon tetrachloride	140	60	Hypochlorous acid	120	49
Carbonic acid	180	82	Iodine solution 10%		
Cellosolve	80	27	Ketones, general	90	32
Chloracetic acid, 50% water	120	49	Lactic acid 25%		
Chloracetic acid	120	49	Lactic acid, concentrated	80	27
Chlorine gas, dry	80	27	Magnesium chloride	180	82

Table 2-73 *Continued*

Chemical	°F	°C	Chemical	°F	°C
Malic acid	80	27	Sodium hydroxide 50%	150	66
Manganese chloride			Sodium hydroxide, con-	x	x
Methyl chloride	80	27	centrated		
Methyl ethyl ketone	x	x	Sodium hypochlorite 10%	130	54
Methyl isobutyl ketone	80	27	Sodium hypochlorite, con-	120	49
Muriatic acid	180	82	centrated		
Nitric acid 5%	90	32	Sodium sulfide to 50%	140	60
Nitric acid 20%	150	66	Stannic chloride	180	82
Nitric acid 70%	x	x	Stannous chloride	180	82
Nitric acid, anhydrous	x	x	Sulfuric acid 10%	120	49
Nitrous acid, concentrated			Sulfuric acid 50%	x	x
Oleum	x	x	Sulfuric acid 70%	x	x
Perchloric acid 10%	130	54	Sulfuric acid 90%	x	x
Perchloric acid 70%	120	49	Sulfuric acid 98%	x	x
Phenol	x	x	Sulfuric acid 100%	x	x
Phosphoric acid 50–80%	130	54	Sulfuric acid, fuming	x	x
Picric acid	120	49	Sulfurous acid	80	27
Potassium bromide 30%	110	43	Thionyl chloride	x	x
Salicylic acid	130	54	Toluene	80	27
Silver bromide 10%			Trichloroacetic acid	80	27
Sodium carbonate	180	82	White liquor		
Sodium chloride	180	82	Zinc chloride	170	77
Sodium hydroxide 10%	90	32			

Source: Schweitzer, Philip A. (1991). *Corrosion Resistance Tables*, Marcel Dekker, Inc., New York, Vols. 1 and 2.

Table 2-74 Physical and Mechanical Properties of Chem-Aire ABS Pipe

Specific gravity	1.03
Water absorption 24 hr at 73°F (23°C), %	0.2–0.4
Tensile strength at 73°F (23°C), psi	5,350
Modulus of elasticity in tension at 73°F (23°C) $\times 10^5$	2.4
Flexural strength, psi	9400
Izod impact strength, notched at 73°F (23°C)	8.5
Coefficient of thermal expansion	
in./in.-°F $\times 10^{-5}$	5.6
in./10°F/100 ft	0.056
Thermal conductivity, Btu/hr/sq ft/°F/in.	1.7
Heat distortion temperature at 66 psi, °F/°C	204/94
Resistance to heat at continuous drainage, °F/°C	140/60
Limiting oxygen index, %	19
Flame spread	not applicable
Underwriters lab rating (Sub 94)	94 HB

Table 2-75 Pressure Rating of Chem-Aire ABS Pipe at Various Temperatures

Temperature (°F/°C)	Maximum pressure rating (psi)
100/38	185
110/43	185
115/46	171
120/49	157
125/52	143
130/54	128
135/57	114
140/60	100

Table 2-76 Support Spacing for Chem-Aire ABS Pipe[a]

Nominal pipe size (in.)	Support spacing (ft)		
	Dry	Wet	
		Sp. Gr. 1	Sp. Gr. 1.35
½	4.0	3.3	3.0
¾	4.6	3.8	3.2
1	5.0	4.2	4.0
1 ¼	5.7	4.8	4.5
1 ½	6.0	5.0	4.8
2	6.8	5.7	5.4

[a]Based on lines being uninsulated and all accessory equipment being independently supported.

Table 2-77 Compatibility of Chem-Aire ABS with Selected Corrodents.
The chemicals listed are in the pure state or in a saturated solution unless otherwise indicated. Compatibility is shown to the maximum allowable temperature for which data are available. Incompatibility is shown by an x. A blank space indicates that data are unavailable.

Chemical	Maximum temp.		Chemical	Maximum temp.	
	°F	°C		°F	°C
Acetaldehyde	x	x	Adipic acid	140	60
Acetamide			Allyl alcohol	x	x
Acetic acid 10%	100	38	Allyl chloride	x	x
Acetic acid 50%	130	54	Alum	140	60
Acetic acid 80%	x	x	Aluminum acetate		
Acetic acid, glacial	x	x	Aluminum chloride, aqueous	140	60
Acetic anhydride	x	x	Aluminum chloride, dry		
Acetone	x	x	Aluminum fluoride	140	60
Acetyl chloride	x	x	Aluminum hydroxide	140	60
Acrylic acid			Aluminum nitrate		
Acrylonitrile			Aluminum oxychloride	140	60

Table 2-77   *Continued*

Chemical	Maximum temp. °F	°C	Chemical	Maximum temp. °F	°C
Aluminum sulfate	140	60	Calcium carbonate	100	38
Ammonia gas dry	140	60	Calcium chlorate	140	60
Ammonium bifluoride	140	60	Calcium chloride	140	60
Ammonium carbonate	140	60	Calcium hydroxide 10%		
Ammonium chloride 10%			Calcium hydroxide, sat.	140	60
Ammonium chloride 50%			Calcium hypochlorite	140	60
Ammonium chloride, sat.	140	60	Calcium nitrate	140	60
Ammonium fluoride 10%	x	x	Calcium oxide	140	60
Ammonium fluoride 25%	x	x	Calcium sulfate 25%	140	60
Ammonium hydroxide 25%	90	32	Caprylic acid		
Ammonium hydroxide, sat.	80	27	Carbon bisulfide	x	x
Ammonium nitrate	140	60	Carbon dioxide, dry	90	32
Ammonium persulfate	140	60	Carbon dioxide, wet	140	60
Ammonium phosphate	140	60	Carbon disulfide	x	x
Ammonium sulfate 10–40%	140	60	Carbon monoxide	140	60
Ammonium sulfide	140	60	Carbon tetrachloride	x	x
Ammonium sulfite			Carbonic acid	140	60
Amyl acetate	x	x	Cellosolve	x	x
Amyl alcohol	80	27	Chloracetic acid, 50% water		
Amyl chloride	x	x	Chloracetic acid	x	x
Aniline	x	x	Chlorine gas, dry	140	60
Antimony trichloride	140	60	Chlorine gas, wet	140	60
Aqua regia 3:1	x	x	Chlorine, liquid	x	x
Barium carbonate	140	60	Chlorobenzene	x	x
Barium chloride	140	60	Chloroform	x	x
Barium hydroxide	140	60	Chlorosulfonic acid	x	x
Barium sulfate	140	60	Chromic acid 10%	90	32
Barium sulfide	140	60	Chromic acid 50%	x	x
Benzaldehyde	x	x	Chromyl chloride		
Benzene	x	x	Citric acid 15%	140	60
Benzene sulfonic acid 10%	80	27	Citric acid, 25%	140	60
Benzoic acid	140	60	Copper acetate		
Benzyl alcohol	x	x	Copper carbonate		
Benzyl chloride	x	x	Copper chloride	140	60
Borax	140	60	Copper cyanide	140	60
Boric acid	140	60	Copper sulfate	140	60
Bromine gas, dry			Cresol	x	x
Bromine gas, moist			Cupric chloride 5%		
Bromine liquid	x	x	Cupric chloride 50%		
Butadiene	x	x	Cyclohexane	80	27
Butyl acetate	x	x	Cyclohexanol	80	27
Butyl alcohol	x	x	Dichloroacetic acid	x	x
n-Butylamine			Dichloroethane (ethylene di-chloride)	x	x
Butyl phthalate					
Butyric acid	x	x	Ethylene glycol	140	60
Calcium bisulfide			Ferric chloride	140	60
Calcium bisulfite	140	60	Ferric chloride 50% in water		

Table 2-77 *Continued*

Chemical	Maximum temp. °F	Maximum temp. °C	Chemical	Maximum temp. °F	Maximum temp. °C
Ferric nitrate 10–50%	140	60	Perchloric acid 70%	x	x
Ferrous chloride	140	60	Phenol	x	x
Ferrous nitrate			Phosphoric acid 50–80%	130	54
Fluorine gas, dry	90	32	Picric acid	x	x
Fluorine gas, moist			Potassium bromide 30%	140	60
Hydrobromic acid, dilute			Salicylic acid		
Hydrobromic acid 20%	140	60	Silver bromide 10%		
Hydrobromic acid 50%			Sodium carbonate	140	60
Hydrochloric acid 20%	90	32	Sodium chloride	140	60
Hydrochloric acid 38%	140	60	Sodium hydroxide 10%	140	60
Hydrocyanic acid 10%			Sodium hydroxide 50%	140	60
Hydrofluoric acid 30%	x	x	Sodium hydroxide, con-	140	60
Hydrofluoric acid 70%	x	x	centrated		
Hydrofluoric acid 100%	x	x	Sodium hypochlorite 20%	140	60
Hypochlorous acid	140	60	Sodium hypochlorite, con-	140	60
Iodine solution 10%			centrated		
Ketones, general	x	x	Sodium sulfide to 50%	140	60
Lactic acid 25%	140	60	Stannic chloride	140	60
Lactic acid, concentrated			Stannous chloride	100	38
Magnesium chloride	140	60	Sulfuric acid 10%	140	60
Malic acid	140	60	Sulfuric acid 50%	130	54
Manganese chloride			Sulfuric acid 70%	x	x
Methyl chloride	x	x	Sulfuric acid 90%	x	x
Methyl ethyl ketone	x	x	Sulfuric acid 98%	x	x
Methyl isobutyl ketone	x	x	Sulfuric acid 100%	x	x
Muriatic acid	140	60	Sulfuric acid, fuming	x	x
Nitric acid 5%	140	60	Sulfurous acid	140	60
Nitric acid 20%	130	54	Thionyl chloride	x	x
Nitric acid 70%	x	x	Toluene	x	x
Nitric acid, anhydrous	x	x	Trichloroacetic acid		
Nitrous acid, concentrated			White liquor	140	60
Oleum	x	x	Zinc chloride	140	60
Perchloric acid 10%	x	x			

Source: Schweitzer, Philip A. (1991). *Corrosion Resistance Tables*, Marcel Dekker, Inc., New York, Vols. 1 and 2.

3

Thermoset Plastic Piping Systems

3.1 DATA AND DESIGN INFORMATION

Thermoset piping materials are the result of a careful selection of base resin backbone, reactive end group, catalyst, accelerator, fillers, and other additives. By varying these ingredients, specific properties can be imparted to the finished product. For thermoset piping systems, the four primary resin groups used are vinyl esters, unsaturated polyesters, epoxies, and furans.

Vinyl esters have a wide range of corrosion resistance, particularly to strong corrosive acids, bases, and salt solutions, up to temperatures of 200°F (93°C). These are produced from the reaction of epoxy resins with ethylene-unsaturated carboxylic acids. The epoxy resin backbone imparts toughness and superior tensile elongation properties. The vinyl ester generally used for chemical process piping systems is derived from a diglycidyl ether of bisphenol-A epoxy resin with the addition of methacrylate end groups to provide maximum corrosion resistance.

Unsaturated polyesters are alkyd thermosetting resins with vinyl unsaturation on the polyester backbone. Because they are simple, versatile, and economical, they are the most widely used resin family. They can be compounded to resist most chemicals at temperatures of 75°F (24°C) and many chemicals to 160°F (71°C) or higher.

As with the vinyl esters, the unsaturated polyesters can be formulated to yield specific properties. A typical unsaturated polyester resin includes a base polyester resin, reactive diluents, catalyst, accelerator, fillers and inhibitors. Fillers are selected to improve flame retardancy, shrinkage control, and impact resistance. Catalyst systems are formulated to extend resin shelf life and control reaction and cure times.

Epoxies are thermosetting matrix resins that cure to cross-linked, insoluble, infusible matrix resins with or without the addition of heat. These resins are specified when continuous operation at elevated temperatures [up to 437°F (225°C)] is anticipated. They can be specially formulated through the selection of a wide variety of base resins, curing agents, catalysts, and additives to meet specific applications.

The furan polymer is a derivative of furfuryl alcohol and furfural. Although these resins cost approximately 30% more than other thermosetting resins, in many cases they are the most economical choice when:

1. Solvents are present in a combination of acids and bases
2. They are an alternate choice to high nickel alloys
3. Process changes may occur that would result in exposure of solvents in oxidizing atmospheres

The greatest single advantage of the furan resins is their extremely good resistance to solvents in combination with acids and bases.

3.1.1   Reinforcing Materials

Because of its low cost, high tensile and impact strength, light weight, and good corrosion resistance, fiberglass is the most widely used reinforcement. The predominant fiberglass used is E-glass (aluminum borosilicate), while S-glass (magnesium aluminoborosilicate) is selected for higher tensile strength, modulus, and temperature requirements. For extremely corrosive applications, C or ECR glass is used.

Although resistant to most chemicals, fiberglass can be attacked by alkalies and a few acids. To overcome this, synthetic veils, most commonly thermoplastic or polyester, are used. For increased resistance to chemical attack a veil of C-glass may be specified.

When chemical resistance to hydrofluoric acid is required, graphite (carbon) fibers are supplied. These fibers also possess high tensile strengths and moduli, low density, and excellent fatigue and creep resistance and can be engineered to yield an almost zero coefficient of thermal expansion. These fibers will also impart some degree of electrical conductivity to the pipe.

Other fibers are finding increasing applications. Among these are aramid, polyester, and oriented polyethylene. Although these fibers have excellent corrosion resistance and high strength-to-weight ratios, they have some drawbacks. The polyethylene has reduced fiber-resin adhesion properties, and the aramid has much lower compressive properties than tensile properties.

Extremely effective is a synthetic nexus surfacing veil. This material has been used in particularly severe applications.

3.1.2   Working Pressure

Thermoset piping systems are not manufactured to a specific standard, therefore there is no generalization as to allowable operating pressures. Some manufacturers supply their piping systems to a true inside diameter, for example, a 1-inch pipe would have a 1-inch inside diameter. Other manufacturers furnish pipe with outside diameters equal to IPS diameters and varying inside diameters depending upon pressure ratings. Therefore, each manufacturer's system must be treated independently as to operating pressures, operating temperatures, and dimensions.

3.1.3   Burial of Thermoset Pipe

This section deals with the design and installation of fiberglass-reinforced polyester, epoxy ester, and vinyl ester piping. Formulas are provided for the calculation of pipe stiffness and allowable deflection. For operating pressures below 100 psi, normally the vertical pressure on the pipe from ground cover and vehicle line loading will dictate the required wall thickness. At operating pressures above 100 psi, the internal pressure usually dictates the required wall thickness. In all cases the wall thickness should be checked for both conditions.

Soil conditions should be checked prior to design since these conditions affect the design calculations. Once these are determined the design calculations can proceed.

The minimum pipe stiffness can be calculated using the following equation:

$$PS = \frac{E_F I}{0.149 \, r_M^3}$$

where:

PS = pipe stiffness (psi)
E_F = hoop flexural modulus of elasticity (psi)
  I = moment of inertia of structural wall (in.^{4}/in. = $t^{-3}/12$, where t = wall thickness)
r_M = mean radius of wall = $D_M/2$ where D_M = mean diameter of wall in.

Table 3-1 shows the minimum stiffness requirements by pipe diameter. If the PS calculated is less than that shown in Table 3-1, a thicker pipe wall must be used. The wall thickness must also be checked to be sure that it is adequate for the internal pressure.

The allowable deflection is calculated by:

$$\Delta Y = \frac{[DW_C + W_L] \, K_X \, r^3}{E_F I + 0.061 \, K_a E' r^3} + a$$

where:

ΔY = vertical pipe deflection (in.) (ΔY should not exceed a value of 0.05 D_M; Y = $D_M/2$.)
  D = deflection lag factor (see Table 2-8)
W_C = vertical soil load
r_M = mean radius wall (in.)
K_X = bedding factor (see Table 2-9)
$E_F I$ = stiffness factor of structural wall (in.^{2} lb/in.)
K_a = deflection coefficient (see Table 3-2)
E' = soil modulus (see Table 2-6)
  a = deflection coefficient (see Table 3-2)

and

$$W_C = \frac{Y_S \, HD_o}{144}$$

where:

Y_S = specific weight of soil (lb/ft^3) (If a specific weight of soil is unknown, use a soil density of 120 lb/ft^3.)
  H = burial depth to top of pipe (ft)
D_o = outside diameter of pipe
W_E = live load on pipe (lb in.)

and

$$W_L = \frac{C_L P (1 + I_F)}{12}$$

where:

C_L = live load coefficient (see Table 3-3A and 3-3B)
P = wheel load (lb)
L_F = impact factor = $0.766 - 0.133H$ $(0 \leq I_F \leq 0.50)$

The trench should be excavated to a depth of approximately 12 inches below the pipe grade, which will permit a 6-inch foundation and a 6-inch layer of pipe bedding material. The bottom of the trench should be as uniform and continuous as possible. High spots in the trench will cause uneven bearing on the pipe, which will cause stress on the pipe during backfill and unnecessary wear at these points. Sharp bends and changes in elevation of the line should be avoided.

All backfill material should be free of stones above ¾ inch in size, vegetation, and hard clods of earth. An ideal type of backfill is pea gravel or crushed stone corresponding to ASTM C33 graduation 67 (grain size ¾–³⁄₁₆ inch).

Pea gravel has some advantages: When poured directly into the trench it will compact to 90% or more of its maximum density, eliminating the need for equipment and labor for compaction. The pea gravel will not retain water, eliminating the problem of clumping. Above the 70% height of the pipe, good native soil, free of clods, stones over ¾ inch, organic matter, and other foreign material can be used. This layer should be placed 6–16 inches above the pipe using any standard compaction method other than hydraulic compaction. Above this level the trench can be backfilled without compaction as long as there are no voids present. The backfilling should be done as soon as possible after testing of the pipe to eliminate the possibility of damage to the pipe, floating of the pipe due to flooding, and shifting of the line due to cave-ins.

If the pipe is to be laid under a road crossing, it is good practice to lay the pipe in a conduit, taking care to see that the pipe is properly bedded at the entrance and exit of the conduit. If not properly bedded, excessive wear and/or stress can be imposed on the pipe.

3.1.4 Supporting Thermoset Pipe

The basic principles used for the support of metallic pipe can be followed for thermoset pipe, with a few modifications. In general, follow the support requirements given for thermoplastic pipe systems starting on page 29. At all anchor points the pipe wall should be built up to provide a sleeve on either side of the anchor with a thickness at least equal to that of the pipe wall (see Fig. 3-1).

Do not exceed the recommended support spacing given for each specific piping system. All valves, instruments, or other accessories in the piping system are to be independently supported. Supports should be installed on either side of a flanged connector regardless of support spacing requirements.

3.1.5 Joining of Thermoset Pipe

As with thermoplastic piping systems, there are several methods available for joining thermoset piping systems. If properly made, the joint will be as strong as the pipe itself, but when improperly made it becomes the weak link in the system. The methods of joining are:

1. butt and strap joint
2. socket-type adhesive joint
3. flanged joints
4. bell and spigot joint
5. threaded joint

Butt and Strap Joint

This joint is made by butting together two sections of pipe and/or fittings and overwrapping the joint with successive layers of resin impregnated mat or mat and roving (see Fig. 3-2). It is the standard method used with polyester and vinyl ester piping systems, and is so stated in the commercial standards, which stipulate that all pipe 20 inches in diameter and larger shall be overlayed both inside and outside. Pipe having diameters of less than 20 inches shall be overlaid on the outside only.

The butt joint provides a mechanical or adhesive bond and not a chemical bond, therefore it is important that the surface be free from any contamination. In addition, the width of the strapping materials is also important since they must be long enough to withstand the shear stress of the pipe. As the strapping material shrinks around the pipe during the cure, it forms a very tight joint.

The width of the strapping material will vary as successive layers are applied. It is important that each layer be long enough to completely surround the pipe and provide approximately a 2-inch overlap. Table 3-4 indicates the material requirements for a butt and strap joint on 50 psi rated pipe, while Table 3-5 provides the same information for 100 psi rated pipe and Table 3-6 for 150 psi rated pipe.

Detailed procedures for making a butt and strap joint can be obtained from any reliable vendor of the pipe. Listed below are the general steps that must be followed:

1. Clean all pipe surfaces in the areas of the joint.
2. Roughen the surface of the pipe or fitting with a file or sander.
3. Coat all raw edges of the pipe with resin to prevent penetration by the fluid to be handled.
4. Cut all ends of pipe straight.
5. Align the two edges of the pipe and check to see that they are square.

Cure time for each joint is approximately 45 minutes.

Socket-Type Adhesive Joint

These joints are used with polyester and vinyl ester piping systems as well as reinforced epoxy systems. The joint is made by utilizing a coupling into which the fitting end or pipe end is placed and cemented. Figure 3-3 illustrates a typical joint.

The adhesive for these joints is furnished in either liquid or paste form in two parts—the cement and the catalyst—which must be mixed together. Since the adhesive only has a pot life of 15–30 minutes, the pipe and fitting surfaces should be cleaned and ready to be joined before the adhesive is mixed. The general steps to follow are:

1. Prepare the pipe and fittings by cleaning and sanding.
2. Mix the cement and catalyst.
3. Apply the prepared adhesive to the pipe and fitting to be joined.
4. Assemble the fitting to the pipe and rotate 180° to distribute the cement and to eliminate air pockets.
5. Wipe a fillet of cement around the fitting and remove any excess.

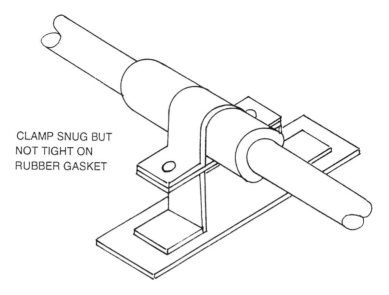

CLAMP SNUG BUT
NOT TIGHT ON
RUBBER GASKET

Figure 3-1   Build-up of pipe wall at anchor point.

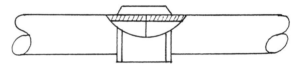

Figure 3-2   Butt and strap joint.

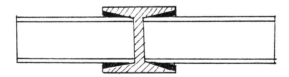

Figure 3-3   Socket-type adhesive joint.

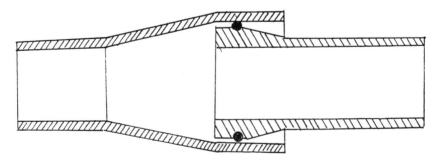

Figure 3-4   Bell and spigot O-ring joint.

6. To speed up the hardening of an epoxy cemented joint, apply heat but do not exceed 180°F (82°C). Heat must be applied slowly. Hold the joint at this temperature for 10 minutes or until the adhesive fillet is no longer tacky, then slowly increase the temperature to 200–230°F (93–110°C) for 10 minutes. If bubbling occurs at the edges of the joint, remove the heat for a few minutes, then reapply.
7. Let the joint cool to 100°F (38°C) before handling.

The application of heat in no way affects the strength or corrosion resistance—it only shortens the curing time. It is not necessary to apply heat to effect a cure. A hot air gun can be used to apply the heat.

Flanged Joints

Thermoset pipe systems can be joined by means of flanged joints, however, a complete system is not normally flanged. Flanges are used when connecting to equipment, e.g., pumps, tanks, etc., or accessory items such as valves or instruments. They are also used when sections of the piping system may have to be removed.

Flanges are generally provided press molded or hand-laid-up stub ends. The press-molded flanges can be furnished with a short length of pipe as a stub end.

Full-face gaskets having a Shore durometer of 50 to 70 should be used. Table 3-7 shows the dimensions of full-face gaskets. Care must be taken in the selection of the gasket material to ensure its compatibility with the material being handled. (Table 1-1 provides the compatibility of selected elastomeric gasket materials with a variety of corrodents.) When connecting flanged joints, a washer should be placed under both the nut and bolt head. Nuts and bolts must also be selected to resist any atmospheric corrosion that may be present.

In order to effect a liquid-tight seal, all bolts should be tightened to the proper torque. Maximum torque ratings are given in Table 3-8. The bolts should be gradually tightened in an alternating 180° pattern so that the maximum pressure is not achieved on one bolt while others have little or no pressure applied.

Bell and Spigot O-Ring Joints

The bell and spigot joint is widely used for buried lines and long straight runs. Sealing of the joint is effected by means of an elastomeric O-ring. Care must be taken in the selection of the O-ring to ensure that it is compatible to the fluid being handled. (Table 1-1 provides the fluid compatibility of various types of elastomeric materials in contact with a variety of corrodents.) Figure 3-4 illustrates a typical joint.

The advantages of this type of joint lie in its ease and speed of assembly regardless of weather conditions, which results in the lowest installation cost for long straight runs. By virtue of the nature of the joint, several degrees of misalignment can be tolerated. The joint also will act as an expansion joint to a limited degree. If installed above ground, provision must be made to restrain thrust.

3.1.6   Fire Hazards

Potential fire hazards should also be considered when selecting the specific thermoset to be used. Although the fluid being conveyed may not be combustible, there is always the potential danger of an external fire.

In the Physical and Mechanical Properties Table found under the heading of each specific resin system are two entries relating to this topic. The first is the limiting oxygen index percent. This is a measure of the minimum oxygen level required to support combustion of the thermoset resin. The second is the flame spread classification. These ratings are based on common tests as outlined by the Underwriters Laboratories and are defined as follows:

Flame Spread Rating	Classification
0–25	noncombustible
25–50	fire retardant
50–200	combustible
over 200	highly combustible

3.2   REINFORCED POLYESTER PIPING SYSTEMS

Polyester does not describe a specific resin, but rather a family of resins. There are three basic types of polyester resins which are used for corrosion-resistant piping. With each of these types there are subtypes.

The isophthalic polyesters are the least expensive of the resins used for piping systems. They have a maximum operating temperature in the range of 140–160°F (60–71°C). These resins have FDA approval for use in the repeated handling of foodstuffs and potable water.

Among the first high-performance polyester resins were the bisphenol polyesters, which have been in service for 25–30 years. These resins have better corrosion resistance than the isophthalic resins and cost approximately 30% more. However, only a small fraction of this cost carries over into the final product. There is a bisphenol A-fumarate and a hydrogenated bisphenol A–bisphenol A. These resins have an operating temperature range of 180–280°F (82–138°C) depending upon the specific formulation.

A third category of polyester resins are the halogenated series—either chlorinated or brominated. These resins can be used at temperatures up to 250°F (121°C). Their cost is approximately the same as the bisphenol polyesters. The chlorinated polyesters have a very high heat distortion point, and the laminates show a very high retention of physical strength at elevated temperatures. These resins are inherently fire retardant. If antimony trioxide is added to these resins, they will burn with extreme difficulty. The other polyesters will support combustion, but burn slowly.

Table 3-9 lists some of the average physical and mechanical properties of the polyester resins. Since specific physical properties are determined by the type of laminate construction, resin selection, selection of reinforcing material, and ratio of resin to reinforcing material, the values shown in Table 3-9 can only be averages and ranges, since these factors will vary between manufacturers.

3.2.1   Pipe Data

Reinforced polyester pipe systems are available in diameters of 1 inch to 24 inches, with larger diameters being available on special order. Standard lengths are 10 and 20 feet. Unlike other piping systems, polyester pipe systems have not been standardized dimensionally. Some manufacturers designate their pipe with a nominal pipe size. For example, a nominal 6-inch-diameter pipe would have an inside diameter of 6.129–6.295 inches depending upon the pressure rating. Other manufacturers would supply a pipe with a true 6-inch inside diameter. Usually, pipe manufactured by the filament-wound process will have a true inside diameter, while pipe manufactured by the centrifugally cast process will have a constant outside diameter and a varying inside diameter.

Table 3-10 shows the minimum wall thicknesses for reinforced polyester pipe at various operating pressures, and Table 3-12 shows the external collapsing pressure of reinforced polyester pipe. This latter set of data is of particular importance when the piping is to be used under negative pressure, such as in vacuum systems or as pump sections.

Piping is usually furnished as a standard with pressure ratings of 50, 100, and 150 psig. Other pressure ratings are easily obtained because of the method of manufacture. Table 3-11 shows the minimum flange thicknesses required for reinforced polyester pipe at various operating pressures.

The life of the pipe in service is affected by more than the type of resin and type of reinforcement. Equally important is the method of manufacture employed, the resin cure, and the reliability and experience of the manufacturer. If the piping is purchased from a reliable and experienced polyester pipe manufacturer, these conditions should be met. In order to guarantee good performance it must be determined that the material being used and the method of manufacture are satisfactory for the specific application.

3.2.2 Joining, Installation, and Support

Reinforced polyester pipe can be joined by any of the methods given on pages 98–101. Of the methods given, the butt and strap technique is the most widely used. Pipe less than 20 inches in diameter is overlayed on the outside only, while pipe exceeding 20 inches in diameter should be overlaid both inside and outside.

Installation should follow the general principles for thermoplastic pipe given on pages 24–32.

Support of the piping is critical, as is the design of the pipe hangers. It is important that there be no point loading on the piping, therefore a relatively wide hanger with a ⅛-inch-thick rubber or elastomeric cushion should be used. Table 3-13 provides suggested hanger widths for both hand-laid-up and filament wound pipe. Support spacing is given in Table 3-14.

3.2.3 Corrosion Resistance

Isophthalic esters show good resistance to a limited range of solvents, including gasoline, kerosene, and naphtha. They are also capable of handling low concentrations of acids and exhibit a broad range of resistance to inorganic salts. Table 3-15 provides the compatibility of isophthalic polyesters with selected corrodents. These resins are unsatisfactory with materials such as acetone, high concentrations of oxidizing acids, alkaline salts of potassium and sodium, benzene, and carbon disulfide.

The bisphenol resins show good performance with moderate alkaline solutions and excellent resistance to bleaching agents. They can also handle a broader range of acids than the isophthalic esters. These resins have a limited range of application with solvents. Higher concentrations of acids are also not recommended. Table 3-16 and Table 3-17 list the compatibility of two of the bisphenol esters with selected solvents.

The halogenated polyesters (chlorinated or brominated) have unique chemical-resistant properties. They are resistant to many of the oxidizing acids and solutions, straight chain solvents of the aliphatic group, and many gases, both wet and dry, at elevated temperatures. Table 3-18 provides the compatibility of halogenated polyesters with selected corrodents.

These compatibility tables provide the corrosion resistance of the polyester resin. Except in the case of hydrofluoric acid and fluoride compounds, glass mat can be used as a reinforcing agent. Carbon fibers or some other resistant reinforcing material must be used for these applications.

In addition, these are general compilations. Since each manufacturer's formulation may change slightly, their recommendations should be requested for any specific application.

3.3 REINFORCED VINYL ESTER PIPING SYSTEMS

The vinyl ester resins became commercially available in the mid-1960s. There are several basic advantages to the vinyl esters.

1. The basic structure of the vinyl ester molecule is such that it is more resistant to some types of chemical attack, such as hydrolysis and oxidation or halogenation.
2. The vinyl esters have better impact resistance and greater tolerance to temperature and pressure fluctuations and mechanical shock than the polyesters.
3. Laminate tests have shown that the vinyl ester resins have strengths somewhat higher than the polyesters but not equal to the heat-cured epoxies.
4. Because of their molecular structure, they cure rapidly and give high early strength and superior creep resistance.
5. They provide excellent fiber wet-out and good adhesion to the glass fiber, in many cases similar to the amine-cured epoxies but less than the heat-cured epoxies.

At elevated temperatures the flexural modulus must be checked, since at 225°F (107°C) the vinyl esters have lost up to half of their flexural strength. This is an important consideration in negative pressure applications.

In general this family of resins possess good physical and mechanical properties, as can be seen from Table 3-19. The values shown in the table are averages since each manufacturer's resin may exhibit different properties.

3.3.1 Pipe Data

Vinyl ester piping systems vary in allowable operating pressures and maximum allowable operating temperatures from manufacturer to manufacturer. This is due to the particular resin used and the method of manufacture. Because of this it is necessary to check with the supplier whose pipe is to be used as to the allowable design conditions.

The Fibercast Company produces a pipe in sizes of 1½ through 14 inches with a maximum allowable operating temperature of 175°F (80°C). Table 3-20 gives the properties of Cl-2030 pipe, including the recommended operating external pressure at 75°F (24°C). As the operating temperature increases, the allowable external pressure decreases (see Table 3-21). Standard fittings to be used with the Cl-2030 pipe have a somewhat lower pressure rating than the pipe (see Table 3-22).

Smith Fiberglass Products produces a line of vinyl ester piping in sizes 2 through 8 inches rated for 150 psi at 225°F (107°C). There are numerous other manufacturers, but these two examples have been chosen to illustrate the differences between manufacturers.

3.3.2   Joining, Installation, and Support

Reinforced vinyl ester piping can be joined by butt and strap joints, socket-type adhesive joints, flanged joints, or bell and spigot joints (see pages 98–101). The specific joining method to be used will depend upon the manufacturer. It is important that the manufacturer's recommendations be strictly adhered to.

Installation of reinforced vinyl ester piping should be done as described on pages 24–32 for thermoplastic pipe. These general principles do not vary between manufacturers.

Support spacing for reinforced vinyl ester piping will depend upon the design of the pipe, which means that it will not be the same for all manufacturers. For example, Table 3-23 specifies the support spacing for Fibercast/s Cl-2030 pipe at various temperatures when handling a fluid having a specific gravity of 1.0. Table 3-24 provides the correction factors when handling fluids of other specific gravities. The support spacing requirements for pipe manufactured by Reinforced Plastic Systems, Inc. are given in Table 3-25.

Regardless of the manufacturer, all valves and other accessories installed in the piping system are to be individually supported. If flanged connections are used, a support should be installed on either side of the flanged joint.

Pipe hanger designs should be the same as those for reinforced polyester pipe and sized as given in Table 3-13. These piping systems may also be buried.

3.3.3   Corrosion Resistance

The specific corrosion resistance of a reinforced vinyl ester piping system will be dependent upon the resin used, the reinforcing material, and the method of manufacture. As a result, there can be a difference in the compatibility of a specific piping system with a specific corrodent. Therefore the manufacturer should be asked for recommendations about the corrodent to be handled.

If the piping system is to be exposed to sunlight, a UV inhibitor for a protection against "glass blooming" caused by ultraviolet radiation from sunlight should be incorporated with the resin.

In general vinyl esters can be used to handle most hot, highly chlorinated, and acidic mixtures at elevated temperatures. They also provide excellent resistance to strong mineral acids and bleaching solutions. The vinyl esters excel in alkaline and bleach environments and are extensively used in the very corrosive conditions found in the pulp and paper industry. Table 3-26 lists the compatibility of vinyl ester with selected corrodents. Remember, a satisfactory rating shown in the table does not mean that all vinyl esters will be satisfactory, but that there is at least one resin that is capable of handling the corrodent.

3.4   REINFORCED EPOXY PIPING SYSTEMS

The family of epoxy resins dominated the pipe market for many years, until the introduction of the vinyl esters. They are still a major factor. In small-diameter high-pressure pipe (through 12 inches) the epoxy resins predominate, although vinyl esters are available. The epoxy resins can provide outstanding service in the chemical processing industry. Virtually all pipe is produced by the heat-cured process, which uses aromatic amines and acid anhydrides, in order to obtain the best physical properties.

There are many epoxy formulations. A formulation can be tailor-made to meet a specific corrosive and/or mechanical requirement. Because of this, the values shown in Table 3-27 are average physical and mechanical properties of reinforced epoxy pipe.

The heat-cured epoxies have higher strengths than the vinyl esters and provide a superior fiber wet-out and adhesion to the glass fiber than the vinyl esters.

3.4.1   Pipe Data

Reinforced epoxy pipe systems are not standardized as to dimensions and/or pressure ratings. They vary depending upon the resin used, the reinforcing material, and the manufacturing techniques employed. The Fibercast Company produces an RB-2530 pipe system in sizes 1 inch through 14 inches, which is nominally rated up to 450 psi in the smaller sizes up to 250°F (121°C). Table 3-28 lists the recommended operating data for the pipe, and Table 3-29 provides the maximum allowable pressure ratings for the epoxy fittings. Since these ratings are lower than the ratings for straight pipe, the fitting ratings will control the overall rating of the system.

Table 3-30 supplies the maximum allowable external pressure for Fibercast RB-2530 epoxy pipe systems at different temperatures.

Reinforced Plastic Systems, Inc. manufactures two epoxy piping systems designated Techstrand 1000 and Techstrand 2000. The Techstrand 1000 system is available in sizes 2 through 16 inches and is rated for 150 psi at 210°F (99°C). Table 3-31 lists the pressure ratings of the Techstrand 1000 piping system.

Techstrand 2000 epoxy piping system is available in sizes 2 through 16 inches and is rated for 150 psi at 225°F (107°C). Table 3-32 provides the pressure ratings of the Techstrand 2000 piping system.

These three piping systems are typical of those available. Many other manufacturers also produce epoxy piping systems.

3.4.2   Joining, Installation, and Support

Reinforced epoxy piping systems can be joined by butt and strap joints, socket-type adhesive joints, and flanged joints as described on pages 98–101. The specific system to be used will be based on the manufacturer's design.

Installation should be in accordance with the general principles described for thermoplastic piping systems, as given on pages 24–32.

Support spacings for epoxy piping systems are specified by each manufacturer. Table 3-33 lists the recommended support spacing for Fibercast RB-2530 epoxy pipe, while Tables 3-34 and 3-35 provides the spacings for Techstrand 1000 and Techstrand 2000, respectively.

Regardless of the manufacturer, all valves and other accessories installed in the piping system must be supported independently. When flanged connections are used, a support should be placed on either side of the flanged joint. Designs of pipe hangers should be the same as those for reinforced polyester pipe (described on page 103). Recommended hanger dimensions are shown in Table 3-13. Reinforced epoxy piping systems can also be buried.

3.4.3   Corrosion Resistance

In general, reinforced epoxy pipe is recommended for handling hot caustics, solvents, acids, and corrosive combinations at elevated temperatures. Specific recommendations should be obtained from the manufacturer whose pipe is to be used since there are differences in the properties of the various epoxy resins. Table 3-36 lists the compatibility of epoxy resins in contact with selected corrodents. The chemical compatibility of epoxy resins varies with each formulation. Consequently Table 3-36 is general. It indicates when there is an epoxy formulation that is suitable to handle a specific corrodent. However, all formulations may not give satisfactory performance, therefore the manufacturer of the piping system must be consulted for recommendations.

3.5 REINFORCED PHENOL-FORMALDEHYDE PIPING SYSTEMS

Reinforced phenol-formaldehyde pipe is manufactured from a proprietary resin produced by the Haveg Division of Ametek, Inc. and sold under the trade name of Haveg 41NA. It is reinforced with silicate fibers. When necessary to provide corrosion resistance against hydrofluoric acid or certain fluoride salts, graphite is substituted for the silicate fillers. This is one of the oldest synthetic piping materials available, having been in existance for more than 50 years. In general it does not have the impact resistance of the polyesters or epoxies. Refer to Table 3-37 for the physical and mechanical properties of Haveg 41NA.

3.5.1 Pipe Data

Pipe is available in sizes ½ inch through 12 inches in lengths of 4 feet in the ½-inch and ¾-inch sizes and in lengths of 10 feet in all other sizes. The pressure rating varies with diameter and operating temperature. Refer to Table 3-38 for operating pressures of each size pipe at different operating temperatures.

Insulation is seldom required on Haveg 41NA because of its low thermal conductivity. The pipe may be steam traced providing the steam temperature does not exceed 300°F (149°C). All sizes of pipe are suitable for use under full vacuum.

3.5.2 Joining, Installation, and Support

Haveg 41NA pipe may be joined by means of either flanged joints or cemented joints. The cemented joint eliminates metal flanges, bolts, and gaskets, as well as the maintenance they entail. The use of flanges is best restricted to make connections to equipment, pumps, expansion joints, or valves that are constructed of a material other than Haveg 41NA. Bell and spigot or threaded joints are available by special order. Flanged connections are made by using cast iron split flanges set in tapered grooves which are machined near the pipe ends.

A cemented joint consists of a machined sleeve that has been split longitudinally and cemented onto the butted ends of two sanded pipes. It is important that the manufacturer's directions for the mixing of the cement, preparation of the pipe ends, and installation of the sleeve be followed exactly if a leak-free joint is to be obtained.

Haveg piping can be installed using essentially the same installation techniques that apply to other piping materials. Although Haveg 41NA pipe and fittings are tough and resilient, care should be taken during handling to prevent mechanical abuse, which might cause chipping or cracking. It is important that the piping be installed stress free. Avoid spring fits, which cause mechanical stress.

Haveg 41NA pipe should be supported every 10 feet of horizontal pipe. Rod hangers, rigid clamps, or U-bolts are all satisfactory for support of the pipe. Hangers must provide definite vertical support but allow movement along the axis of the pipe. On long vertical runs, support the weight of the pipeline between expansion joints by means of brackets, straps, or clamps. Support only the pipe itself. Do not attach supports to pipe flanges or expansion joints.

3.5.3   Corrosion Resistance

Haveg 41NA is generally recommended for service with mineral acids, salts, and chlorinated aromatic hydrocarbons. Haveg 46 has the same resin as Haveg 41NA, but the silicate fibers have been replaced by graphite. This permits the material to be used to handle hydrofluoric acid, fluosilicic acid, and other related fluoride-bearing compounds. Refer to Table 3-39 for the compatibility of phenol-formaldehyde pipe with selected corrodents. Since these resins possess little alkaline and bleach resistance, application in such services should be avoided.

3.6   REINFORCED FURAN RESIN PIPING SYSTEMS

Furan resins are produced from furfuryl alcohol and furfural. These resins are more expensive than other thermoset resins but are the most economical choice when the presence of solvents exists in a combination with acids and bases or when process changes may occur that result in exposure to solvents in oxidizing atmospheres.

The furan laminates have the ability to retain their physical properties at elevated temperatures. Pipe produced from these resins can be used at a temperature of 300°F/149°C.

Reinforced furfuryl alcohol-formaldehyde pipe is produced from a proprietary resin with silicate fillers manufactured by the Haveg Division of Ametek, Inc. and sold under the trade name of Haveg 61NA. It is considered a furan resin. Graphite filler is substituted when the pipe is to be used to convey hydrofluoric acid, or certain fluoride salts. The material is tough, durable, light weight, and has been available for over 50 years. Refer to Table 3-40 for the average physical and mechanical properties of the Furan resins.

3.6.1   Pipe Data

Haveg 61NA is available in sizes ½ inch through 12 inches. Standard lengths of the ½-inch and ¾-inch sizes are 4 feet. All other diameters are available in 10 foot lengths. The pressure rating varies with temperature and pipe diameter. Table 3-41 provides the maximum allowable operating pressure at different temperatures for each diameter of pipe. All sizes are suitable for full vacuum at 300°F (149°C).

Because of the low thermal conductivity of the pipe, insulation is seldom required. The pipe may be steam traced providing the steam temperature does not exceed 300°F (149°C).

3.6.2   Joining, Installation, and Support

Haveg 61NA pipe may be connected by either flanged or cemented joints. The cemented joint eliminates metal flanges, bolts, and gaskets, as well as the maintenance they require. It is a good policy to limit the use of flanges to making connections to pumps, equipment, expansion joints, valves, or other accessory equipment to be installed in the pipe line. Flanges can also be used to connect Haveg 61NA to dissimilar materials. Bell and spigot or threaded connections are available by special order.

Flanged connections are made by using cast iron split flanges set in tapered grooves, which are machined near the end of the pipe.

Cemented joints are made by using a machined sleeve, which has been split longitudinally and cemented on to the butted, sanded ends of two pipe lengths. It is important that the manufacturer's directions for the mixing of the cement, preparation of the pipe ends, and installation of the sleeve be followed exactly if a leak-free joint is to be obtained. Haveg piping can be installed using essentially the same techniques that apply to other piping materials. Although the Haveg material is tough and resilient, care should be taken during handling to prevent mechanical abuse, which might cause chipping or cracking. When installing the pipe, do not spring it into place. It is important that the pipe be installed stress free.

Supports for Haveg 61NA pipe should be spaced every 10 feet of horizontal pipe. Rod hangers, rigid clamps, or U-bolts are all satisfactory for the support of the pipe. Hangers must provide positive vertical support but allow movement along the axis of the pipe. On long vertical runs, support the weight of the pipeline between expansion joints by means of brackets, straps, or clamps. Support only the pipe itself. Do not attach supports to pipe flanges or expansion joints.

3.6.3   Corrosion Resistance

The strong point of the furans is their excellent resistance to solvents in combination with acids and alkalies. The furans are not resistant to bleaches, such as peroxides and hypochlorites, concentrated sulfuric acid, phenol, and free chlorine, or to higher concentrations of chromic or nitric acids. Since there are different formulations of furans, the supplier should be checked as to the compatibility of a particular resin with the corrodents to be encountered.

Haveg 61NA is recommended for use with alcohols, ketones, chlorinated hydrocarbons, dilute acids, and alkaline reactions. When graphite is substituted for the silicate filler, the material known as Haveg 66 can also be used to handle hydrofluoric acid and related compounds as well as all of the corrodents that can be handled by Haveg 61A.

Refer to Table 3-42 for a listing of the compatibility of furan with selected corrodents. This is a general tabulation of different furan formulations. It is important to verify with the supplier the compatibility of any particular resin with the corrodents to be encountered. The allowable operating temperature and pressure of the piping system should also be checked, particularly at elevated temperatures. Although the resin resists corrosion when in contact with a specific corrodent at an elevated temperature, it does not necessarily mean that the piping system can operate under these conditions.

3.7   REINFORCED PHENOLIC PIPING SYSTEM

Reinforced phenolic pipe consists of a proprietary phenolic resin produced by Haveg Division of Ametek, Inc. and silica filaments and fillers. It is sold under the trade name of Haveg SP. See Table 3-43 for the physical and mechanical properties.

3.7.1   Pipe Data

Haveg SP pipe and fittings are made to nominal IPS schedule 40 OD and are available in sizes 1 inch through 8 inches. Piping in diameters of 1, 1½, 2, 3, and 4 inches are designed for pressure applications and are furnished with threaded ends in nominal

10-foot lengths. The 6- and 8-inch-diameter pipe is designed for low-pressure drainage operations. This pipe is supplied with plain nonthreaded ends. The maximum operating temperature for all sizes is 300°F (149°C). Refer to Table 3-44 for maximum allowable operating pressures at 70°F (21°C) and 300°F (149°C).

3.7.2 Joining, Installation, and Support

The pressure-rated piping system, sizes 1 inch through 4 inches, is assembled using threaded joints and cement. The purpose of the threads is to align and immobilize the joint until the cement hardens.

Low-pressure drain piping, sizes 6 and 8 inches, are assembled by flanging and/or cementing. Since it is somewhat difficult to field thread 6- and 8-inch pipe, a spigot-and-socket cemented joint is used instead of a threaded joint. If a flange connection is desired, it can be made with a slip-on cemented construction rather than a threaded cemented construction.

Care should be exercised in the handling of the pipe to protect it from mechanical damage, which can cause a stress in the pipe. Normal installation procedures should be followed as for any piping system. Alignment is critical. The pipe should not be sprung into place, nor should there be any bending pressures transmitted to the pipe, as these cause stress in the pipe.

Pipe supports should be installed so that there is complete vertical support, but the pipe should be free to move along its longitudinal axis. Spacing of hangers for the different diameter pipes is as follows:

Pipe size (in.)	1	1½	2	3	4	6	8
Spacing (ft)	7	10	12	15	20	20	20

All vales or other equipment installed in the pipe line should be supported separately. Haveg SP piping is suitable for underground installation.

3.7.3 Corrosion Resistance

Haveg SP piping is suitable for indoor, outdoor, or underground installation. It is not subject to degradation by acid or soils and is not affected by corrosive spills or fumes.

In general, Haveg SP is suitable for use with chlorinated hydrocarbons, inorganic chlorides, and weak organic acids. It is not recommended for use with alkalies, or strong acids. Table 3-45 compares the compatibility of Haveg SP with selected corrodents.

Table 3-1 Minimum Pipe Stiffness Requirements

Nominal diameter (in.)	Minimum pipe stiffness at 5% deflection (psi)
1–8	35
10	20
12–144	10

Table 3-2 Deflection Coefficients

Installed condition	K_a	Δa
For all installation conditions with burial depths of 16 ft or less	0.75	0
For burial depths greater than 16 ft and installation conditions as follows:		
Dumped or slight degree of compaction (Proctor less than 85% or relative density less than 40%)	1.00	$0.02\ D_M$
Moderate degree of compaction (Proctor of 85–95% or relative density of 40–70%)	1.00	$0.01\ D_M$
High degree of compaction (Proctor greater than 95% or relative density greater than 70%)	1.00	$0.005\ D_M$

Table 3-3A Live Load Coefficient C_L-Single Wheel Load

Pipe diameter (in.)	C_L				
	4[a]	6	8	12	16
8	0.020	0.010	0.006	0.003	0.001
10	0.025	0.012	0.007	0.003	0.002
12	0.029	0.014	0.008	0.004	0.002
14	0.034	0.016	0.009	0.004	0.002
16	0.038	0.018	0.010	0.005	0.003
18	0.042	0.020	0.012	0.005	0.003
20	0.046	0.022	0.013	0.006	0.003
24	0.055	0.026	0.015	0.007	0.004
30	0.066	0.032	0.019	0.007	0.005
36	0.076	0.038	0.022	0.010	0.006
42	0.085	0.044	0.026	0.012	0.007
48	0.094	0.049	0.029	0.014	0.008
54	0.101	0.053	0.032	0.015	0.009
60	0.104	0.055	0.033	0.016	0.009

[a]Height of cover over pipe in feet.

Table 3-3B Live Load Coefficient C_L for Two Passing Trucks

Pipe diameter (in.)	C_L				
	4[a]	6	8	12	16
8	0.0294	0.0169	0.0112	0.0062	0.0039
10	0.0367	0.0210	0.0139	0.0077	0.0049
12	0.0443	0.0253	0.0167	0.0092	0.0059
14	0.0517	0.0295	0.0195	0.0108	0.0069
16	0.0589	0.0336	0.0223	0.0123	0.0078
18	0.0664	0.0379	0.0251	0.0139	0.0088
20	0.0740	0.0422	0.0280	0.0155	0.0098
24	0.0886	0.0506	0.0335	0.0185	0.0118
30	0.1108	0.0632	0.0419	0.0232	0.0147
36	0.1329	0.0759	0.0503	0.0278	0.0177
42	0.1551	0.0886	0.0586	0.0325	0.0207
48	0.1773	0.1012	0.0670	0.0371	0.0236
54	0.1994	0.1139	0.0754	0.0417	0.0266
60	0.2216	0.1265	0.0838	0.0464	0.0295

[a]Height of cover over pipe in feet.

Table 3-4 Strapping Material Required for Butt and Strap Joint of 50 psi Rated Pipe

Nominal pipe size (in.)	Mat width (in.)			Roving width (in.), layer sequence 4
	3[a]	4[a]	6[a]	
2	1,2	3	4	4
3	1,2	3	4	4
4	1,2	3	4	4
6	1,2	3	4	4
8	1,2	3,5,6	4	4
10	1,2	3,5,6	4	4
12	1,2	3,5,6	4	4
14		1,2,3	4,6,7	5
16		1,2,3	4,6,7	5
18		1,2,3	5,7,8	4,6
20		1,2,3	5,7,8	4,6
24		1,2,3	4,6,8,9	5,7

[a]Layer sequence.

Table 3-5 Strapping Material Required for Butt and Strap Joint of 100 psi Rated Pipe

Nominal pipe size (in.)	Mat width (in.)					Roving width (in.)		
	3[a]	4[a]	6[a]	8[a]	12[a]	4[a]	6[a]	8[a]
2	1,2	3				4		
3	1,2	3				4		
4	1,2	3,5,6				4		
6	1,2	3,5,6				4		
8		1,2,3	4,6,7			5		
10		1,2,3	5,7,8			4,6		
12		1,2,3	4,6,8			5,7		
14		1,2,3	4,5,6	8,10,11			7,9	
16			1,2,3	4,5,6	7,9,11,12			8,10
18			1,2,3,4	5,6,7,8	10,12,13,14			9,11
20			1,2,3,4	5,6,7,8,9	11,13,14,15			10,12
24			1,2,3,4,5	6,7,8,9,10	11,13,15,16			12,14

[a]Layer sequence.

Table 3-6 Strapping Material Required for Butt and Strap Joint of 150 psi Rated Pipe

Nominal pipe size (in.)	Mat width (in.)					Roving width (in.)		
	3[a]	4[a]	6[a]	8[a]	12[a]	4[a]	6[a]	8[a]
2	1,2	3				4		
3	1,2	3,5,6				4		
4	1,2	3,5,6				4		
6		1,2,3	5,7,8			4,6		
8		1,2,3	4,6,8,9			5,7		
10		1,2,3	4,5,6	8,10,11			7,9	
12			1,2,3,4	5,6,8	10,12,13,14			9,11
14			1,2,3,4,5	6,7,8,9,10	11,13,15,16			12,14

[a]Layer sequence.

Table 3-7 Standard Full-Face Gasket Dimensions

Nominal pipe size (in.)	Gasket I.D. (in.)	Gasket O.D. (in.)
1	$1\frac{5}{16}$	$4\frac{1}{4}$
$1\frac{1}{2}$	$1\frac{29}{32}$	5
2	$2\frac{3}{8}$	6
$2\frac{1}{2}$	$2\frac{7}{8}$	7
3	$3\frac{1}{2}$	$7\frac{1}{2}$
4	$4\frac{1}{2}$	9
6	$6\frac{5}{8}$	11
8	$8\frac{5}{8}$	$13\frac{1}{2}$
10	$10\frac{3}{4}$	16
12	$12\frac{3}{4}$	19
14	14	21
16	16	$23\frac{1}{2}$
18	18	25
20	20	$27\frac{1}{2}$
24	24	32

Table 3-8 Maximum Bolt Torque for Planged Joints on Reinforced Thermoset Pipe at Different Pressure Ratings

Nominal pipe size (in.)	Torque (ft-lb)					
	25 psi	50 psi	75 psi	100 psi	125 psi	150 psi
2	25	25	25	25	25	25
3	25	25	25	25	25	25
4	25	25	25	25	25	25
6	25	25	25	25	35	40
8	25	25	30	40	50	60
10	25	25	30	40	50	70
12	25	25	35	45	60	80
14	25	30	40	60	75	100
16	25	30	50	70	80	
18	30	35	50	80	100	
20	30	35	60	90		
24	35	40	70			

Table 3-9 Average Physical and Mechanical Properties of Reinforced Polyester Pipe[a]

Specific gravity	1.3–1.7
Water absorption 24 hr at 73°F (23°C), %	0.2
Tensile strength at 73° (23°C), psi	9000–15000
Modulus of elasticity in tension at 73°F (23°C) $\times 10^5$	6–10
Compressive strength, psi	18000–24000
Flexural strength, psi	16000–22000
Izod impact strength, notched at 73°F (23°C) lb/in.	30–40
Coefficient of thermal expansion	
in./in.°F $\times 10^{-5}$	1.5–1.9
in./10°F/100 ft	0.015–0.019
Thermal conductivity, Btu/hr/ft^2/°F/in.	1.2–1.5
Heat distortion temperature at 264 psi, °F/°C	252/122–350/177
Limiting oxygen index, %	20–43
Flame spread	15–30

[a]Values depend upon the formulation of the manufacturer.

Table 3-10 Reinforced Polyester Pipe Wall Thicknesses at Different Pressure Ratings

Pipe size (in.)	Minimum pipe wall thicknesses (in.)					
	25 psi	50 psi	75 psi	100 psi	125 psi	150 psi
2	0.1875	0.1875	0.1875	0.1875	0.1875	0.1875
3	0.1875	0.1875	0.1875	0.1875	0.25	0.25
4	0.1875	0.1875	0.1875	0.25	0.25	0.25
6	0.1875	0.1875	0.25	0.25	0.3125	0.375
8	0.1875	0.25	0.25	0.3125	0.375	0.4375
10	0.1875	0.25	0.3125	0.375	0.4375	0.50
12	0.1875	0.25	0.375	0.4375	0.50	0.625
14	0.25	0.3125	0.375	0.50	0.625	0.75
16	0.25	0.3125	0.4375	0.5625	0.6875	
18	0.25	0.375	0.50	0.625	0.75	
20	0.25	0.375	0.50	0.6875		
24	0.25	0.4375	0.625	0.8125		
30	0.3125	0.50	0.75			
36	0.375	0.625				
42	0.375	0.75				

Table 3-11 Minimum Flange Thicknesses Required for Reinforced Polyester Pipe at Various Operating Pressures

Pipe size (in.)	Minimum flange thicknesses (in.)					
	25 psi	50 psi	75 psi	100 psi	125 psi	150 psi
2	0.5	0.5	0.5	0.5625	0.625	0.6875
3	0.5	0.5	0.625	0.6875	0.75	0.8125
4	0.5	0.5625	0.6875	1.1875	0.875	1.3125
6	0.5	0.625	0.75	0.875	1.0	1.0625
8	0.5625	0.75	0.875	1.0	1.125	1.25
10	0.6875	0.875	1.0625	1.1875	1.3125	1.4375
12	0.75	1.0	1.25	1.4375	1.625	1.75
14	0.8125	1.0625	1.3125	1.5	1.75	1.875
16	0.875	1.1875	1.4375	1.625	1.875	
18	0.9375	1.25	1.5	1.75	2.0	
20	1.0	1.3125	1.625	1.875		
24	1.125	1.5	1.875			
30	1.375	1.875				
36	1.75					
42	2.0					

Table 3-12 Collapsing Pressure of Reinforced Polyester Pipe, Based on 20-Foot Lengths, at Different Wall Thicknesses (in.)

Pipe size (in.)	Collapsing pressure (psig)									
	0.1875	0.25	0.3125	0.375	0.4375	0.50	0.5625	0.625	0.6875	0.75
6	38.5	100[b]	207	378	570	800	1040	1410	1780	2200
8	17.3	44.7	94	173	264	378	515	680	865	1040
10	9.23	23.7	50	93	143	206	284	378	485	610
12	5.36	14.2	30	56	86	125	173	230	298	370
14	3.42	9	19.3	36	56	81	113	151	196	249
16	2.3	6.1	13	21.6	38	56	78	104	136	173
18	1.62	4.3	9.3	17.4	27	39.7	56	75	98	125
20	1.2	3.2	6.88	13	20.7	29.7	41.5	56	73	99
24	0.89	2.1	4.1	7.65	12	17.6	24.6	33.4	43.6	56
30	0.63	1.5	2.97	5.2	7.6	10.4	13.8	18	23	29.7
36	0.48	1.15	2.28	3.99	5.78	8	10.7	13.8	17.4	21.6
42	0.39	0.91	1.8	3.17	4.6	6.4	8.5	11	13.9	17.4
48	0.31	0.75	1.48	2.59	3.75	5.2	7	9.18	11.4	14.3
54	0.26	0.63	1.24	2.18	3.16	4.4	5.9	7.7	9.7	12
60	0.22	0.54	1.06	1.85	2.07	3.8	5.06	6.6	8.3	10.3

[a]Any reading above 14.7 will withstand full vacuum. If stiffeners are added every 10 feet, the above figures will increase.

[b]Wall thicknesses above the horizontal lines are capable of withstanding full vacuum with a 5:1 safety factor, minimum.

Table 3-13 Minimum Hanger Widths for Hand-Laid-Up and Filament Wound Pipe

Nominal pipe size (in.)	Hanger widths (in.)	
	Hand-laid-up pipe	Filament wound pipe
2	2	2
3	2	2
4	2	2
6	3	2
8	4	2
10	5	2
12	6	3
14	7	3
16	8	3
18	9	3
20	10	4
24	12	4

Table 3-14 Maximum Support Spacing for Reinforced Polyester Pipe at Various Pressure Ratings

Nominal pipe size (in.)	Hanger spacing (ft)					
	25 psi	50 psi	75 psi	100 psi	125 psi	150 psi
2	6.0	6.0	6.0	6.0	6.0	6.0
3	6.5	6.5	6.5	6.5	8.0	8.0
4	7.0	7.0	7.0	8.5	8.5	8.5
6	8.0	8.0	9.0	9.0	10.0	10.5
8	8.5	10.0	10.0	10.5	11.0	11.5
10	9.5	10.5	11.5	12.0	12.5	13.0
12	10.0	11.5	12.5	13.0	13.5	14.0
14	11.5	12.5	13.0	14.0	15.0	15.5
16	12.0	13.0	14.0	15.5	16.5	17.0
18	12.5	14.5	15.0	16.0	16.5	17.5
20	12.5	15.0	15.0	17.0	18.0	18.5
24	8.5	15.0	17.0	18.5	19.0	
30	9.5	17.5	19.5	21.0		
36	10.5	19.5	21.0			
42	8.0	21.0	22.5			

Table 3-15 Compatibility of Isophthalic Polyester with Selected Corrodents
The chemicals listed are in the pure state or in a saturated solution unless otherwise indicated. Compatibility is shown to the maximum allowable temperature for which data are available. Incompatibility is shown by an x. A blank space indicates that data are unavailable.

Chemical	Maximum temp. °F	Maximum temp. °C	Chemical	Maximum temp. °F	Maximum temp. °C
Acetaldehyde	x	x	Barium carbonate	190	88
Acetamide			Barium chloride	140	60
Acetic acid 10%	180	82	Barium hydroxide	x	x
Acetic acid 50%	110	43	Barium sulfate	160	71
Acetic acid 80%	x	x	Barium sulfide	90	32
Acetic acid, glacial	x	x	Benzaldehyde	x	x
Acetic anhydride	x	x	Benzene	x	x
Acetone	x	x	Benzene sulfonic acid 10%	180	82
Acetyl chloride	x	x	Benzoic acid	180	82
Acrylic acid	x	x	Benzyl alcohol	x	x
Acrylonitrile	x	x	Benzyl chloride	x	x
Adipic acid	220	104	Borax	140	60
Allyl alcohol	x	x	Boric acid	180	82
Allyl chloride	x	x	Bromine gas, dry	x	x
Alum	250	121	Bromine gas, moist	x	x
Aluminum acetate			Bromine liquid	x	x
Aluminum chloride, aqueous	180	82	Butadiene		
Aluminum chloride, dry	170	77	Butyl acetate	x	x
Aluminum fluoride 10%	140	60	Butyl alcohol	80	27
Aluminum hydroxide	160	71	n-Butylamine	x	x
Aluminum nitrate	160	71	Butyl phthalate		
Aluminum oxychloride			Butyric acid 25%	120	49
Aluminum sulfate	180	82	Calcium bisulfide	160	71
Ammonia gas	90	32	Calcium bisulfite	150	66
Ammonium bifluoride			Calcium carbonate	160	71
Ammonium carbonate	x	x	Calcium chlorate	160	71
Ammonium chloride 10%	160	71	Calcium chloride	180	82
Ammonium chloride 50%	160	71	Calcium hydroxide 10%	160	71
Ammonium chloride, sat.	180	82	Calcium hydroxide, sat.	160	71
Ammonium fluoride 10%	90	32	Calcium hypochlorite 10%	120	49
Ammonium fluoride 25%	90	32	Calcium nitrate	140	60
Ammonium hydroxide 25%	x	x	Calcium oxide	160	71
Ammonium hydroxide, sat.	x	x	Calcium sulfate	160	71
Ammonium nitrate	160	71	Caprylic acid	160	71
Ammonium persulfate	160	71	Carbon bisulfide	x	x
Ammonium phosphate	160	71	Carbon dioxide, dry	160	71
Ammonium sulfate 10%	180	82	Carbon dioxide, wet	160	71
Ammonium sulfide	x	x	Carbon disulfide	x	x
Ammonium sulfite	x	x	Carbon monoxide	160	71
Amyl acetate	x	x	Carbon tetrachloride	x	x
Amyl alcohol	160	71	Carbonic acid	160	71
Amyl chloride	x	x	Cellosolve	x	x
Aniline	x	x	Chloracetic acid, 50% water	x	x
Antimony trichloride	160	71	Chloracetic acid to 25%	150	66
Aqua regia 3:1	x	x	Chlorine gas, dry	160	71

Table 3-15 *Continued*

Chemical	Maximum temp. °F	Maximum temp. °C	Chemical	Maximum temp. °F	Maximum temp. °C
Chlorine gas, wet	160	71	Magnesium chloride	180	82
Chlorine, liquid	x	x	Malic acid	90	32
Chlorobenzene	x	x	Manganese chloride		
Chloroform	x	x	Methyl chloride		
Chlorosulfonic acid	x	x	Methyl ethyl ketone	x	x
Chromic acid 10%	x	x	Methyl isobutyl ketone	x	x
Chromic acid 50%	x	x	Muriatic acid	160	71
Chromyl chloride	140	60	Nitric acid 5%	120	49
Citric acid 15%	160	71	Nitric acid 20%	x	x
Citric acid, concentrated	200	93	Nitric acid 70%	x	x
Copper acetate	160	71	Nitric acid, anhydrous	x	x
Copper carbonate			Nitrous acid, concentrated	120	49
Copper chloride	180	82	Oleum	x	x
Copper cyanide	160	71	Perchloric acid 10%	x	x
Copper sulfate	200	93	Perchloric acid 70%	x	x
Cresol	x	x	Phenol	x	x
Cupric chloride 5%	170	77	Phosphoric acid 50–80%	180	82
Cupric chloride 50%	170	77	Picric acid	x	x
Cyclohexane	80	27	Potassium bromide 30%	160	71
Cyclohexanol			Salicylic acid	100	38
Dichloroacetic acid	x	x	Silver bromide 10%		
Dichloroethane (ethylene di-chloride)	x	x	Sodium carbonate 20%	90	32
			Sodium chloride	200	93
Ethylene glycol	120	49	Sodium hydroxide 10%	x	x
Ferric chloride	180	82	Sodium hydroxide 50%	x	x
Ferric chloride 50% in water	160	71	Sodium hydroxide, concen-trated	x	x
Ferric nitrate 10–50%	180	82	Sodium hypochlorite 20%	x	x
Ferrous chloride	180	82	Sodium hypochlorite, concentrated	x	x
Ferrous nitrate	160	71			
Fluorine gas, dry	x	x	Sodium sulfide to 50%	x	x
Fluorine gas, moist	x	x	Stannic chloride	180	82
Hydrobromic acid, dilute	120	49	Stannous chloride	180	82
Hydrobromic acid 20%	140	60	Sulfuric acid 10%	160	71
Hydrobromic acid 50%	140	60	Sulfuric acid 50%	150	66
Hydrochloric acid 20%	160	71	Sulfuric acid 70%	x	x
Hydrochloric acid 38%	160	71	Sulfuric acid 90%	x	x
Hydrocyanic acid 10%	90	32	Sulfuric acid 98%	x	x
Hydrofluoric acid 30%	x	x	Sulfuric acid 100%	x	x
Hydrofluoric acid 70%	x	x	Sulfuric acid, fuming	x	x
Hydrofluoric acid 100%	x	x	Sulfurous acid	x	x
Hypochlorous acid	90	32	Thionyl chloride	x	x
Iodine solution 10%			Toluene	110	43
Ketones, general	x	x	Trichloroacetic acid 50%	170	77
Lactic acid 25%	160	71	White liquor	x	x
Lactic acid, concentrated	160	71	Zinc chloride	180	82

Source: Schweitzer, Philip A. (1991). *Corrosion Resistance Tables*, Marcel Dekker, Inc., New York, Vols. 1 and 2.

Table 3-16 Compatibility of Bisphenol A-Fumarate Polyester with Selected Corrodents
The chemicals listed are in the pure state or in a saturated solution unless otherwise indicated. Compatibility is shown to the maximum allowable temperature for which data are available. Incompatibility is shown by an x. A blank space indicates that data are unavailable.

Chemical	Maximum temp. °F	Maximum temp. °C	Chemical	Maximum temp. °F	Maximum temp. °C
Acetaldehyde	x	x	Barium carbonate	200	93
Acetamide			Barium chloride	220	104
Acetic acid 10%	220	104	Barium hydroxide	150	66
Acetic acid 50%	160	171	Barium sulfate	220	104
Acetic acid 80%	160	171	Barium sulfide	140	60
Acetic acid, glacial	x	x	Benzaldehyde	x	x
Acetic anhydride	110	43	Benzene	x	x
Acetone	x	x	Benzene sulfonic acid 10%	200	93
Acetyl chloride	x	x	Benzoic acid	180	82
Acrylic acid	100	38	Benzyl alcohol	x	x
Acrylonitrile	x	x	Benzyl chloride	x	x
Adipic acid	220	104	Borax	220	104
Allyl alcohol	x	x	Boric acid	220	104
Allyl chloride	x	x	Bromine gas, dry	90	32
Alum	220	104	Bromine gas, moist	100	38
Aluminum acetate			Bromine liquid	x	x
Aluminum chloride, aqueous	200	93	Butadiene		
Aluminum chloride, dry			Butyl acetate	80	27
Aluminum fluoride 10%	90	32	Butyl alcohol	80	27
Aluminum hydroxide	160	71	*n*-Butylamine	x	x
Aluminum nitrate	200	93	Butyl phthalate		
Aluminum oxychloride			Butyric acid	220	93
Aluminum sulfate	200	93	Calcium bisulfide		
Ammonia gas	200	93	Calcium bisulfite	180	82
Ammonium bifluoride			Calcium carbonate	210	99
Ammonium carbonate	90	32	Calcium chlorate	200	93
Ammonium chloride 10%	200	93	Calcium chloride	220	104
Ammonium chloride 50%	220	104	Calcium hydroxide 10%	180	82
Ammonium chloride, sat.	220	104	Calcium hydroxide, sat.	160	71
Ammonium fluoride 10%	180	82	Calcium hypochlorite 10%	80	27
Ammonium fluoride 25%	120	49	Calcium nitrate	220	93
Ammonium hydroxide 25%	100	38	Calcium oxide		
Ammonium hydroxide, 20%	140	60	Calcium sulfate	220	93
Ammonium nitrate	220	104	Caprylic acid	160	71
Ammonium persulfate	180	82	Carbon bisulfide	x	x
Ammonium phosphate	80	27	Carbon dioxide, dry	350	177
Ammonium sulfate 10–40%	220	104	Carbon dioxide, wet	210	99
Ammonium sulfide	110	43	Carbon disulfide	x	x
Ammonium sulfite	80	27	Carbon monoxide	350	177
Amyl acetate	80	27	Carbon tetrachloride	110	43
Amyl alcohol	200	93	Carbonic acid	90	32
Amyl chloride	x	x	Cellosolve	140	60
Aniline	x	x	Chloracetic acid, 50% water	140	60
Antimony trichloride	220	104	Chloracetic acid to 25%	80	27
Aqua regia 3 : 1	x	x	Chlorine gas, dry	200	93

Table 3-16 *Continued*

Chemical	Maximum temp. °F	Maximum temp. °C	Chemical	Maximum temp. °F	Maximum temp. °C
Chlorine gas, wet	200	93	Magnesium chloride	220	104
Chlorine, liquid	x	x	Malic acid	160	71
Chlorobenzene	x	x	Manganese chloride		
Chloroform	x	x	Methyl chloride		
Chlorosulfonic acid	x	x	Methyl ethyl ketone	x	x
Chromic acid 10%	x	x	Methyl isobutyl ketone	x	x
Chromic acid 50%	x	x	Muriatic acid	130	54
Chromyl chloride	150	66	Nitric acid 5%	160	71
Citric acid 15%	220	104	Nitric acid 20%	100	38
Citric acid, concentrated	220	104	Nitric acid 70%	x	x
Copper acetate	180	82	Nitric acid, anhydrous	x	x
Copper carbonate			Nitrous acid, concentrated		
Copper chloride	220	104	Oleum	x	x
Copper cyanide	220	104	Perchloric acid 10%		
Copper sulfate	220	104	Perchloric acid 70%		
Cresol	x	x	Phenol	x	x
Cupric chloride 5%			Phosphoric acid 50–80%	220	104
Cupric chloride 50%			Picric acid	110	43
Cyclohexane	x	x	Potassium bromide 30%	200	93
Cyclohexanol			Salicylic acid	150	66
Dichloroacetic acid	100	38	Silver bromide 10%		
Dichloroethane (ethylene di-chloride)	x	x	Sodium carbonate	160	71
			Sodium chloride	220	104
Ethylene glycol	220	104	Sodium hydroxide 10%	130	54
Ferric chloride	220	104	Sodium hydroxide 50%	220	104
Ferric chloride 50% in water	220	104	Sodium hydroxide, concen-trated	200	93
Ferric nitrate 10–50%	220	104	Sodium hypochlorite 20%	x	x
Ferrous chloride	220	104	Sodium hypochlorite, concentrated		
Ferrous nitrate	220	104	Sodium sulfide to 50%	210	99
Fluorine gas, dry			Stannic chloride	200	93
Fluorine gas, moist			Stannous chloride	220	104
Hydrobromic acid, dilute	220	104	Sulfuric acid 10%	220	104
Hydrobromic acid 20%	220	104	Sulfuric acid 50%	220	104
Hydrobromic acid 50%	160	71	Sulfuric acid 70%	160	71
Hydrochloric acid 20%	190	88	Sulfuric acid 90%	x	x
Hydrochloric acid 38%	x	x	Sulfuric acid 98%	x	x
Hydrocyanic acid 10%	200	93	Sulfuric acid 100%	x	x
Hydrofluoric acid 30%	90	32	Sulfuric acid, fuming	x	x
Hydrofluoric acid 70%			Sulfurous acid	110	43
Hydrofluoric acid 100%			Thionyl chloride	x	x
Hypochlorous acid 20%	90	32	Toluene	x	x
Iodine solution 10%	200	104	Trichloroacetic acid 50%	180	82
Ketones, general			White liquor	180	82
Lactic acid 25%	210	99	Zinc chloride	250	121
Lactic acid, concentrated	220	104			

Source: Schweitzer, Philip A. (1991). *Corrosion Resistance Tables*, Marcel Dekker, Inc., New York, Vols. 1 and 2.

Table 3-17 Compatibility of Hydrogenated Bisphenol A–Bisphenol A Polyester with Selected Corrodents

The chemicals listed are in the pure state or in a saturated solution unless otherwise indicated. Compatibility is shown to the maximum allowable temperature for which data are available. Incompatibility is shown by an x. A blank space indicates that data are unavailable.

Chemical	Maximum temp. °F	Maximum temp. °C	Chemical	Maximum temp. °F	Maximum temp. °C
Acetaldehyde			Aqua regia 3:1	x	x
Acetamide			Barium carbonate	180	82
Acetic acid 10%	200	93	Barium chloride	200	93
Acetic acid 50%	160	71	Barium hydroxide		
Acetic acid 80%			Barium sulfate		
Acetic acid, glacial			Barium sulfide		
Acetic anhydride	x	x	Benzaldehyde	x	x
Acetone	x	x	Benzene	x	x
Acetyl chloride	x	x	Benzene sulfonic acid 10%		
Acrylic acid			Benzoic acid	210	99
Acrylonitrile	x	x	Benzyl alcohol	x	x
Adipic acid			Benzyl chloride	x	x
Allyl alcohol			Borax		
Allyl chloride			Boric acid	210	99
Alum			Bromine gas, dry		
Aluminum acetate			Bromine gas, moist		
Aluminum chloride, aqueous	200	93	Bromine liquid	x	x
Aluminum chloride, dry			Butadiene		
Aluminum fluoride	x	x	Butyl acetate	x	x
Aluminum hydroxide			Butyl alcohol		
Aluminum nitrate			n-Butylamine	x	x
Aluminum oxychloride			Butyl phthalate		
Aluminum sulfate	200	93	Butyric acid	x	x
Ammonia gas			Calcium bisulfide	120	49
Ammonium bifluoride			Calcium bisulfite		
Ammonium carbonate			Calcium carbonate		
Ammonium chloride 10%			Calcium chlorate	210	99
Ammonium chloride 50%			Calcium chloride	210	99
Ammonium chloride, sat.	200	93	Calcium hydroxide 10%		
Ammonium fluoride 10%			Calcium hydroxide, sat.		
Ammonium fluoride 25%			Calcium hypochlorite 10%	180	82
Ammonium hydroxide 25%			Calcium nitrate		
Ammonium hydroxide, sat.			Calcium oxide		
Ammonium nitrate	200	93	Calcium sulfate		
Ammonium persulfate	200	93	Caprylic acid		
Ammonium phosphate			Carbon bisulfide	x	x
Ammonium sulfate 10–40%			Carbon dioxide, dry		
Ammonium sulfide	100	38	Carbon dioxide, wet		
Ammonium sulfite			Carbon disulfide	x	x
Amyl acetate	x	x	Carbon monoxide		
Amyl alcohol	200	93	Carbon tetrachloride	x	x
Amyl chloride	90	32	Carbonic acid		
Aniline	x	x	Cellosolve		
Antimony trichloride	80	27	Chloracetic acid, 50% water	90	32

Table 3-17 *Continued*

Chemical	Maximum temp. °F	Maximum temp. °C	Chemical	Maximum temp. °F	Maximum temp. °C
Chloracetic acid			Lactic acid, concentrated	210	99
Chlorine gas, dry	210	99	Magnesium chloride	210	99
Chlorine gas, wet	210	99	Malic acid		
Chlorine, liquid			Manganese chloride		
Chlorobenzene			Methyl chloride		
Chloroform	x	x	Methyl ethyl ketone	x	x
Chlorosulfonic acid			Methyl isobutyl ketone	x	x
Chromic acid 10%			Muriatic acid	190	88
Chromic acid 50%	x	x	Nitric acid 5%	90	32
Chromyl chloride			Nitric acid 20%		
Citric acid 15%	200	93	Nitric acid 70%		
Citric acid, concentrated	210	99	Nitric acid, anhydrous		
Copper acetate	210	99	Nitrous acid, concentrated		
Copper carbonate			Oleum	x	x
Copper chloride	210	99	Perchloric acid 10%	x	x
Copper cyanide	210	99	Perchloric acid 70%	x	x
Copper sulfate	210	99	Phenol	x	x
Cresol	x	x	Phosphoric acid 50–80%	210	99
Cupric chloride 5%			Picric acid		
Cupric chloride 50%			Potassium bromide 30%		
Cyclohexane	210	99	Salicylic acid		
Cyclohexanol			Silver bromide 10%		
Dichloroacetic acid			Sodium carbonate 10%	100	38
Dichloroethane (ethylene di-chloride)	x	x	Sodium chloride	210	99
Ethylene glycol			Sodium hydroxide 10%	100	38
Ferric chloride	210	99	Sodium hydroxide 50%	x	x
Ferric chloride 50% in water	200	93	Sodium hydroxide, concentrated	x	x
Ferric nitrate 10–50%	200	93	Sodium hypochlorite 10%	160	71
Ferrous chloride	210	99	Sodium hypochlorite, concentrated		
Ferrous nitrate	210	99	Sodium sulfide to 50%		
Fluorine gas, dry			Stannic chloride		
Fluorine gas, moist			Stannous chloride		
Hydrobromic acid, dilute			Sulfuric acid 10%	210	99
Hydrobromic acid 20%	90	32	Sulfuric acid 50%	210	99
Hydrobromic acid 50%	90	32	Sulfuric acid 70%	90	32
Hydrochloric acid 20%	180	82	Sulfuric acid 90%	x	x
Hydrochloric acid 38%	190	88	Sulfuric acid 98%	x	x
Hydrocyanic acid 10%	x	x	Sulfuric acid 100%	x	x
Hydrofluoric acid 30%	x	x	Sulfuric acid, fuming	x	x
Hydrofluoric acid 70%	x	x	Sulfurous acid 25%	210	99
Hydrofluoric acid 100%	x	x	Thionyl chloride		
Hypochlorous acid 50%	210	99	Toluene	90	32
Iodine solution 10%			Trichloroacetic acid	90	32
Ketones, general			White liquor		
Lactic acid 25%	210	99	Zinc chloride	200	93

Source: Schweitzer, Philip A. (1991). *Corrosion Resistance Tables*, Marcel Dekker, Inc., New York, Vols. 1 and 2.

Table 3-18 Compatibility of Halogenated Polyester with Selected Corrodents
The chemicals listed are in the pure state or in a saturated solution unless otherwise indicated. Compatibility is shown to the maximum allowable temperature for which data are available. Incompatibility is shown by an x. A blank space indicates that data are unavailable.

Chemical	Maximum temp. °F	Maximum temp. °C	Chemical	Maximum temp. °F	Maximum temp. °C
Acetaldehyde	x	x	Barium carbonate	250	121
Acetamide			Barium chloride	250	121
Acetic acid 10%	140	60	Barium hydroxide	x	x
Acetic acid 50%	90	32	Barium sulfate	180	82
Acetic acid 80%			Barium sulfide	x	x
Acetic acid, glacial	110	43	Benzaldehyde	x	x
Acetic anhydride	100	38	Benzene	90	32
Acetone	x	x	Benzene sulfonic acid 10%	120	49
Acetyl chloride	x	x	Benzoic acid	250	121
Acrylic acid	x	x	Benzyl alcohol	x	x
Acrylonitrile	x	x	Benzyl chloride	x	x
Adipic acid	220	104	Borax	190	88
Allyl alcohol	x	x	Boric acid	180	82
Allyl chloride	x	x	Bromine gas, dry	100	38
Alum 10%	200	93	Bromine gas, moist	100	38
Aluminum acetate			Bromine liquid	x	x
Aluminum chloride, aqueous	120	49	Butadiene		
Aluminum chloride, dry			Butyl acetate	80	27
Aluminum fluoride 10%	90	32	Butyl alcohol	100	38
Aluminum hydroxide	170	77	n-Butylamine	x	x
Aluminum nitrate	160	71	Butyl phthalate	100	38
Aluminum oxychloride			Butyric acid 20%	200	93
Aluminum sulfate	250	121	Calcium bisulfide	x	x
Ammonia gas	150	66	Calcium bisulfite	150	66
Ammonium bifluoride			Calcium carbonate	210	99
Ammonium carbonate	140	60	Calcium chlorate	250	121
Ammonium chloride 10%	200	93	Calcium chloride	250	121
Ammonium chloride 50%	200	93	Calcium hydroxide 10%		
Ammonium chloride, sat.	200	93	Calcium hydroxide, sat.	x	x
Ammonium fluoride 10%	140	60	Calcium hypochlorite 20%	80	27
Ammonium fluoride 25%	140	60	Calcium nitrate	220	104
Ammonium hydroxide 25%	90	32	Calcium oxide	150	66
Ammonium hydroxide, sat.	90	32	Calcium sulfate	250	121
Ammonium nitrate	200	93	Caprylic acid	140	60
Ammonium persulfate	140	60	Carbon bisulfide	x	x
Ammonium phosphate	150	66	Carbon dioxide, dry	250	121
Ammonium sulfate 10–40%	200	93	Carbon dioxide, wet	250	121
Ammonium sulfide	120	49	Carbon disulfide	x	x
Ammonium sulfite	100	38	Carbon monoxide	170	77
Amyl acetate	190	85	Carbon tetrachloride	120	49
Amyl alcohol	200	93	Carbonic acid	160	71
Amyl chloride	x	x	Cellosolve	80	27
Aniline	120	49	Chloracetic acid, 50% water	100	38
Antimony trichloride 50%	200	93	Chloracetic acid 25%	90	32
Aqua regia 3:1	x	x	Chlorine gas, dry	200	93

Table 3-18 *Continued*

Chemical	°F	°C	Chemical	°F	°C
Chlorine gas, wet	220	104	Magnesium chloride	250	121
Chlorine, liquid	x	x	Malic acid 10%	90	32
Chlorobenzene	x	x	Manganese chloride		
Chloroform	x	x	Methyl chloride	80	27
Chlorosulfonic acid	x	x	Methyl ethyl ketone	x	x
Chromic acid 10%	180	82	Methyl isobutyl ketone	80	27
Chromic acid 50%	140	60	Muriatic acid	190	88
Chromyl chloride	210	99	Nitric acid 5%	210	99
Citric acid 15%	250	121	Nitric acid 20%	80	27
Citric acid, concentrated	250	121	Nitric acid 70%	80	27
Copper acetate	210	99	Nitric acid, anhydrous		
Copper carbonate			Nitrous acid, concentrated	90	32
Copper chloride	250	121	Oleum	x	x
Copper cyanide	250	121	Perchloric acid 10%	90	32
Copper sulfate	250	121	Perchloric acid 70%	90	32
Cresol	x	x	Phenol 5%	90	32
Cupric chloride 5%			Phosphoric acid 50–80%	250	121
Cupric chloride 50%			Picric acid	100	38
Cyclohexane	140	60	Potassium bromide 30%	230	110
Cyclohexanol			Salicylic acid	130	54
Dichloroacetic acid	100	38	Silver bromide 10%		
Dichloroethane (ethylene di-chloride)	x	x	Sodium carbonate 10%	190	88
			Sodium chloride	250	121
Ethylene glycol	250	121	Sodium hydroxide 10%	110	43
Ferric chloride	250	121	Sodium hydroxide 50%	x	x
Ferric chloride 50% in water	250	121	Sodium hydroxide, concen-trated	x	x
Ferric nitrate 10–50%	250	121	Sodium hypochlorite 20%	x	x
Ferrous chloride	250	121	Sodium hypochlorite, concentrated	x	x
Ferrous nitrate	160	71			
Fluorine gas, dry			Sodium sulfide to 50%	x	x
Fluorine gas, moist			Stannic chloride	80	27
Hydrobromic acid, dilute	200	93	Stannous chloride	250	121
Hydrobromic acid 20%	160	71	Sulfuric acid 10%	260	127
Hydrobromic acid 50%	200	93	Sulfuric acid 50%	200	93
Hydrochloric acid 20%	230	110	Sulfuric acid 70%	190	88
Hydrochloric acid 38%	180	82	Sulfuric acid 90%	x	x
Hydrocyanic acid 10%	150	66	Sulfuric acid 98%	x	x
Hydrofluoric acid 30%	120	49	Sulfuric acid 100%	x	x
Hydrofluoric acid 70%			Sulfuric acid, fuming	x	x
Hydrofluoric acid 100%			Sulfurous acid 10%	80	27
Hypochlorous acid 10%	100	38	Thionyl chloride	x	x
Iodine solution 10%			Toluene	110	43
Ketones, general			Trichloroacetic acid 50%	200	93
Lactic acid 25%	200	93	White liquor	x	x
Lactic acid, concentrated	200	93	Zinc chloride	200	93

Source: Schweitzer, Philip A. (1991). *Corrosion Resistance Tables*, Marcel Dekker, Inc., New York, Vols. 1 and 2.

Table 3-19 Average Physical and Mechanical Properties of Reinforced Vinyl Ester Pipe[a]

Specific gravity	1.58
Tensile strength at 73°F (23°C), psi	12000
Modulus of elasticity in tension at 73°F (23°C) × 10^5	4.9
Flexural strength, psi	18000
Coefficient of thermal expansion	
in./in. °F × 10^{-5}	1.3
in./10°F/100 ft	0.013
Thermal conductivity, Btu/hr/ft^2/°F/in.	0.8712
Heat distortion temperature at 264 psi, °F/°C	220/104
Resistance to heat at continuous drainage, °F/°C	200–280/93–140
Flame spread	350–500

[a]These are average values since manufacturers resins vary in properties.

Table 3-20 Recommended Operating Data of Fibercast Cl-2030 Pipe

Nominal pipe size (in.)	Internal pressure at 175°F (80°C) (psi)	External pressure at 75°F (24°C) (psi)	Axial load at 75°F (24°C) (lb)
1½	875	432	3000
2	800	511	3700
3	525	302	6500
4	500	206	10200
6	500	141	19400
8	375	80	25500
10	300	26	32000
12	250	16	38100
14	175	12	41900

Source: Courtesy of Fibercast Company.

Table 3-21 Operating External Pressure for Cl-2030 Pipe at Various Temperatures

Nominal pipe size (in.)	Maximum allowable external pressure (psi) at various temperatures (°F/°C)			
	75/24	150/66	175/80	200/93
1½	432	384	327	261
2	511	455	387	309
3	302	269	229	183
4	206	183	156	125
6	141	125	107	85
8	80	71	61	48
10	26	23	20	16
12	16	14	12	9
14	12	11	9	7

Source: Courtesy of Fibercast Company.

Table 3-22 Pressure Ratings (psi) of Fibercast Vinyl Ester Fittings at 175°F (80°C) Using Weldfast 200 Adhesive

| Nominal pipe size (in.) | Elbows, tees, reducers, couplings, flanges, socket and threaded nipples | | Laterals, crosses, saddles, and grooved nipples |
	Socket fittings	Flanged fittings	
1½	300	300	
2	275	200	125
3	200	150	125
4	150	150	100
6	150	150	100
8	150	150	100
10	150	150	75
12	150	150	75
14	125	150	

Source: Courtesy of Fibercast Company.

Table 3-23 Maximum Support Spacing for Cl-2030 Reinforced Vinyl Ester Pipe[a]

| Nominal pipe size (in.) | Maximum span (ft) at various temperatures (°F/°C) | | | |
	75/24	150/66	175/80	200/93
1½	11.2	11.1	11.1	10.7
2	12.1	11.9	11.9	11.6
3	14.1	13.9	13.9	13.6
4	15.9	15.7	15.7	15.3
6	18.8	18.6	18.6	18.1
8	20.3	20.1	20.1	19.6
10	21.7	21.4	21.4	20.9
12	22.7	22.5	22.5	21.9
14	23.4	23.1	23.1	22.5

[a]Based on fluid having a specific gravity of 1.0 and lines being uninsulated.
Source: Courtesy of Fibercast Company.

Table 3-24 Correction Factors for Maximum Support Spacings of Cl-2030 Reinforced Vinyl Ester Pipe When Handling Various Fluids

Specific gravity	Multiplier
1.25	0.95
1.50	0.90
2.0	0.84
3.0	0.76
Air	1.40

Source: Courtesy of Fibercast Company.

Table 3-25 Maximum Support Spacing for P-150 Pipe at 75°F (24°C) Manufactured by Reinforced Plastic Systems Inc.

Nominal pipe size (in.)	Continuous Span (ft)	
	Specific gravity 1.0	Specific gravity 1.3
1½	9.3	8.7
2	9.8	9.2
3	10.3	9.5
4	10.7	9.8
6	13.3	12.2
8	15.6	14.2
10	17.5	16.0
12	20.0	18.0

Table 3-26 Compatibility of Vinyl Ester with Selected Corrodents
The chemicals listed are in the pure state or in a saturated solution unless otherwise indicated. Compatibility is shown to the maximum allowable temperature for which data are available. Incompatibility is shown by an x. A blank space indicates that data are unavailable.

Chemical	Maximum temp.		Chemical	Maximum temp.	
	°F	°C		°F	°C
Acetaldehyde	x	x	Ammonium carbonate	150	66
Acetamide			Ammonium chloride 10%	200	93
Acetic acid 10%	200	93	Ammonium chloride 50%	200	93
Acetic acid 50%	180	82	Ammonium chloride, sat.	200	93
Acetic acid 80%	150	66	Ammonium fluoride 10%	140	60
Acetic acid, glacial	150	66	Ammonium fluoride 25%	140	60
Acetic anhydride	100	38	Ammonium hydroxide 25%	100	38
Acetone	x	x	Ammonium hydroxide, sat.	130	54
Acetyl chloride	x	x	Ammonium nitrate	250	121
Acrylic acid	100	38	Ammonium persulfate	180	82
Acrylonitrile	x	x	Ammonium phosphate	200	93
Adipic acid	180	82	Ammonium sulfate 10–40%	220	104
Allyl alcohol	90	32	Ammonium sulfide	120	49
Allyl chloride	90	32	Ammonium sulfite	220	104
Alum	240	116	Amyl acetate	110	38
Aluminum acetate	210	99	Amyl alcohol	210	99
Aluminum chloride, aqueous	260	127	Amyl chloride	120	49
Aluminum chloride, dry	140	60	Aniline	x	x
Aluminum fluoride	100	38	Antimony trichloride	160	71
Aluminum hydroxide	200	93	Aqua regia 3:1	x	x
Aluminum nitrate	200	93	Barium carbonate	260	127
Aluminum oxychloride			Barium chloride	200	93
Aluminum sulfate	250	121	Barium hydroxide	150	66
Ammonia gas	100	38	Barium sulfate	200	93
Ammonium bifluoride	150	66	Barium sulfide	180	82

Table 3-26 *Continued*

Chemical	Maximum temp. °F	Maximum temp. °C	Chemical	Maximum temp. °F	Maximum temp. °C
Benzaldehyde	x	x	Chromyl chloride	210	99
Benzene	x	x	Citric acid 15%	210	99
Benzene sulfonic acid 10%	200	93	Citric acid, concentrated	210	99
Benzoic acid	180	82	Copper acetate	210	99
Benzyl alcohol	100	38	Copper carbonate		
Benzyl chloride	90	32	Copper chloride	220	104
Borax	210	99	Copper cyanide	210	99
Boric acid	200	93	Copper sulfate	240	116
Bromine gas, dry	100	38	Cresol	x	x
Bromine gas, moist	100	38	Cupric chloride 5%	260	127
Bromine liquid	x	x	Cupric chloride 50%	220	104
Butadiene			Cyclohexane	150	66
Butyl acetate	80	27	Cyclohexanol	150	66
Butyl alcohol	120	49	Dichloroacetic acid	100	38
n-Butylamine	x	x	Dichloroethane (ethylene di-chloride)	110	43
Butyl phthalate	200	93			
Butyric acid	130	54	Ethylene glycol	210	99
Calcium bisulfide			Ferric chloride	210	99
Calcium bisulfite	180	82	Ferric chloride 50% in water	210	99
Calcium carbonate	180	82	Ferric nitrate 10–50%	200	93
Calcium chlorate	260	127	Ferrous chloride	200	93
Calcium chloride	180	82	Ferrous nitrate	200	93
Calcium hydroxide 10%	180	82	Fluorine gas, dry	x	x
Calcium hydroxide, sat.	180	82	Fluorine gas, moist	x	x
Calcium hypochlorite	180	82	Hydrobromic acid, dilute	180	82
Calcium nitrate	210	99	Hydrobromic acid 20%	180	82
Calcium oxide	160	71	Hydrobromic acid 50%	200	93
Calcium sulfate	250	116	Hydrochloric acid 20%	220	104
Caprylic acid	220	104	Hydrochloric acid 38%	180	82
Carbon bisulfide	x	x	Hydrocyanic acid 10%	160	71
Carbon dioxide, dry	200	93	Hydrofluoric acid 30%	x	x
Carbon dioxide, wet	220	104	Hydrofluoric acid 70%	x	x
Carbon disulfide	x	x	Hydrofluoric acid 100%	x	x
Carbon monoxide	350	177	Hypochlorous acid	150	66
Carbon tetrachloride	180	82	Iodine solution 10%	150	66
Carbonic acid	120	49	Ketones, general	x	x
Cellosolve	140	60	Lactic acid 25%	210	99
Chloracetic acid, 50% water	150	66	Lactic acid, concentrated	200	93
Chloracetic acid	200	93	Magnesium chloride	260	127
Chlorine gas, dry	250	121	Malic acid 10%	140	60
Chlorine gas, wet	250	121	Manganese chloride	210	99
Chlorine, liquid	x	x	Methyl chloride		
Chlorobenzene	110	43	Methyl ethyl ketone	x	x
Chloroform	x	x	Methyl isobutyl ketone	x	x
Chlorosulfonic acid	x	x	Muriatic acid	180	82
Chromic acid 10%	150	66	Nitric acid 5%	180	82
Chromic acid 50%	x	x	Nitric acid 20%	150	66

Table 3-26 *Continued*

Chemical	Maximum temp. °F	Maximum temp. °C	Chemical	Maximum temp. °F	Maximum temp. °C
Nitric acid 70%	x	x	Sodium hypochlorite, concentrated	100	38
Nitric acid, anhydrous	x	x	Sodium sulfide to 50%	220	104
Nitrous acid 10%	150	66	Stannic chloride	210	99
Oleum	x	x	Stannous chloride	200	93
Perchloric acid 10%	150	66	Sulfuric acid 10%	200	93
Perchloric acid 70%	x	x	Sulfuric acid 50%	210	99
Phenol	x	x	Sulfuric acid 70%	180	82
Phosphoric acid 50–80%	210	99	Sulfuric acid 90%	x	x
Picric acid	200	93	Sulfuric acid 98%	x	x
Potassium bromide 30%	160	71	Sulfuric acid 100%	x	x
Salicylic acid	150	66	Sulfuric acid, fuming	x	x
Silver bromide 10%			Sulfurous acid 10%	120	49
Sodium carbonate	180	82	Thionyl chloride	x	x
Sodium chloride	180	82	Toluene	120	49
Sodium hydroxide 10%	170	77	Trichloroacetic acid 50%	210	99
Sodium hydroxide 50%	220	104	White liquor	180	82
Sodium hydroxide, concentrated			Zinc chloride	180	82
Sodium hypochlorite 15%	180	82			

Source: Schweitzer, Philip A. (1991). *Corrosion Resistance Tables*, Marcel Dekker, Inc., New York, Vols. 1 and 2.

Table 3-27 Average Physical and Mechanical Properties of Reinforced Epoxy Pipe[a]

Specific gravity	1.58–1.9
Modulus of elasticity in tension at 73°F (23°C) $\times 10^5$	16.9–28
Coefficient of thermal expansion	
in./in. °F $\times 10^{-5}$	1.1–1.3
in./10°F/100 ft	0.011–0.013
Thermal conductivity, Btu/hr/ft^2/°F/in.	2.8
Resistance to heat at continuous drainage, °F/°C	250/121

[a]These are average values since each manufacturer's resins vary in properties.

Table 3-28 Recommended Operating Data for Fibercast RB-2530 Epoxy Pipe

Nominal pipe size (in.)	Internal pressure at 270°F (132°C) (psi)	Internal pressure at 250°F (121°C) (psi)	Internal pressure at 75°F (24°C) (psi)	Axial load at 75°F (24°C) (lb)
1	200	300	1469	1680
1½	850	1275	1454	3700
2	670	1000	837	4700
3	430	650	678	7100
4	370	500	274	9200
6	370	500	160	17,700
8	250	375	90	23,200
10	200	300	34	29,100
12	165	250	20	34,600
14	150	225	15	38,000

Source: Courtesy of Fibercast Company.

Table 3-29 Pressure Rating of Fibercast Epoxy Fittings Up to 225°F (107°C) Using Weldfast 440 Adhesive

Nominal pipe size (in.)	Maximum allowable operating pressure (psi)			
	Elbows, tees, reducers, couplings flanges, sockets, and threaded nipples		Flanges	Laterals, crosses, saddles and grooved nipples
	Socket flanges	Flanged fitting		
1	300	300	300	
1½	450	150	450	
2	450	150	450	125
3	300	150	300	125
4	225	150	225	100
6	225	150	225	100
8	225	150	225	100
10	225	150	225	75
12	225	150	225	75
14	125	150	150	

Source: Courtesy of Fibercast Company.

Table 3-30 Allowable Operating External Pressures for Fibercast RB-2530 Epoxy Pipe at Various Temperatures

Nominal pipe size (in.)	Allowable pressure (psi) at °F/°C						
	75/24	150/66	175/80	200/93	225/107	250/121	270/132
1	1469	1439	1378	1343	1160	940	812
1½	1454	1425	1364	1329	1149	931	804
2	837	820	785	765	661	536	463
3	678	654	636	619	535	434	375
4	274	269	257	251	217	176	152
6	160	157	150	146	126	102	89
8	90	88	84	82	71	58	50
10	34	33	32	31	27	22	19
12	20	20	19	19	16	13	11
14	15	15	14	14	12	10	9

Source: Courtesy of Fibercast Company.

Table 3-31 Pressure Rating of Techstrand 1000 Epoxy Piping System

Nominal pipe size (in.)	Short-term burst pressure[a] 210°F (99°C) (psi)	External pressure at 75°F (24°C) (psi)
2	2200	100
3	1800	70
4	1600	40
6	1500	30
8	1300	17
10	1300	17
12	1300	17
14	1300	17
16	1300	17

[a]Failure by weeping.

Table 3-32 Pressure Rating of Techstrand 2000 Epoxy Piping System

Nominal pipe size (in.)	Short-term burst pressure[a] 225°F (107°C) (psi)	External pressure at 75°F (24°C) (psi)
2	3000	260
3	2000	120
4	1600	75
6	1400	31
8	1300	18
10	1300	18
12	1300	18
14	1300	18
16	1300	18

[a]Failure by weeping.

Table 3-33 Support Spacing for Fibercast RB-2530 Pipe at Various Temperatures

Nominal pipe size (in.)	Maximum support spacing (ft) at °F/°C						
	75/24	150/66	175/80	200/93	225/107	250/121	270/132
1	9.6	9.5	9.4	9.1	9.0	8.3	7.0
1½	11.2	11.1	10.9	10.6	10.5	9.7	8.2
2	12.6	12.5	12.3	12.0	11.8	11.0	9.2
3	14.3	14.1	13.9	13.6	13.4	12.4	10.4
4	15.5	15.3	15.2	14.7	14.5	13.5	11.4
6	18.6	18.3	18.1	17.6	17.3	16.1	13.6
8	20.0	19.8	19.6	19.0	18.7	17.4	14.7
10	21.4	21.1	20.8	20.3	20.0	18.6	15.6
12	22.4	22.1	21.9	21.3	20.9	19.5	16.4
14	23.0	22.8	22.5	21.9	21.5	20.0	16.9

Source: Courtesy of Fibercast Company.

Table 3-34 Support Spacing for Techstrand 1000 Epoxy Pipe at 75°F (24°C)[a]

| Nominal pipe size (in.) | Spacing (ft) | | | |
| | Single span | | Continuous span | |
	Specific gravity 1.0	Specific gravity 1.3	Specific gravity 1.0	Specific gravity 1.3
2	7.5	7.0	11.2	10.5
3	8.6	8.1	12.8	12.0
4	9.8	9.2	14.6	13.7
6	11.7	11.0	17.4	15.0
8	14.5	13.6	18.0	16.5
10	16.2	15.2	20.0	18.5
12	17.6	16.5	22.0	20.5
14	19.4	18.2	24.0	22.5
16	20.7	19.4	25.0	23.5

[a]For 210°F (99°C) operation, reduce spans by 0.85.

Table 3-35 Support Spacing for Techstrand 2000 Epoxy Pipe at 75°F (24°C)[a]

| Nominal pipe size (in.) | Spacing (ft) | | | |
| | Single span | | Continuous span | |
	Specific gravity 1.0	Specific gravity 1.3	Specific gravity 1.0	Specific gravity 1.3
2	9.4	8.8	14.0	13.1
3	10.6	10.0	15.8	14.0
4	11.4	10.7	16.2	14.5
6	13.7	12.9	17.0	15.8
8	15.3	14.3	18.5	17.5
10	17.1	16.1	20.5	19.8
12	18.6	17.5	22.5	21.5
14	20.1	18.9	24.5	23.0
16	21.4	20.1	26.0	24.5

Table 3-36 Compatibility of Epoxy with Selected Corrodents
The chemicals listed are in the pure state or in a saturated solution unless otherwise indicated. Compatibility is shown to the maximum allowable temperature for which data are available. Incompatibility is shown by an x. A blank space indicates that data are unavailable.

Chemical	Maximum temp.		Chemical	Maximum temp.	
	°F	°C		°F	°C
Acetaldehyde	150	66	Aqua regia 3:1	x	x
Acetamide	90	32	Barium carbonate	240	116
Acetic acid 10%	190	88	Barium chloride	250	121
Acetic acid 50%	110	43	Barium hydroxide 10%	200	93
Acetic acid 80%	110	43	Barium sulfate	250	121
Acetic acid, glacial			Barium sulfide	300	149
Acetic anhydride	x	x	Benzaldehyde	x	x
Acetone	110	43	Benzene	160	71
Acetyl chloride	x	x	Benzene sulfonic acid 10%	160	71
Acrylic acid	x	x	Benzoic acid	200	93
Acrylonitrile	90	32	Benzyl alcohol	x	x
Adipic acid	250	121	Benzyl chloride	60	16
Allyl alcohol	x	x	Borax	250	121
Allyl chloride	140	60	Boric acid 4%	200	93
Alum	300	149	Bromine gas, dry	x	x
Aluminum acetate			Bromine gas, moist	x	x
Aluminum chloride, aqueous 1%	300	149	Bromine liquid	x	x
			Butadiene	100	38
Aluminum chloride, dry	90	32	Butyl acetate	170	77
Aluminum fluoride	180	82	Butyl alcohol	140	60
Aluminum hydroxide	180	82	*n*-Butylamine	x	x
Aluminum nitrate	250	121	Butyl phthalate		
Aluminum oxychloride			Butyric acid	210	99
Aluminum sulfate	300	149	Calcium bisulfide		
Ammonia gas dry	210	99	Calcium bisulfite	200	93
Ammonium bifluoride	90	32	Calcium carbonate	300	149
Ammonium carbonate	140	60	Calcium chlorate	200	93
Ammonium chloride 10%			Calcium chloride 37.5%	190	88
Ammonium chloride 50%			Calcium hydroxide 10%		
Ammonium chloride, sat.	180	82	Calcium hydroxide, sat.	180	82
Ammonium fluoride 10%			Calcium hypochlorite 70%	150	66
Ammonium fluoride 25%	150	66	Calcium nitrate	250	121
Ammonium hydroxide 25%	140	60	Calcium oxide		
Ammonium hydroxide, sat.	150	66	Calcium sulfate	250	121
Ammonium nitrate 25%	250	121	Caprylic acid	x	x
Ammonium persulfate	250	121	Carbon bisulfide	100	38
Ammonium phosphate	140	60	Carbon dioxide, dry	200	93
Ammonium sulfate 10–40%	300	149	Carbon dioxide, wet		
Ammonium sulfide			Carbon disulfide	100	38
Ammonium sulfite	100	38	Carbon monoxide	80	27
Amyl acetate	80	27	Carbon tetrachloride	170	77
Amyl alcohol	140	60	Carbonic acid	200	93
Amyl chloride	80	27	Cellosolve	140	60
Aniline	150	66	Chloracetic acid, 92% water	150	66
Antimony trichloride	180	82	Chloracetic acid	x	x

Table 3-36 *Continued*

Chemical	Maximum temp. °F	Maximum temp. °C	Chemical	Maximum temp. °F	Maximum temp. °C
Chlorine gas, dry	150	66	Magnesium chloride	190	88
Chlorine gas, wet	x	x	Malic acid		
Chlorine, liquid			Manganese chloride		
Chlorobenzene	150	66	Methyl chloride	x	x
Chloroform	110	43	Methyl ethyl ketone	90	32
Chlorosulfonic acid	x	x	Methyl isobutyl ketone	140	60
Chromic acid 10%	110	43	Muriatic acid	140	60
Chromic acid 50%	x	x	Nitric acid 5%	160	71
Chromyl chloride			Nitric acid 20%	100	38
Citric acid 15%	190	88	Nitric acid 70%	x	x
Citric acid, 32%	190	88	Nitric acid, anhydrous	x	x
Copper acetate	200	93	Nitrous acid, concentrated	x	x
Copper carbonate	150	66	Oleum	x	x
Copper chloride	250	121	Perchloric acid 10%	90	32
Copper cyanide	150	66	Perchloric acid 70%	80	27
Copper sulfate 17%	210	99	Phenol	x	x
Cresol	100	38	Phosphoric acid 50–80%	110	43
Cupric chloride 5%	80	27	Picric acid	80	27
Cupric chloride 50%	80	27	Potassium bromide 30%	200	93
Cyclohexane	90	32	Salicylic acid	140	60
Cyclohexanol	80	27	Silver bromide 10%		
Dichloroacetic acid	x	x	Sodium carbonate	300	149
Dichloroethane (ethylene di-chloride)	x	x	Sodium chloride	210	99
			Sodium hydroxide 10%	190	88
Ethylene glycol	300	149	Sodium hydroxide 50%	200	93
Ferric chloride	300	149	Sodium hydroxide, con-centrated		
Ferric chloride 50% in water	250	121			
			Sodium hypochlorite 20%	x	x
Ferric nitrate 10–50%	250	121	Sodium hypochlorite, con-centrated	x	x
Ferrous chloride	250	121			
Ferrous nitrate			Sodium sulfide to 10%	250	121
Fluorine gas, dry	90	32	Stannic chloride	200	93
Fluorine gas, moist			Stannous chloride	160	71
Hydrobromic acid, dilute	180	82	Sulfuric acid 10%	140	60
Hydrobromic acid 20%	180	82	Sulfuric acid 50%	110	43
Hydrobromic acid 50%	110	43	Sulfuric acid 70%	110	43
Hydrochloric acid 20%	200	93	Sulfuric acid 90%	x	x
Hydrochloric acid 38%	140	60	Sulfuric acid 98%	x	x
Hydrocyanic acid 10%	160	71	Sulfuric acid 100%	x	x
Hydrofluoric acid 30%	x	x	Sulfuric acid, fuming	x	x
Hydrofluoric acid 70%	x	x	Sulfurous acid 20%	240	116
Hydrofluoric acid 100%	x	x	Thionyl chloride	x	x
Hypochlorous acid	200	93	Toluene	150	66
Iodine solution 10%			Trichloroacetic acid	x	x
Ketones, general	x	x	White liquor	90	32
Lactic acid 25%	220	104	Zinc chloride	250	121
Lactic acid, concentrated	200	93			

Source: Schweitzer, Philip A. (1991). *Corrosion Resistance Tables*, Marcel Dekker, Inc., New York, Vols. 1 and 2.

Table 3-37 Physical and Mechanical Properties of Haveg 41NA

Specific gravity	1.8
Tensile strength at 73°F (23°C), psi	5540
Modulus of elasticity in tension at 73°F (23°C) $\times$ 10^5	13.7
Compressive strength, psi	19300
Flexural strength, psi	9840
Izod impact strength, notched at 73°F (23°C)	0.60
Coefficient of thermal expansion	
in./in. °F $\times$ 10^{-5}	0.83
in./10°F/100 ft	0.0083
Thermal conductivity, Btu/hr/ft^2/°F/in.	1.96
Resistance to heat at continuous drainage, °F/°C	300/149
Flame spread	<50

Source: Courtesy of Ametek Inc.

Table 3-38 Operating Pressure Versus Temperature for Haveg 41NA Pipe Systems

Nominal pipe size (in.)	Maximum operating pressure (psi) at °F/°C					
	50/10	100/38	150/66	200/93	250/121	300/149
½	100	100	100	100	100	100
¾	100	100	100	100	100	100
1	100	100	100	100	100	100
1½	100	100	100	100	100	100
2	100	100	100	100	100	100
3	95	90	85	82	80	75
4	67	65	61	59	55	52
6	41	40	39	37	35	30
8	35	34	32	31	30	29
10	29	28	27	26	25	24
12	24	23	22	20	19	18

Source: Courtesy Ametek Inc.

Table 3-39 Compatibility of Haveg 41NA with Selected Corrodents
The chemicals listed are in the pure state or in a saturated solution unless otherwise indicated. Compatibility is shown to the maximum allowable temperature for which data are available. Incompatibility is shown by an x. A blank space indicates that data are unavailable.

Chemical	Maximum temp. °F	Maximum temp. °C	Chemical	Maximum temp. °F	Maximum temp. °C
Acetaldehyde	x	x	Barium carbonate		
Acetamide			Barium chloride		
Acetic acid 10%	212	100	Barium hydroxide		
Acetic acid 50%	160	71	Barium sulfate		
Acetic acid 80%	120	49	Barium sulfide		
Acetic acid, glacial	120	49	Benzaldehyde		
Acetic anhydride			Benzene	160	71
Acetone	x	x	Benzene sulfonic acid 10%	160	71
Acetyl chloride	x	x	Benzoic acid		
Acrylic acid 90%	80	27	Benzyl alcohol		
Acrylonitrile	x	x	Benzyl chloride	160	71
Adipic acid			Borax		
Allyl alcohol			Boric acid	300	149
Allyl chloride			Bromine gas, dry		
Alum			Bromine gas, moist		
Aluminum acetate			Bromine liquid 3% max.	300	149
Aluminum chloride, aqueous	300	149	Butadiene		
Aluminum chloride, dry	300	149	Butyl acetate		
Aluminum fluoride			Butyl alcohol		
Aluminum hydroxide			n-Butylamine		
Aluminum nitrate			Butyl phthalate		
Aluminum oxychloride			Butyric acid	260	127
Aluminum sulfate	300	149	Calcium bisulfide		
Ammonia gas			Calcium bisulfite		
Ammonium bifluoride			Calcium carbonate		
Ammonium carbonate			Calcium chlorate		
Ammonium chloride 10%			Calcium chloride	300	149
Ammonium chloride 50%			Calcium hydroxide 10%	x	x
Ammonium chloride, sat.			Calcium hydroxide, sat.	x	x
Ammonium fluoride 10%			Calcium hypochlorite	x	x
Ammonium fluoride 25%			Calcium nitrate		
Ammonium hydroxide 25%	x	x	Calcium oxide		
Ammonium hydroxide, sat.	x	x	Calcium sulfate		
Ammonium nitrate			Caprylic acid		
Ammonium persulfate			Carbon bisulfide	160	71
Ammonium phosphate			Carbon dioxide, dry		
Ammonium sulfate 10–40%			Carbon dioxide, wet		
Ammonium sulfide			Carbon disulfide		
Ammonium sulfite			Carbon monoxide		
Amyl acetate			Carbon tetrachloride	212	100
Amyl alcohol	160	71	Carbonic acid		
Amyl chloride			Cellosolve		
Aniline	x	x	Chloracetic acid, 50% water		
Antimony trichloride			Chloracetic acid		
Aqua regia 3:1	x	x	Chlorine gas, dry	160	71

Table 3-39 *Continued*

Chemical	Maximum temp. °F	Maximum temp. °C	Chemical	Maximum temp. °F	Maximum temp. °C
Chlorine gas, wet	160	71	Magnesium chloride		
Chlorine, liquid			Malic acid 10%		
Chlorobenzene			Manganese chloride		
Chloroform	160	71	Methyl chloride	300	149
Chlorosulfonic acid	80	27	Methyl ethyl ketone	x	x
Chromic acid 10%	x	x	Methyl isobutyl ketone	x	x
Chromic acid 50%	x	x	Muriatic acid	300	149
Chromyl chloride			Nitric acid 5%	x	x
Citric acid 15%			Nitric acid 20%	x	x
Citric acid, concentrated			Nitric acid 70%	x	x
Copper acetate			Nitric acid, anhydrous	x	x
Copper carbonate			Nitrous acid, concentrated		
Copper chloride			Oleum		
Copper cyanide			Perchloric acid 10%		
Copper sulfate	300	149	Perchloric acid 70%		
Cresol			Phenol	x	x
Cupric chloride 5%	300	149	Phosphoric acid 50	212	100
Cupric chloride 50%	300	149	Picric acid		
Cyclohexane			Potassium bromide 30%		
Cyclohexanol			Salicylic acid		
Dichloroacetic acid			Silver bromide 10%		
Dichloroethane (ethylene di- chloride)			Sodium carbonate	x	x
			Sodium chloride		
Ethylene glycol	80	27	Sodium hydroxide 10%	x	x
Ferric chloride			Sodium hydroxide 50%	x	x
Ferric chloride 50% in water	300	149	Sodium hydroxide, con- centrated	x	x
Ferric nitrate 10–50%	300	149	Sodium hypochlorite 15%	x	x
Ferrous chloride 40%	300	149	Sodium hypochlorite, con- centrated	x	x
Ferrous nitrate					
Fluorine gas, dry			Sodium sulfide to 50%		
Fluorine gas, moist			Stannic chloride		
Hydrobromic acid, dilute	212	100	Stannous chloride		
Hydrobromic acid 20%	212	100	Sulfuric acid 10%	300	149
Hydrobromic acid 50%			Sulfuric acid 50%	300	149
Hydrochloric acid 20%	300	149	Sulfuric acid 70%	250	121
Hydrochloric acid 38%	300	149	Sulfuric acid 90%	100	38
Hydrocyanic acid 10%	160	71	Sulfuric acid 98%		
Hydrofluoric acid 30%	x	x	Sulfuric acid 100%		
Hydrofluoric acid 70%	x	x	Sulfuric acid, fuming		
Hydrofluoric acid 100%	x	x	Sulfurous acid	160	71
Hypochlorous acid			Thionyl chloride	80	27
Iodine solution 10%	x	x	Toluene	212	100
Ketones, general			Trichloroacetic acid 30%	80	27
Lactic acid 25%	160	71	White liquor		
Lactic acid, concentrated	160	71	Zinc chloride	300	149

Source: Schweitzer, Philip A. (1991). *Corrosion Resistance Tables*, Marcel Dekker, Inc., New York, Vols. 1 and 2.

Table 3-40   Average Physical and Mechanical Properties of Furan Resins

Specific gravity	1.8
Water absorption 24 hr at 73°F (23°C), %	2.65
Tensile strength at 73°F (23°C), psi	5040
Modulus of elasticity in tension at 73°F (23°C) $\times 10^5$	9.8
Compressive strength, psi	20,300
Flexural strength, psi	9,260
Izod impact strength, notched at 73°F (23°C) ft-lb/in.	0.67
Coefficient of thermal expansion	
in./in. °F $\times 10^{-5}$	1.35
in./10°F/100 ft	0.0135
Thermal conductivity, Btu/hr/ft^2/°F/in.	1.96
Resistance to heat at continuous drainage, °F/°C	300/149
Flame spread	<25

Source: Courtesy of Ametek Inc.

Table 3-41   Operating Pressure Versus Temperature for Haveg 61NA Pipe System

Nominal pipe size (in.)	Maximum operating pressure (psi) at °F/°C					
	50/10	100/38	150/66	200/93	250/121	300/149
½	100	100	100	100	100	100
¾	100	100	100	100	100	100
1	100	100	100	100	100	100
1 ½	100	100	100	100	100	100
2	100	100	100	100	100	100
3	95	90	85	82	80	75
4	67	65	61	59	55	52
6	41	40	39	37	35	30
8	35	34	32	31	30	29
10	29	28	27	26	25	24
12	24	23	22	20	19	18

Source: Courtesy Ametek Inc.

Table 3-42 Compatibility of Furan Resins with Selected Corrodents.
The chemicals listed are in the pure state or in a saturated solution unless otherwise indicated.
Compatibility is shown to the maximum allowable temperature for which data are available.
Incompatibility is shown by an x. A blank space indicates that data are unavailable.

Chemical	Maximum temp. °F	Maximum temp. °C	Chemical	Maximum temp. °F	Maximum temp. °C
Acetaldehyde	x	x	Barium carbonate	240	116
Acetamide			Barium chloride	260	127
Acetic acid 10%	212	100	Barium hydroxide	260	127
Acetic acid 50%	160	71	Barium sulfate		
Acetic acid 80%	80	27	Barium sulfide	260	127
Acetic acid, glacial	80	27	Benzaldehyde	80	27
Acetic anhydride	80	27	Benzene	160	71
Acetone	80	27	Benzene sulfonic acid 10%	160	71
Acetyl chloride	200	93	Benzoic acid	260	127
Acrylic acid	80	27	Benzyl alcohol	80	27
Acrylonitrile	80	27	Benzyl chloride	140	60
Adipic acid 25%	280	138	Borax	140	60
Allyl alcohol	300	149	Boric acid	300	149
Allyl chloride	300	149	Bromine gas, dry	x	x
Alum 5%	140	60	Bromine gas, moist	x	x
Aluminum acetate			Bromine liquid 3% max.	300	149
Aluminum chloride, aqueous	300	149	Butadiene		
Aluminum chloride, dry	300	149	Butyl acetate	260	127
Aluminum fluoride	280	138	Butyl alcohol	212	100
Aluminum hydroxide	260	127	*n*-Butylamine	x	x
Aluminum nitrate			Butyl phthalate		
Aluminum oxychloride			Butyric acid	260	127
Aluminum sulfate	160	71	Calcium bisulfide		
Ammonia gas			Calcium bisulfite	260	127
Ammonium bifluoride			Calcium carbonate		
Ammonium carbonate	240	116	Calcium chlorate		
Ammonium chloride 10%			Calcium chloride	160	71
Ammonium chloride 50%			Calcium hydroxide 10%		
Ammonium chloride, sat.			Calcium hydroxide, sat.	260	127
Ammonium fluoride 10%			Calcium hypochlorite	x	x
Ammonium fluoride 25%			Calcium nitrate	260	127
Ammonium hydroxide 25%	250	121	Calcium oxide		
Ammonium hydroxide, sat.	200	93	Calcium sulfate	260	127
Ammonium nitrate	250	121	Caprylic acid	250	121
Ammonium persulfate	260	127	Carbon bisulfide	160	71
Ammonium phosphate	260	127	Carbon dioxide, dry	90	32
Ammonium sulfate 10–40%	260	127	Carbon dioxide, wet	80	27
Ammonium sulfide	260	127	Carbon disulfide	260	127
Ammonium sulfite	240	116	Carbon monoxide		
Amyl acetate	260	127	Carbon tetrachloride	212	100
Amyl alcohol	278	137	Carbonic acid		
Amyl chloride	x	x	Cellosolve	240	116
Aniline	80	27	Chloracetic acid, 50% water	100	38
Antimony trichloride	250	121	Chloracetic acid	240	116
Aqua regia 3:1	x	x	Chlorine gas, dry	260	127

Table 3-42 *Continued*

Chemical	Maximum temp. °F	Maximum temp. °C	Chemical	Maximum temp. °F	Maximum temp. °C
Chlorine gas, wet	260	127	Magnesium chloride	260	127
Chlorine, liquid	x	x	Malic acid 10%	260	127
Chlorobenzene	260	127	Manganese chloride	200	93
Chloroform	x	x	Methyl chloride	120	49
Chlorosulfonic acid	260	127	Methyl ethyl ketone	80	27
Chromic acid 10%	x	x	Methyl isobutyl ketone	160	71
Chromic acid 50%	x	x	Muriatic acid	80	27
Chromyl chloride	250	121	Nitric acid 5%	x	x
Citric acid 15%	250	121	Nitric acid 20%	x	x
Citric acid, concentrated	250	121	Nitric acid 70%	x	x
Copper acetate	260	127	Nitric acid, anhydrous	x	x
Copper carbonate			Nitrous acid, concentrated	x	x
Copper chloride	260	127	Oleum	190	88
Copper cyanide	240	116	Perchloric acid 10%	x	x
Copper sulfate	300	149	Perchloric acid 70%	260	127
Cresol	260	127	Phenol	x	x
Cupric chloride 5%	300	149	Phosphoric acid 50	212	100
Cupric chloride 50%	300	149	Picric acid		
Cyclohexane	140	60	Potassium bromide 30%	260	127
Cyclohexanol			Salicylic acid	260	127
Dichloroacetic acid	x	x	Silver bromide 10%		
Dichloroethane (ethylene di-chloride)	250	121	Sodium carbonate	212	100
			Sodium chloride	260	127
Ethylene glycol	160	71	Sodium hydroxide 10%	x	x
Ferric chloride	260	127	Sodium hydroxide 50%	x	x
Ferric chloride 50% in water	160	71	Sodium hydroxide, con-centrated	x	x
Ferric nitrate 10–50%	160	71	Sodium hypochlorite 15%	x	x
Ferrous chloride	160	71	Sodium hypochlorite, con-centrated	x	x
Ferrous nitrate					
Fluorine gas, dry	x	x	Sodium sulfide to 10%	260	127
Fluorine gas, moist	x	x	Stannic chloride	260	127
Hydrobromic acid, dilute	212	100	Stannous chloride	250	121
Hydrobromic acid 20%	212	100	Sulfuric acid 10%	160	71
Hydrobromic acid 50%	212	100	Sulfuric acid 50%	80	27
Hydrochloric acid 20%	212	100	Sulfuric acid 70%	80	27
Hydrochloric acid 38%	80	27	Sulfuric acid 90%	x	x
Hydrocyanic acid 10%	160	71	Sulfuric acid 98%	x	x
Hydrofluoric acid 30%	230	110	Sulfuric acid 100%	x	x
Hydrofluoric acid 70%	140	60	Sulfuric acid, fuming	x	x
Hydrofluoric acid 100%	140	60	Sulfurous acid	160	71
Hypochlorous acid	x	x	Thionyl chloride	x	x
Iodine solution 10%	x	x	Toluene	212	100
Ketones, general	100	38	Trichloroacetic acid 30%	80	27
Lactic acid 25%	212	100	White liquor	140	60
Lactic acid, concentrated	160	71	Zinc chloride	160	71

Source: Schweitzer, Philip A. (1991). *Corrosion Resistance Tables*, Marcel Dekker, Inc., New York, Vols. 1 and 2.

Table 3-43 Physical and Mechanical Properties of Reinforced Phenolic Pipe

Specific gravity	1.75
Modulus of elasticity in tension at 70°F (21°C) $\times 10^5$	9.6
Flexural strength at 70°F (21°C), psi	9200
Coefficient of thermal expansion	
in./in. °F $\times 10^{-5}$	0.5–0.7
in./10°F/100 ft	0.005–0.007
Thermal conductivity, Btu/hr/ft²/°F/in.	2–3
Resistance to heat at continuous drainage, °F/°C	300/149
Flame spread	Nonflammable

Source: Courtesy of Haveg Division of Ametek Inc.

Table 3-44 Maximum Allowable Working Pressure of Haveg SP Pipe

Nominal pipe size (in.)	Maximum allowable working pressure (psi)	
	70°F (21°C)	300°F (149°C)
1	150	100
1½	150	100
2	150	100
3	100	75
4	100	75
6	25	15
8	15	10

Source: Courtesy of Haveg Div. of Ametek Inc.

Table 3-45 Compatibility of Haveg SP with Selected Corrodents.
The chemicals listed are in the pure state or in a saturated solution unless otherwise indicated. Compatibility is shown to the maximum allowable temperature for which data are available. Incompatibility is shown by an x. A blank space indicates that data are unavailable.

Chemical	Maximum temp.		Chemical	Maximum temp.	
	°F	°C		°F	°C
Acetaldehyde			Barium carbonate		
Acetamide			Barium chloride		
Acetic acid 10%	212	100	Barium hydroxide		
Acetic acid 50%			Barium sulfate		
Acetic acid 80%			Barium sulfide		
Acetic acid, glacial	70	21	Benzaldehyde	70	21
Acetic anhydride	70	21	Benzene	160	71
Acetone	x	x	Benzene sulfonic acid 10%	70	21
Acetyl chloride			Benzoic acid		
Acrylic acid			Benzyl alcohol		
Acrylonitrile			Benzyl chloride	70	21
Adipic acid			Borax		
Allyl alcohol			Boric acid		
Allyl chloride			Bromine gas, dry		
Alum			Bromine gas, moist		
Aluminum acetate			Bromine liquid		
Aluminum chloride, aqueous	90	32	Butadiene		
Aluminum chloride, dry			Butyl acetate	x	x
Aluminum fluoride			Butyl alcohol		
Aluminum hydroxide			*n*-Butylamine		
Aluminum nitrate			Butyl phthalate		
Aluminum oxychloride			Butyric acid	160	71
Aluminum sulfate	300	149	Calcium bisulfide		
Ammonia gas	90	32	Calcium bisulfite		
Ammonium bifluoride			Calcium carbonate		
Ammonium carbonate	90	32	Calcium chlorate		
Ammonium chloride 10%	80	27	Calcium chloride	300	149
Ammonium chloride 50%	80	27	Calcium hydroxide 10%		
Ammonium chloride, sat.	80	27	Calcium hydroxide, sat.		
Ammonium fluoride 10%			Calcium hypochlorite	x	x
Ammonium fluoride 25%			Calcium nitrate		
Ammonium hydroxide 25%	x	x	Calcium oxide		
Ammonium hydroxide, sat.	x	x	Calcium sulfate		
Ammonium nitrate	160	71	Caprylic acid		
Ammonium persulfate			Carbon bisulfide		
Ammonium phosphate			Carbon dioxide, dry	300	149
Ammonium sulfate 10–40%	300	149	Carbon dioxide, wet	300	149
Ammonium sulfide			Carbon disulfide		
Ammonium sulfite			Carbon monoxide		
Amyl acetate			Carbon tetrachloride	200	93
Amyl alcohol			Carbonic acid	200	93
Amyl chloride			Cellosolve		
Aniline	x	x	Chloracetic acid, 50% water		
Antimony trichloride			Chloracetic acid		
Aqua regia 3 : 1			Chlorine gas, dry		

Table 3-45 *Continued*

Chemical	Maximum temp. °F	Maximum temp. °C	Chemical	Maximum temp. °F	Maximum temp. °C
Chlorine gas, wet	x	x	Magnesium chloride		
Chlorine, liquid	x	x	Malic acid 10%		
Chlorobenzene	260	127	Manganese chloride		
Chloroform	160	71	Methyl chloride	160	71
Chlorosulfonic acid			Methyl ethyl ketone	x	x
Chromic acid 10%	x	x	Methyl isobutyl ketone	160	71
Chromic acid 50%	x	x	Muriatic acid	300	149
Chromyl chloride			Nitric acid 5%	x	x
Citric acid 15%	160	71	Nitric acid 20%	x	x
Citric acid, concentrated	160	71	Nitric acid 70%	x	x
Copper acetate			Nitric acid, anhydrous	x	x
Copper carbonate			Nitrous acid, concentrated		
Copper chloride			Oleum		
Copper cyanide			Perchloric acid 10%		
Copper sulfate	300	149	Perchloric acid 70%		
Cresol			Phenol	x	x
Cupric chloride 5%			Phosphoric acid 50–80%	212	100
Cupric chloride 50%			Picric acid		
Cyclohexane			Potassium bromide 30%		
Cyclohexanol			Salicylic acid		
Dichloroacetic acid			Silver bromide 10%		
Dichloroethane (ethylene di-chloride)			Sodium carbonate		
			Sodium chloride	300	149
Ethylene glycol	70	21	Sodium hydroxide 10%	x	x
Ferric chloride	300	149	Sodium hydroxide 50%	x	x
Ferric chloride 50% in water	300	149	Sodium hydroxide, con-centrated	x	x
Ferric nitrate 10–50%			Sodium hypochlorite 15%	x	x
Ferrous chloride 40%			Sodium hypochlorite, con-centrated	x	x
Ferrous nitrate			Sodium sulfide to 50%		
Fluorine gas, dry			Stannic chloride		
Fluorine gas, moist			Stannous chloride		
Hydrobromic acid, dilute	200	93	Sulfuric acid 10%	250	121
Hydrobromic acid 20%	200	93	Sulfuric acid 50%	250	121
Hydrobromic acid 50%	200	93	Sulfuric acid 70%	200	93
Hydrochloric acid 20%	300	149	Sulfuric acid 90%	70	21
Hydrochloric acid 38%	300	149	Sulfuric acid 98%	x	x
Hydrocyanic acid 10%			Sulfuric acid 100%	x	x
Hydrofluoric acid 30%	x	x	Sulfuric acid, fuming		
Hydrofluoric acid 70%	x	x	Sulfurous acid	80	27
Hydrofluoric acid 100%	x	x	Thionyl chloride	200	93
Hypochlorous acid			Toluene		
Iodine solution 10%			Trichloroacetic acid 30%		
Ketones, general			White liquor		
Lactic acid 25%	160	71	Zinc chloride	300	149
Lactic acid, concentrated					

Source: Schweitzer, Philip A. (1991). *Corrosion Resistance Tables*, Marcel Dekker, Inc., New York, Vols. 1 and 2.

4

Thermoplastic-Lined Piping Systems

Neither metallic nor nonmetallic piping systems can completely meet the needs of today's industries in solving corrosion problems. The least expensive of the metallic materials, and the most common, is carbon steel. It is strong, easy to work with, and readily available. However, even mildly corrosive materials can destroy the inner surface, thereby limiting its safe or economic use in handling corrosive materials.

Stainless steel and other alloys are able to provide improved corrosion resistance, but at a much greater cost. The ability of metals to resist corrosion is the result of the formation of a thin protective oxide film. While the film is present, additional corrosion does not take place. This film can be quickly eroded by high velocities, turbulence, or mildly abrasive fluids. Stainless steel pipe is susceptible to corrosion at the weld areas, more commonly referred to as the "heat-affected zone." It is also subject to chemically induced stress cracking, particularly in the presence of chlorides.

Other metal alloys, while they may have an improved resistance to corrosion relative to stainless steel, must maintain their protective film in order to be effective. At the same time, these materials are more expensive than stainless steel. Even when a metal is said to resist corrosive attack, there still may be some corrosion. Every metal experiences a corrosion rate in the presence of a specific corrodent. When this rate is low enough (below 5 mil/year), the material is said to be resistant.

Nonmetallic materials such as the thermoplastics, thermosets, and glass do not have a corrosion rate like metals. These materials will either be completely resistant to a corrodent or be completely destroyed. When resistant, there is no "corrosion rate." Therefore these materials possess a decided advantage over the metallics in resisting corrosion.

However, the thermoplasts do not have the physical strength or ruggedness of metals. In addition, special pipe supports are often required and safety considerations may limit not only allowable pressures and temperatures, but also system locations. The majority of the thermoplasts are affected by ultraviolet light and oxygen, requiring UV stabilizers and antioxidants to be added to the formulation.

The thermosets increase the range of temperatures and pressures but still require special supporting techniques, protection from external loading, and allowances for expansion and contraction. As operating temperature increases, the allowable operating pressure decreases.

Glass piping, although extremely corrosion resistant, must be carefully installed and protected when used in pressurized applications.

Lined piping has the advantages of the physical strength of the metallics and the corrosion resistance of the nonmetallics. Inside the pipe is a corrosion-resistant liner best suited for the application. It can be a thermoplastic material, glass, or even another metal. Outside the pipe, where structural strength and shock resistance are required, the metal shell provides these features. Lined piping systems also increase the allowable operating pressure over that of the solid plastic pipe material.

4.1   LINER SELECTION

With a wide choice of liners available, consideration should be given to their selection. In order to make the best selection, several broad categories must be considered:

1. Materials being handled
2. Operating conditions
3. Conditions external to the pipe

First, the following questions must be asked about the materials being handled:

1. What are the primary chemicals being handled and at what concentrations?
2. Are there any secondary chemicals, and if so, at what concentrations?
3. Are there any trace impurities or chemicals?
4. Are there any solids present, and if so, what is the particle size and concentration?
5. What are the flow rates, maximum and minimum?
6. What are the fluid purity requirements?

The correct answers to these questions are critical for proper liner selection.
    Second, questions about the operating conditions must be asked:

1. What is the normal operating temperature and temperature range?
2. What peak temperatures can be reached during shutdown, start-up, process upset, etc.?
3. Will any mixing areas exist where exothermic or heat of mixing temperatures can develop?
4. What is the normal operating pressure?
5. What vacuum conditions and range are possible during operation, start-up, shutdown, or upset conditions?
6. Will there be temperature cycling?
7. What are the line-cleaning methods (chemical, steam, high pressure water, pigging, etc.)?

Finally, the following conditions external to the pipe should be examined:

1. Ambient temperature conditions
2. Maximum surface temperature during static operation
3. Insulation requirements

4. Exterior finish requirements
5. Heat-tracing requirements
6. Need for grounding

With this information an intelligent liner selection can be made.

4.2 DESIGN CONSIDERATIONS

All piping systems are designed to operate under specific fluid pressure, temperature, and flow rate specifications. In order to maximize the life of a piping system, the system needs to be engineered properly in order to meet the specified process conditions. Proper design of the system requires more than sizing the pipe to meet the flow rate and allowable pressure drop. Other factors must be taken into account.

4.2.1 Permeation

Physical properties of the liner determine the reaction of the lining material to such physical actions as permeation and absorption. If a lining material is subject to permeation by a corrosive chemical, it is possible for the base metal to be attacked and corroded even though the lining material itself is unaffected. Because of this, permeation and absorption are two factors that must be taken into account when specifying a lining material. All materials are somewhat permeable to chemical molecules, but plastic materials tend to be an order of magnitude greater in their permeability rates than metals. Gases, vapors, or liquids will permeate polymers. Permeation is a molecular migration either through microvoids in the polymer (if the structure is more or less porous) or between polymer molecules. In neither case is there any chemical attack on the polymer. This action is strictly a physical phenomenon.

Permeation can result in:

1. Bond failure and blistering, resulting from the accumulation of fluids at the bond when the substrate is less permeable than the liner, or from corrosion/reaction products if the substrate is attacked by the permeant.
2. Failure of the substrate from corrosive attack.
3. Loss of contents through substrate and liner as a result of eventual failure of the substrate. In unbonded linings it is important that the space between the liner and support member be vented to the atmosphere, not only to allow minute quantities of permeant vapors to escape, but to prevent expansion of entrapped air from collapsing the liner.

Permeation is a function of two variables: one relating to diffusion between molecular chains and the other to the solubility of the permeant in the polymer. The partial pressure gradient for gases and the concentration gradient of liquids are the driving forces of diffusion. Solubility is a function of the affinity of the permeant for the polymer.

A distinction must be made between permeation and the passage of materials through cracks and continuous voids. Even though in both cases migrating chemicals travel through the polymer from one side to the other, the two phenomena are in no way related.

Permeation is affected by the following factors, over which the user has some control:

1. Temperature and pressure
2. Permeant concentration
3. Thickness of the polymer

Increasing the temperature will increase the permeation rate, since the solubility of the permeant in the polymer will increase, and as the temperature rises the polymer chain movement is stimulated, permitting more permeants to diffuse among the chain more easily. For many gases the permeation rates increase linearly with the partial pressure gradient, and the same effect is experienced with concentration gradients of liquids. If the permeant is highly soluble in the polymer, the permeability increase may be nonlinear. The thickness of the polymer affects permeation. An increase in the thickness will generally decrease permeation by the square of the thickness. For general corrosion resistance, thicknesses of 0.010–0.020 in. are usually satisfactory.

In addition to thickness, the density of the polymer will also have an effect on the permeation rate. The higher the specific gravity of the polymer liner, the fewer the voids through which permeation can take place.

A comparison of the specific gravity of liners produced from different polymers does not provide any indication as to the relative rates of permeation. However, a comparison of liners produced from the same polymer having different specific gravities will provide an indication of the relative permeation rates, the denser liner having the lowest rate.

Other factors affecting permeation consisting of chemical and physicochemical properties are:

1. Ease of condensation of the permeant: chemicals that condense readily will permeate at higher rates.
2. The higher the intermolecular chain forces (e.g., van der Waals hydrogen bonding) of the polymer, the lower the permeation rate.
3. The higher the level of crystallinity in the polymer, the lower the permeation rate, since these ordered regions reduce the molecular volume available for passage. Amorphous regions are easier to penetrate.
4. The greater the degree of cross-linking within the polymer, the lower the permeation rate.
5. Chemical similarity between the polymer and the permeant: when the polymer and permeant both have similar functional groups, the permeation rate will increase.
6. The smaller the molecule of the permeant, the greater the permeation rate.

The magnitude of any of the effects will be a function of the combination of polymer and permeant in actual service. All piping products lined with fluoro-polymers, such as PTFE, FEP, PFA, ETFE, and ECTFE, are subject to some permeation from certain fluids. For example, in nitrate-containing fluids the nitrates can slowly permeate the liner. However, ECTFE liners can be designed to prevent permeation. If the outer pipe shell is of carbon steel construction, these permeant vapors can stress crack the interior of the steel shell. This can lead to an intergranular phenomena known as nitrate stress cracking. Because of its nearly universal chemical resistance and high temperature capability, PTFE-lined pipe is generally considered to be the universal solution to chemical resistance problems. If a corrodent to be handled is known or suspected of being able to permeate PTFE, FEP, PFA, ETFE, or ECTFE, it is advisable to consider an alternate lining material. In many cases it may be advisable to conduct tests beforehand based on the actual combination of conditions.

4.2.2 Absorption

Polymers, unlike metals, absorb varying quantities of the corrodents they come into contact with, especially organic liquids. This can result in swelling, cracking, and penetration to the substrate. Swelling can cause softening of the polymer. In a lined pipe this can introduce high stresses and failure of the bond. If the polymer has a high absorption rate, permeation will probably occur. An approximation of the expected permeation and/or absorption of a polymer can be based on the absorption of water. These data are usually available. Table 4-1 provides the water absorption rates for the more common polymers used as pipe liners.

The failure due to absorption can best be understood by considering the "steam cycle" test described in the ASTM standards for lined pipe. A section of lined pipe is subjected to 125 psi(0.8 M Pa) steam, alternating with low-pressure cold water, causing very severe thermal and pressure fluctuations. This is repeated for 100 cycles. The steam creates a pressure and temperature gradient through the liner, causing absorption of a small quantity of steam, which condenses to water within the inner wall. Upon pressure release, or on reintroduction of steam, the entrapped water can expand to vapor causing an original micropore. The repeated pressure and thermal cycling enlarges the micropores, ultimately producing visible water-filled blisters within the liner.

In an actual process the polymer may absorb process fluids, and repeated temperature or pressure cycling can cause blisters. Eventually corrodent may find its way to the substrate. Related effects can occur when process chemicals are absorbed, which may later react, decompose, or solidify within the structure of the plastic. Prolonged retention of absorbed chemicals may lead to their decomposition within the polymer. Although unusual, it is possible for absorbed monomers to polymerize.

In order to reduce absorption, several steps can be taken. Thermal insulation of the lined pipe will reduce the temperature gradient in the pipe liner, thereby preventing condensation and subsequent expansion of the absorbed fluids. This also reduces the rate and magnitude of temperature changes, keeping blistering to a minimum. The use of operating procedures or devices that limit the ratio of process pressure reductions or temperature increases will provide additional protection.

4.2.3 Environmental Stress Cracking

When a tough polymer is stressed for an extended period of time under loads that are small relative to the polymers yield point, stress cracks develop. Cracking occurs with little elongation of the material, even though it may exhibit more than 300% elongation before failure under other circumstances. The higher the molecular weight of the polymer, the less the likelihood of environmental stress cracking, other things being equal. Molecular weight is a function of the length of individual chains that make up the polymer. Longer-chain polymers tend to crystallize less than polymers of lower molecular weight or shorter chains, and also have a greater load-bearing capacity.

Crystallinity is an important factor affecting stress corrosion cracking. The less the crystallization that takes place, the less the likelihood of stress cracking. Unfortunately, the lower the crystallinity, the greater the likelihood of permeation.

Resistance to stress cracking can be reduced by the absorption of substances that chemically resemble the polymer and will plasticize it. In addition the mechanical strength will also be reduced. Halogenated chemicals, particularly those consisting of small molecules containing fluorine or chlorine, are especially likely to be similar to the fluoropolymers and should be tested for their effect.

Improper installation, fabrication, and support of a lined pipe system can aggravate the stress cracking problem. Stress crack resistance is lower under tensile stress than under compressive stress. The steel pipe in a lined piping system which supports the lining handles most of the stress. However, differing rates of shrinkage or expansion in the lining can induce stresses. In addition, residual fabrication stresses may be present in the liner due to careless fabrication. The latter potential problem can be eliminated by working with an experienced, competent manufacturer.

The presence of contaminants in a fluid may act as an accelerator. For example, while polypropylene can safely handle sulfuric or hydrochloric acids, iron or copper contamination in concentrated sulfuric or hydrochloric acids can result in stress cracking of polypropylene liners.

4.2.4 Plastic-Lined Piping Manufacture

Most of the thermoplastic and elastomeric materials are available in lined piping systems. Some of the materials used do not have sufficient mechanical strength to permit the manufacture of solid piping systems, but they can be utilized as liners. In general, lining thicknesses range from 0.10 to 0.20 inch in the widely used pipe diameters. Several manufacturers offer large-diameter pipe (>8 inches) in some liners.

The outer shell is usually schedule 40 or schedule 80 carbon steel pipe, although one manufacturer produces a shell of 304 ELC as a standard option. Flanges are ANSI class 150 ductile iron or cast steel. Class 300 ANSI cast steel flanges are also available.

Plastic-lined piping systems are actually two separate systems joined to operate as one. Most plastic liners have thermal expansion coefficients 6–15 times greater than steel. The thermal stress that results from this difference could create cracking of the plastic at the flange face.

Several manufacturing techniques have been developed to overcome this problem. By properly securing the liner within the shell, both the shell and liner expand and contract as one. The stresses caused by thermal expansion and contraction are distributed evenly throughout the entire length of the pipe section.

The first method of manufacture is known as swaging. In this method a plastic liner is inserted into an oversized steel tubing, and the loose assembly is fed into a swaging machine, where the tubing is compressed down on the liner under tremendous pressure. Both the tubing and liner are reduced in size to the finished dimension.

One manufacturer prepares the inner surface of the tubing with thousands of tiny barbs prior to swaging. As the tubing is pressed down onto the liner, the tiny barbs are imbedded slightly into the liner surface, while the plastic fills the metal cavities. This inner locking secures the liner to the tubing over the entire length of the pipe.

The second method of manufacture is known as interference fit or reverse swaging. In this method, a slightly oversized plastic liner is compressed with a sizing dye as it is pulled into the steel tubing. After insertion, the liner expands, creating a tight fit between liner and tube.

A third method of manufacture provides a slip fit or loose lining. This type of lining is produced by slipping a liner into the steel housing. It is so loose that it can be easily removed in the field. This type of liner has the advantage of easy field fabrication, but it lacks the tight bond between the liner and housing. Because of its construction little protection is provided against flange face failure resulting from differential thermal expansion of the liner. This method of manufacture is employed when liners are subject to permeation. The space between the outer surface of the liner and the inner surface of the pipe is vented to permit the escape of vapors that have permeated.

The fourth method of manufacture involves the bonding of unformed lining material directly to the pipe. Various techniques can be employed depending upon the plastic selected. In some cases, the molten lining material is applied directly to the inside of the pipe; in other cases a fluidized bed process is used. Other similar methods are used to produce dense, homogeneous linings. When the fluidized bed technique is employed, the metallic pipe is lined internally and coated externally. This provides added protection to piping systems installed in corrosive atmospheres.

4.2.5 Producing Odd Lengths

All of the major lined pipe manufacturers have local stocking distributors capable of producing odd lengths of pipe. These distributors have qualified, trained personnel who can supply odd lengths or closure pieces quickly. Also available from the manufacturers are field fabrication kits, which will permit the production of odd-length pieces in the field. Training in the use of this equipment is provided by the manufacturer and/or distributor.

Swaged and reverse swaged pipe can be fabricated only with threaded flanges. Slip-fit pipe can be fabricated with lap joint, threaded, or welded flanges. Lined pipe produced by fluidized bed techniques cannot be field fabricated. Piping for these systems must be accurately laid out and ordered from the factory.

4.2.6 Fluid Velocity

If abrasives are not present, plastic-lined pipe will not erode due to corrosive fluids. Under these conditions flow through the pipe should be limited to a maximum of 12 ft/sec.

When slurries are being handled, the allowable velocity will vary, depending upon the nature of the slurry. In general solids concentration in the slurry should be kept at least 50% or higher, with flow velocities in the 2–4 ft/sec range. Particle sizes will also have an effect. Sizes greater than 100 mesh may result in significant wear. If particle sizes are less than 250 mesh, little wear will result, while particle sizes between 250 and 100 mesh will result in some wear.

The recommended liners for slurry application are PVDF and PTFE. Compared with PTFE or PVDF, polypropylene wears twice as fast and carbon steel wears 6.5 times faster in corrosive slurries.

Plastic-lined piping systems have a lower friction factor than metallic piping. This is because of its smooth surface. As a result, lined piping has a lower pressure drop throughout the pipe and fittings as compared to equivalent size metallic pipe. This smooth surface also tends to provide a nonstick surface that prevents material from adhering to the pipe.

4.2.7 Heat Tracing

Heat tracing of the pipe line may be required when handling materials having a high freezing point or materials having a high viscosity. Lined piping can be traced successfully, providing certain basic rules are followed. Three heat tracing systems may be used: liquid tracing, electrical tracing, and steam tracing.

Factors to be considered when selecting the tracing system are:

1. The plastic liner
2. Climate

3. Length of heating season
4. Temperature range to be maintained
5. Initial cost
6. Operating costs
7. Need for future expansion

If PTFE-lined pipe is being used, the dew point of the fluid being handled must be considered. The pipe shell temperature must be kept high enough that any material permeating the liner will not condense on the inner metal wall. This will minimize shell corrosion.

Sizing of the tracing system for lined pipe requires the same steps that would be used for a metal pipe. There must be a balance between the design load and available heat input. The only potential difficulty is the temperature sensitivity of the plastic liner. Table 4-2 lists the maximum temperature limits for plastic-lined piping systems. Remember that these temperatures may have to be reduced when handling certain corrosives. The compatibility tables must be checked for these limits.

Liquid Tracing

Hot water and water-glycol systems are preferred when there is a need for close control of operating temperatures or to hold them within the temperature limitations of plastic liners such as Saran or PP. The tracing should be installed along the bottom of the line, imbedded in heat transfer cement, according to the manufacturer's recommendation. When the tracing runs are long, it will be necessary to provide expansion loops in the tracing tubing. Do not apply heat transfer cement to the expansion loops. Flanges, valves, fittings, and other pieces of equipment are sources of concentrated heat loss, therefore make sure that they are sufficiently traced (see Fig. 4-1).

Electrical Tracing

Electrical tracing is available in cable or strip forms, either of which can be used (see Fig. 4-2). Proper engineering of the electrical tracing system is required if damage to the liner is to be prevented. Keep in mind that if the temperature limit of the liner is reduced because of the nature of the fluid being handled, it will be necessary to limit the electrical input to prevent overheating and deterioration of the liner. This is particularly important with Saran, PP, and PVDF linings. Careful design of the electrical heating system is also required. If improperly designed, local overheating can occur, which will cause liner deterioration. In PTFE-lined pipe, localized overheating can cause increased vapor permeation through the liner and may even cause liner buckling.

All electrical tracing has a factor called T-rating. This is the highest temperature that the exposed surface of the cable can attain when drawing maximum wattage in a perfectly insulated environment. For proper design and installation of an electrical tracing system, the following rules should be adhered to:

1. Use cable with a T-rating below the liner's temperature limit for your application and/or use self-limiting electrical tracing.
2. Place the thermostat as close as possible to the electrical tracing to minimize heat loss effects and to decrease reaction time. The heater cable output should be in the range of 4–6 watts per foot or less.
3. Make sure that the cable does not overlap (criss-cross) onto itself. If it does, heat concentration at this point will be greatly increased.

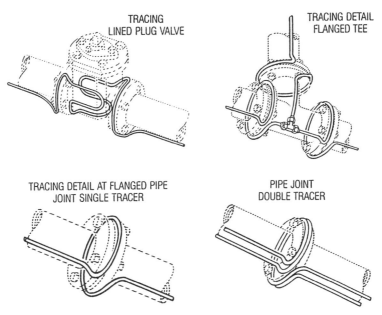

Figure 4-1 Recommended tracing configuration for joints, fittings, and valves. (Courtesy of Dow Chemical.)

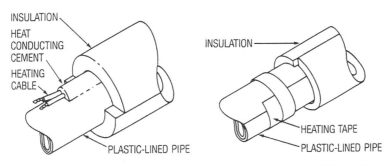

Figure 4-2 Electrical tracing methods and configurations. (Courtesy of Dow Chemical.)

Steam Tracing

Steam tracing should never be used on Saran- or PP-lined pipe. Steam temperature must be controlled so that the temperature limit of the liner is not exceeded (as rated by the fluid being handled). Application of a direct contact steam tracing on PTFE-lined pipe in an axial direction can result in premature failure of the lining. The tracing should be installed in an annular space between the outside shell of the lined pipe and the inside of oversized pipe insulation (see Fig. 4-3). This will alleviate localized overheating of the liner.

4.2.8　Jacketed Pipe

Some manufacturers supply their lined pipe with a jacket. This jacket can be used with the same considerations required for tracing. The temperature of the heating medium must not exceed the maximum allowable temperature of the liner. Adequate controls will be required on the heating medium to ensure this.

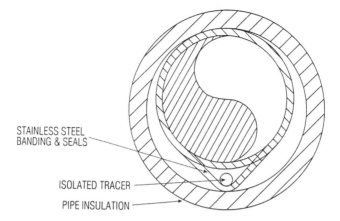

STAINLESS STEEL
BANDING & SEALS

ISOLATED TRACER

PIPE INSULATION

Figure 4-3 Preferred pipe line tracing details. (Courtesy of Dow Chemical.)

4.2.9 Venting

When plastic-lined pipe is insulated, the insulation may block the pipe's vent holes, which allow permeating vapors to escape. To overcome this, vent hole extenders should be provided. These are available from the pipe manufacturers on special order (see Fig. 4-4).

4.2.10 Bolt Torquing

Plastic-lined pipe, with few exceptions, is joined by flanges. The plastic-lined faces of the flange act as a gasket. Proper assembly technique is required in order to ensure a leak-free joint.

When assembling always use new, high-strength B7 machine bolts and a torque wrench for bolt tightening. Tables 4-3 and 4-4 provide average bolt torque ratings. Since recommendations for torque ratings vary between manufacturers, use the rating recom-

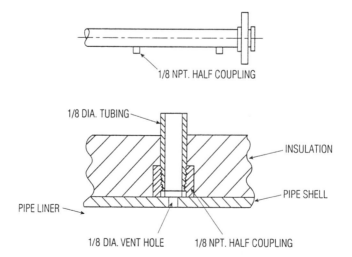

1/8 NPT. HALF COUPLING

1/8 DIA. TUBING

INSULATION

PIPE LINER

PIPE SHELL

1/8 DIA. VENT HOLE   1/8 NPT. HALF COUPLING

Figure 4-4 Venting of permeant vapor. (Courtesy of Dow Chemical.)

mended for the particular lined pipe being installed. Do not overtorque, since this can cause damage to the plastic sealing surfaces.

The torquing process should be done in five steps, bringing the torque up in 20% increments to the final torque. Tightening should be done on a crisscross sequential basis until the final torque is reached. At this point, rotational tightening is required to ensure that bolts are stable at the final values. Usually twice around at this step is all that is required.

After 24 hours the bolts should be checked for proper torque values. During the 24-hour period, the plastic has had an opportunity to seat and the bolts to relax. If the system is to operate at elevated temperatures, it is recommended that hot water be circulated through the system for 24 hours at the operating temperature. After the system has cooled to ambient temperature, the bolts should be retorqued. Never torque the bolts while the system is above ambient temperature.

If a flange leak occurs and the bolts on the leaking side are at the proper torque value, do not tighten further as permanent damage to the seal face may occur. The bolts on the opposite side should be loosened a half turn at a time and the corresponding bolts on the opposite side tightened the same amount. Should the leak persist the joint should be disassembled and the sealing faces checked for scratches or dents across an entire face which could provide a leak path. Any scratches or dents that do not exceed 20% of the liner thickness can be eliminated by hand-polishing with a fine abrasive cloth or paper.

4.2.11 Supporting Lined Piping

Although carbon steel pipe is used as an outer shell, it should not be assumed that support spacing used for carbon steel pipe can be used for lined pipe. It is recommended that a hanger be used to support each spool piece and that it be placed near the flange. This is done to prevent excess deflection at the joint. Additional support should also be provided for valves, clusters of fittings, instruments, or other accessories that may be installed in the pipe line. All supports should permit the pipe to move freely as it expands or contracts due to thermal changes.

In systems that will experience thermal cycling, provision must be made to compensate for expansion and contraction. One way to do this is to install expansion joints, which are available in comparable materials of construction. However, there are some disadvantages to this approach. The expansion joints have a bellows form of construction. This is usually construed as the weakest link in the piping system. In addition, when installed in a horizontal line, corrosive materials will collect in the bellows. If the line is disassembled for any reason, this could pose a safety hazard.

Good engineering design can eliminate the need for expansion joints. On long straight runs either the equipment or the piping should be allowed to move to compensate for thermal expansion and contraction. Lined piping systems can be treated the same as carbon steel piping when installed in a long straight run. The pipe must be allowed to move freely. Figure 4.5 illustrates the correct way to handle this situation.

Many piping systems involve the installation of piping between pumps and equipment. On new installations the equipment can be laid out in such a manner as to permit a proper installation, without having direct runs.

In some instances the layout will be very tight and the installation of a piping loop is not possible. Under these circumstances the pump should be allowed to float freely. Large pump motor base assemblies are mounted on a platform with stilt legs. Several man-

WRONG

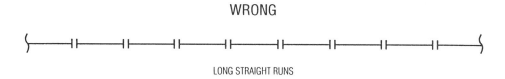

LONG STRAIGHT RUNS

RIGHT

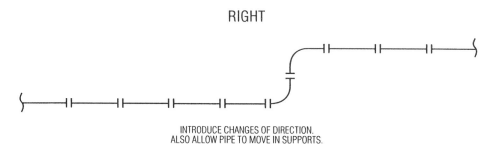

INTRODUCE CHANGES OF DIRECTION.
ALSO ALLOW PIPE TO MOVE IN SUPPORTS.

Figure 4-5   Installation of long straight runs. (Courtesy of Dow Chemical.)

ufacturers are able to provide these. This arrangement allows enough movement to relieve normal stresses in a piping system. Smaller pumping systems can be mounted on a square tubing with a ⅛-inch-thick sheet of PTFE fixed on top of a section of C channel. The pumping system and the square tubing would free float, as a single unit, on top of the channel (see Fig. 4.6).

4.2.12   Friction Factors for Lined Pipe

In order to determine the pressure drop of a piping system, it is necessary to know the friction factor. This factor can be calculated using a Moody's diagram when the Reynolds number and roughness factor are known.

The parametric curve related to the roughness factor for plastic-lined pipe is labeled "smooth pipes." Use this curve along with the Reynolds number to determine the friction factor for a lined piping system.

The basic definition of the Reynolds number is:

$$N_R = \frac{vD\rho}{\mu}$$

where:

N_R = Reynolds number
v = average fluid velocity
D = inside pipe diameter
ρ = fluid density
μ = fluid viscosity

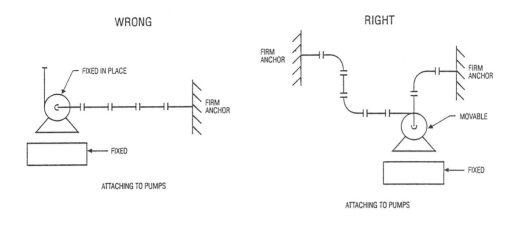

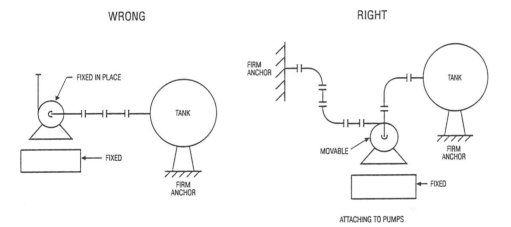

Figure 4-6 Installation of piping between fixed points. (Courtesy of Dow Chemical.)

4.3   HANDLING OF PLASTIC-LINED PIPE

Plastic-lined pipe is shipped with protective end caps in place to protect the flared or molded sealing faces. These should be left in place until the pipe section is installed. When removing the flange covers, do not rotate the flanges (unless rotatable flanges had been specified). Exercise care so as not to damage the sealing faces.

Under no circumstance should any welding be performed on plastic-lined pipe. Heat caused during the welding operation would be detrimental to the plastic liner.

Plastic-lined pipe is usually supplied with a prime coat of paint to protect the carbon steel shell during shipping. Normally the pipe is sandblasted and painted. Both operations should be carried out carefully to protect the plastic faces. Do not remove the end caps during these operations. Direct all sandblasting away from the flange face. If the pipe is supplied with vent holes, make sure that these are not plugged with paint. These vent holes are required to prevent possible gas build-up behind the liner, which could cause liner collapse.

4.4.　GASKETS

Gaskets are not normally required or used when joining plastic-lined pipe to plastic-lined pipe initially. If piping is disconnected and later reinstalled, or if piping is reconnected after maintenance work, a gasket should be used.

Gaskets are required when connecting plastic-lined pipe to non–plastic-lined pipe, valves, fittings, or equipment. Make sure that the gasket material is compatible with the fluid being handled (see Table 1-1).

4.5　LINING MATERIALS

With the exception of Saran, PVC, PE, and PP, lining materials are considered fluoroplastics. These are divided into two groups: fully fluorinated fluorocarbon polymers, such as PTFE, FEP, and PFA, called perfluoropolymers and partially fluorinated polymers, such as ETFE, Kynar PVDF, Solef PVDF, and Halar ECTFE, which are called fluoropolymers. The polymeric characteristics within each group are similar, but there are important differences between the groups.

4.5.1　Fluorocarbon Resins

The fluorocarbon resins consist of PTFE, FEP, and PFA. They are characterized by the following properties:

1. Nonpolarity—the carbon backbone of the linear polymer is completely sheathed by the tightly held electron cloud of fluorine atoms, with electronegatives balanced
2. High C-F and C-C bond strengths
3. Low interchain forces—interactive forces between the two adjacent polymer chains are significantly lower than the bond forces within one chain
4. Crystallinity
5. High degree of polymerization

These properties provide the following advantages to these materials:

1. High melting point
2. High thermal stability
3. High upper service temperatures
4. Inertness to chemical attack by almost all chemicals
5. Low coefficient of friction
6. Low water absorbability
7. Weatherability
8. Flame resistance
9. Toughness

Table 4-5 lists the properties of these resins.

4.5.2　Fluoropolymer Resins

The fluoropolymers are resistant to a broader range of chemicals at higher temperatures than chlorinated or hydrogenated polymers, polyesters, and polyamides. However, their properties are significantly different from those of fully fluorinated resins. Included in this category are ETFE, (Tefzel), PVDF (Kynar, Solef), and ECTFE (Halar).

The polarity of these resins is increased as the result of substituting hydrogen or chlorine, which have different electronegatives relative to fluorine. The length of their bonds to the carbon backbone also differs from those with fluorine. The centers of electronegativity and electropositivity are not held as tightly as with carbon-fluorine bonds. As a result differential separation of charge can be induced chemically between atoms in adjacent chains to permit electrostatic interaction between chains. Higher mechanical properties are produced because of the increased interpolymer chain attraction along with the interlocking of differently sized atoms. In addition, the increased polarity/ interpolymer attraction reduces the permeation of penetrants through the resin. Because of the substitution of hydrogen or of hydrogen and chlorine, for fluorine, chemical and thermal stability are sacrificed.

The chemical stability is affected by the arrangement of the substituting elements along the polymer chain. Solubility can be a leading indicator. For example, ETFE has no known solvent under ordinary circumstances, while PVDF is soluble in common industrial ketones (e.g., methyl ethyl ketone) and ECTFE is soluble in some fluorinated solvents. While the fully fluorinated polymers are resistant to strong acids and alkalies, the substituted polymers are adversely affected. However, these resins do possess very advantageous properties, both mechanical and chemically resistant (Table 4-6).

4.6   PTFE-LINED PIPE

Polytetrafluorethylene (PTFE) was originally developed by DuPont in 1938, made commercially available in 1948, and is sold under the trade name of Teflon PTFE. This resin is also produced by Ausimont USA, Inc. and sold under the trade name of Fluon and by others with their own trade names. It has become the standard against which other fluoroplastics are compared. It is a fully fluorinated thermoplastic with the following formula:

$$
\begin{array}{ccc}
\text{F} & & \text{F} \\
| & & | \\
-\;\text{C} & - & \text{C}\; - \\
| & & | \\
\text{F} & & \text{F}
\end{array}
$$

PTFE-lined pipe has the widest range of operating temperatures of any of the lined piping systems, being capable of operation between –20°F (29°C) and 450°F (232°C). This range is based on the physical/mechanical properties of the liner. When handling certain aggresive chemicals, the upper temperature limit may have to be reduced.

The operating pressure of the lined piping system is based on the steel pipe and cast iron or ductile iron fitting housing. Because of the poor impact strength or shock resistance, cast iron fittings are not normally supplied with PTFE-lined pipe. The standard used is ductile iron. These housings permit operating pressures of 150 or 625 psi, both at 450°F (232°C). The pressure rating is based on the pressure rating of the unlined fitting. Table 4-7 provides the vacuum ratings of PTFE-lined pipe based on temperature and the method of securing the liner in the pipe.

Pipe is available in lengths of up to 10 feet. It is customary to lay out the piping runs and order the appropriate lengths from the manufacturer. Odd lengths of pipe can be field-fabricated using special tools supplied by the piping manufacturer and/or distributor.

PTFE liners are subject to permeation by some corrodents. When this condition exists, vents must be provided on the outer steel pipe shell to permit escape of the permeant vapors. Table 4-8 provides some typical permeation rates for PTFE with a specific gravity greater than 2.2. Table 4-9 provides the physical and mechanical properties of PTFE.

Pipe is available in standard sizes of 1 inch through 12 inches, with larger sizes available on special order. With the exception of Dow, the outer shell is of carbon steel. Dow produces a line of PTFE-lined pipe in sizes of 1 inch through 4 inches with a type 304L stainless steel shell.

PTFE is unique in its corrosion-resistant properties. It is chemically inert in the presence of most materials, which means that no detectable chemical reaction is taking place. There are very few chemicals that will attack PTFE within normal use temperature. These reactants are among the most violent oxidizers and reducing agents known. Elemental sodium in intimate contact with fluorocarbons removes fluorine from the polymer molecule. The other alkali metals (potassium, lithium, etc.) react in a similar manner.

Fluorine and related compounds (e.g., chlorine trifluoride) are absorbed into the PTFE resin with such intimate contact that the mixture becomes sensitive to a source of ignition such as impact. These potent oxidizers should only be handled with great care and a recognition of the potential hazards.

The handling of 80% sodium hydroxide, aluminum chloride, ammonia, and certain amines at high temperatures may produce the same effect as elemental sodium. Also, slow oxidative attack can be produced by 70% nitric acid under pressure and at 480°F (250°C).

Table 4-10 provides the compatibility of PTFE with selected corrodents.

4.7 FEP-LINED PIPE

Fluorinated ethylene propylene (FEP) was introduced in 1960. It is a fully fluorinated thermoplastic with some branching but consists mainly of linear chains having the following formula:

$$
\begin{array}{ccccc}
F & F & F & F & F \\
| & | & | & | & | \\
-C- & C- & C- & C- & C- \\
| & | & | & | & | \\
F & F & | & F & F \\
 & & F-C-F & & \\
 & & | & & \\
 & & F & & N
\end{array}
$$

It was developed to add melt processibility properties to the desirable properties of PTFE. Although FEP has a higher temperature rating than other plastics, it has a lower temperature rating than PTFE. It exhibits changes in physical strength after prolonged exposure above 400°F (204°C), which accounts for the lower temperature rating.

FEP is a relatively soft plastic with lower tensile strength, wear resistance, and creep resistance than other plastics. It is insensitive to notched impact forces and has excellent permeation resistance to most liquids except some chlorinated hydrocarbons. Table 4-11 lists the physical and mechanical properties of FEP.

Permeation of FEP liners can pose a problem. Table 4-12 provides some permeation data relating to the more common chemicals. As with PTFE there is some absorption of chemicals by FEP. This absorption can lead to problems similar to those encountered with PTFE. Table 4-13 is a listing of the absorption of selected liquids in FEP.

FEP-lined pipe has an operating range of –20 to 400°F (–29 to 204°C). However, when used with certain agressive chemicals, the upper temperature limit may have to be lowered. The operating pressure will depend upon the outer pipe shell and the ductile iron or cast steel fitting housings. Maximum pressure/temperature ratings for standard pipe and fittings with class 150 ANS/B16.42 flanges and fittings are as follows:

Temperature (°F/°C)	Pressure (psi)
100/38	250
200/93	235
300/149	215
400/205	200

Operation under negative pressure is a function of both pipe size and operating temperature. Pipe diameters of from 1 inch through 2 inches with an interference fit liner will withstand full vacuum up to 225°F (107°C). The 3-inch-diameter pipe will withstand full vacuum up to 200°F (93°C), while the 4-, 6-, and 8-inch-diameter pipe will only withstand 10 in. Hg at 200°F (93°C). Ten-inch-diameter pipe will withstand 10 in. vacuum at 150°F (66°C).

Pipe having slip fit liners may be used under vacuum in special circumstances. It may be necessary to apply vacuum to one of the vent openings in the pipe to balance the pressure on either side of the liner. When this approach is taken, the manufacturer should be consulted for recommendations.

In general, FEP-lined pipe is available as a standard through 12 inches in diameter. Larger sizes up to 30 inches in diameter are available but not normally stocked.

Pipe is available in lengths up to 20 feet. It is customary to lay out the piping runs and order the appropriate lengths from the manufacturer. Odd lengths of pipe can be field-fabricated using special tools supplied by the pipe manufacturer and/or distributor.

FEP basically exhibits the same corrosion resistance as PTFE, with a few exceptions, but at a lower operating temperature. It is resistant to practically all chemicals, the exceptions being the extremely potent oxidizers, such as chlorine trifluoride and related compounds. Some chemicals will attack FEP when present in high concentrations at or near the service temperature limit of 400°F (204°C).

Table 4-14 lists the compatibility of FEP with selected corrodents.

4.8 PFA-LINED PIPE

Perfluoroalkoxy (PFA) is a fully fluorinated polymer with the following formula:

$$
\begin{array}{ccccc}
F & F & F & F & F \\
| & | & | & | & | \\
-C & -C & -C & -C & -C- \\
 & & | & & \\
 & & O & & \\
 & & | & & \\
 & & R_F & & \\
\end{array}
$$

$$R_F = C_nF_{2n} + 1$$

PFA has somewhat better physical and mechanical properties than FEP above 300°F (149°C) and can be used up to 500°F (260°C), but it lacks the physical strength of PTFE at elevated temperatures. For example, although PFA has reasonable tensile strength at 68°F (20°C), its heat deflection temperature is the lowest of all fluoroplastics. Although PFA matches the hardness and impact strength of PTFE, it sustains only one quarter of the life of PTFE in flexibility tests. Table 4-15 provides the physical and mechanical properties of PFA.

Like PTFE, PFA is subject to permeation by certain gases and will also absorb selected chemicals. Table 4-16 illustrates the permeability of PFA, whereas Table 4-17 lists the absorption of representative liquids in PFA.

PFA also performs well at cryogenic temperatures. Table 4-18 shows the results of cryogenic tests performed with liquid nitrogen on PFA.

PFA-lined pipe has an allowable operating temperature range of from –20°F (29°C) to 450°F (232°C). It is available in diameters of 1 inch to 30 inches, with 1 inch to 12 inches usually stocked. Standard lengths are 20 feet through 16-inch diameter. Above 16-inch diameter, standard lengths are 10 feet. This pipe is also available jacketed for heating or cooling operations. Odd lengths of pipe can be field-fabricated using special equipment furnished by the manufacturer or distributor. Alternatively, odd lengths can be ordered directly from the factory.

As with other lined piping systems, the maximum allowable operating pressure is dependent upon the outer housing. These housings permit operating pressure of 150 and 625 psi, both at 450°F (232°C). Allowable operation under negative pressure is a function of the type of liner construction and the operating temperature. With an interference fit liner, pipe diameters of 1 inch through and including 3 inches can be operated at full vacuum at 450°F (232°C). Above these diameters, full vacuum is permitted but at reduced operating temperatures as follows:

4-inch FV at 300°F (149°C)
6-inch FV at 250°F (121°C)
8-inch FV at 150°F (66°C)

For vacuum service of pipe sizes above 8 inches and pipe with other types of liners, the manufacturer must be consulted.

The compatibility of PFA with corrodents is typical of fully fluorinated polymers. It is not degraded by chemical systems commonly encountered in chemical processes. It is inert to strong mineral acids, inorganic bases, inorganic oxidizers, aromatics, some aliphatic hydrocarbons, alcohols, aldehydes, ketones, ethers, esters, chlorocarbons, fluorocarbons, and mixtures of these mentioned.

PFA will be attacked by certain halogenated complexes containing fluorine. These include chlorine trifluoride, bromine trifluoride, iodine pentafluoride, and fluorine. PFA can also be attacked by such metals as sodium or potassium, particularly in their molten state.

Table 4-19 lists the compatibility of PFA with selected corrodents.

4.9 PVDF-LINED PIPE

Polyvinylidene fluoride (PVDF) is manufactured by several companies under different trade names as follows:

Kynar—Elf Atochem
Hylar—Ausimont USA, Inc.
Solef—Solvay

It is a homopolymer of vinylidene fluoride containing 59% fluorine with the following chemical structure:

```
        H    H
        |    |
    –   C  – C  –
        |    |
        F    H
```

which is linear and similar to PTFE but is not fully fluorinated.

In contact with many fluids, particularly inorganics, PVDF is comparable to the more expensive FEP, with an allowable maximum temperature of operation 25°F (14°C) less than that of FEP. In addition it offers greater resistance to permeation and vacuum collapse and is less likely to "creep" under loads like those imposed on sealing faces. The physical and mechanical properties of PVDF liners are shown in Table 4-20.

PVDF-lined pipe has an operating temperature range of –20°F (–29°C) to 275°F (135°C).

As with other lined piping systems, the maximum allowable operating pressure is a function of the outer metallic shell, which are 150 and 625 psi, both at 275°F (135°C). Operation under negative pressure is dependent upon the type of liner used. Table 4-21 provides the allowable operating conditions.

Lined pipe is available from ½ inch through 10 inches in diameter. Standard lengths are 10 and 20 feet. Intermediate lengths may be ordered from the factory or distributor or may be field-fabricated using equipment supplied by the manufacturer.

Unlike PTFE, FEP, and PFA, PVDF liners can be constructed to prevent permeation. By use of proper thicknesses, permeation is eliminated. Table 4-22 provides the typical liner thicknesses required to prevent permeation.

PVDF is highly resistant to a number of agressive corrosives including halogens, acids, alkalies, and strong oxidizers. It will be attacked by fuming sulfuric acid at room temperature, strong sulfuric acid, and other sulfonating agents at high temperatures. It has been approved by the Food and Drug Administration for repeated use in contact with foodstuffs in food-handling and food-processing equipment. Table 4-23 provides the compatibility of lined PVDF pipe with selected corrodents.

4.10   ECTFE-LINED PIPE

Ethylene chlorotrifluorethylene (ECTFE) is a 1:1 alternating copolymer of ethylene and chlorotrifluoroethylene and is sold under the trade name of Halar by Ausimont USA, Inc. The chemical structure:

```
        H    H    F    F
        |    |    |    |
    –   C  – C  – C  – C  –
        |    |    |    |
        H    H    F    Cl
```

provides the polymer with a unique combination of physical, mechanical, and corrosion resistance properties.

The physical and mechanical properties of ECTFE are shown in Table 2-56. Pipe lined with ECTFE has an allowable operating temperature range of from cryogenic to 300°F (149°C). Allowable operating pressures are a function of the outer metallic shell. Schedule 10 through schedule 80 carbon steel pipe is the usual outer shell component, although stainless steel outer shells are also available. Liner thicknesses are usually 0.160 inch, which resist permeation. It is not necessary to provide weep holes in the outer pipe.

The extruded or molded liner, with integral flange faces, is fusion bonded and locked into the pipe or fitting to prevent stress cracking, creep, and cold flow. Pipe is available in standard sizes of 1 inch through 8 inches. It is necessary to lay out and specify each pipe length for fabrication at the factory. Alternatively, the lengths can be prepared at the job site with field-fabrication tools supplied by the manufacturer.

ECTFE is very similar in its corrosion resistance to PTFE but does not have the permeation problem associated with PTFE, PFA, and FEP. It exhibits excellent resistance to abrasion and is virtually immune to attack by all corrosive chemicals commonly encountered in industry. Among the substances that ECTFE is resistant to are strong mineral and oxidizing acids, alkalies, metal etchants, liquid oxygen, and essentially all organic solvents except hot amines (e.g., aniline, dimethylamine). Typical of the fluoropolymers ECTFE is attacked by are metallic sodium and potassium. Table 2-59 provides additional information on the compatibility of ECTFE with selected corrodents.

4.11 POLYPROPYLENE (PP)-LINED PIPE

Polypropylene-lined pipe is one of the least expensive of the lined piping systems. The corrosion-resistant properties of the solid polypropylene are retained in the lined pipe, but because of the metallic housing the allowable operating temperatures and pressures are increased. Lined PP pipe has an allowable operating temperature range of from 0°F (−18°C) to 225°F (107°C).

Allowable operating pressures are based on the metallic shell being used. Standard systems are available with pressure ratings of 150 and 300 psi. Carbon steel pipe is the most commonly used outer shell material. The vacuum rating of PP pipe depends upon the type of liner construction and operating temperature. Pipe sizes of 1 inch through 12 inches can withstand full vacuum up to a temperature of 225°F (107°C), provided the liner is of swaged construction. Liners of the interference fit type of construction have full vacuum ratings in sizes 1 inch through 8 inches at 225°F (107°C) and in the 10- and 12-inch sizes at 100°F (38°C).

PP-lined pipe is available in diameters of 1 inch through 12 inches. Table 4-24 provides the typical dimensions of PP-lined pipe. The PP liner is not subject to permeation, therefore the outer shell is not fitted with a vent connection. Standard pipe lengths are 10 and 20 feet. Intermediate lengths can be ordered from the factory or provided in the field using equipment supplied by the manufacturer. Gaskets are not required when joining two flanged sections of PP pipe and/or fitting.

The physical and mechanical properties of copolymer polypropylene can be found in Table 2-39. Copolymer material is used to produce the PP liners.

In general polypropylene exhibits good corrosion resistance to strong and weak acids, although it is subject to slow attack by oxidizing acids and strong or weak alkalies. It can be used with most solvents below 180°F (82°C), although it is affected by hot aromatic hydrocarbons, benzene, and chlorosulfonic acid or fuming nitric acid even at room temperature.

The compatibility of polypropylene with selected corrodents may be found in Table 4-25. When handling certain corrosives, the maximum allowable operating temperature of the lined PP pipe may be reduced. In some instances the ratings in Table 4-25 vary slightly from those found in Table 2-48, since the lined pipe can be operated at higher temperatures than the solid pipe.

4.12 SARAN-LINED PIPE

Saran (polyvinylidene chloride) is manufactured by Dow Chemical. The resin is a proprietary product of Dow. It has found wide application in the plating industry and for handling deionized water, pharmaceuticals, food processing, and other applications where stream purity protection is critical. The material complies with FDA regulations for food processing and potable water and also with regulations prescribed by the Meat Inspection Division of the Department of Agriculture for transporting fluids used in meat production. In applications such as plating solutions, chlorines, and certain other chemicals, Saran is superior to polypropylene and finds many applications in the handling of municipal water supplies and waste waters.

The system is rated at 150 or 300 psi, depending upon the outer shell design, with a temperature range of 0–175°F (18–80°C). It can withstand full vacuum at 175°F (80°C) throughout the entire range of pipe sizes. It is available in nominal pipe diameters of 1 inch through 8 inches (Table 4-26). Pipe is usually supplied in 10-foot lengths, but shorter lengths are available. Intermediate pipe lengths can be field-fabricated.

Although the piping system in general is rated to 175°F (80°C), lower operating temperatures may be required when in contact with certain corrosives. Refer to Table 2-73 for conditions of operation when in contact with selected corrodents.

4.13 POLYETHYLENE-LINED PIPE

The system used to produce this pipe is one of molecular bonding. The process coats the pipe both internally and externally, providing corrosion resistance both internally and externally. Excellent corrosion resistance is provided against acids, alkalies, salt solutions, and low and high purity water. The material is approved for potable water use by the National Sanitation Foundation.

Schedule 10 aluminum and/or schedule 5 steel pipe is used as the metallic housing to which the lining is applied. It is available, with a complete line of fittings, in nominal diameters of 2 through 8 inches. Standard factory lengths are 30 feet.

Several alternative methods of joining are available. It may be joined by means of a bell and spigot joint (either) gasketed or epoxied), a flare flange, or a grooved connection. Operating pressure of the system is dependent upon the system of joining used (see Table 4-27). The maximum allowable operating temperature is 150°F (66°C). Refer to Table 4-28 for compatibility of polyethylene-lined pipe with selected corrodents.

Table 4-1 Water Absorption Rates of Polymers
Used for Pipe Lining

Polymer	Water absorption 24 hr at 73°F (23°C) (%)
PVC	0.05
CPVC	0.03
PP (Homo)	0.02
PP (Co)	0.03
PE (EHMW)	<0.01
E-CTFE	<0.1
PVDF	<0.04
Saran	nil
PFA	<0.03
ETFE	0.029
PTFE	<0.01
FEP	<0.01

Table 4-2 Maximum Liner Temperature Limits

Liner material	Maximum temperature (°F/°C)
PTFE	450/232
PFA	500/260
FEP	400/204
ECTFE	300/150
ETFE	300/150
PVDF	275/135
PP	225/107
Saran	175/79
PVC	180/82
PE	150/66

Table 4-3 Average Bolt Torques for Class 125 and 150 Systems

Pipe size (in.)	No. of bolts	Bolt diameter (in.)	Average bolt torque (ft-lb)					
			Saran	PP	PVDF	PTFE	FEP	PFA
1	4	½	30	30–35	40	20	20	15
1½	4	½	40	50	50	40	25	20
2	4	⅝	45	70	80	50	30	30
3	4	⅝	55	90	95	75	40	40
4	8	⅝	70	85	90	65	40	35
6	8	¾	50	140	150	90	50	55
8	8	¾	95	160	160	125	70	75
10	12	⅞	160	200	140	130	70	70
12	12	⅞		240		260	80	
14	12	1		160			110	
16	16	1		165			125	

Values are based on the use of clean, lubricated nuts and bolts.

Table 4-4 Average Bolt Torques for Class 300 Systems

Pipe size (in.)	No. of bolts	Bolt diameter (in.)	Average bolt torque (ft-lb)			
			Saran	PP	PVDF	PTFE
1	4	⅝	35	45	70	30
1½	8	¾	60	75	170	80
2	8	⅝	25	55	85	40
3	8	¾	40	100	155	65
4	8	¾	60	170	225	110
6	12	¾	65	165	225	85
8	12	⅞	125	325	365	150

Values are based on the use of clean, lubricated nuts and bolts.

Table 4-5 Typical Properties of the Fluorocarbon Resins

	ASTM standard	PTFE	FEP	PFA
Specific gravity	D792	2.13–2.22	2.15	2.15
Tensile strength, psi	D638	2500–4000	3400	3600
Elongation, %	D638	200–400	325	300
Flexural modulus, psi	D790	27,000	90,000	90,000
Impact strength, ft-lb/in.	D256	3.5	no break	no break
Hardness, Shore D	D2240	50–65	56	60
Coefficient of friction	D1894	0.1	0.2	0.2
Upper service temperature, °F/°C	UL746B	500/260	400/204	500/260
Flame rating	UL94	VO	VO	VO
Limiting oxygen index, %	D2863	>95	>95	>95
Chemical/Solvent resistance	D543	Outstanding	Outstanding	Outstanding
Water absorption, 24 hr	D570	>0.01	>0.01	>0.03

Table 4-6 Typical Properties of the Fluoropolymer Resins

	ASTM standard	ETFE	PVDF	ECTFE
Specific gravity	D792	1.70	1.78	1.68
Tensile strength, psi	D638	6500	4500	7000
Elongation, %	D638	300	50	210
Flexural modulus, psi $\times 10^5$	D790	1.7	2.5	2.4
Impact strength, ft-lb/in.	D256	no break	2	no break
Hardness, Shore D	D2240	67	78	75
Coefficient of friction	D1894	0.4		
Upper service temperature, °F/°C	UL746B	300/150		300/150
Flame rating	UL94	VO	VO	VO
Limiting oxygen index, %	D2863	30	30	30
Chemical/Solvent resistance	D543	excellent	fair	good
Water absorption, 24 hr	D570	<0.03	<0.03	>0.1

Table 4-7   Vacuum Rating of PTFE-Lined Pipe

Nominal pipe size (in.)		Method of securing liner		
		Interference fit	Swaged	Slip fit
1	°F/°C	450/232	450/232	450/232
	in Hg	FV	FV	FV
1½	°F/°C	450/232	450/232	450/232
	in Hg	FV	FV	FV
2	°F/°C	450/232	450/232	450/232
	in Hg	FV	FV	FV
3	°F/°C	450/232	450/232	450/232
	in Hg	FV	FV	FV
4	°F/°C	450/232	450/232	450/232
	in Hg	FV	FV	FV
6	°F/°C	350/177	450/232	
	in Hg	FV	FV	
8	°F/°C	350/177	450/232	
	in Hg	FV	FV	

FV = full vacuum.
Caution: Although the pipe section may be rated for full vacuum, certain of the piping fittings may not have an equivalent rating. Check with the manufacturer.

Table 4-8   Vapor Permeation of PTFE Liners

Gases	Permeation (g/100 in.^{2}/24 hr/mil) at:	
	73°F (23°C)	86°F (30°C)
Carbon dioxide		0.66
Helium		0.22
Hydrogen chloride, anh.		<.01
Nitrogen		0.11
Acetophenone	0.56	
Benzene	0.36	0.80
Carbon tetrachloride	0.06	
Ethyl alcohol	0.13	
Hydrochloric acid 20%	<0.01	
Piperdine	0.07	
Sodium hydroxide 50%	5×10^{-5}	
Sulfuric acid 98%	1.8×10^{-5}	

Based on PTFE having a specific gravity > 2.2

Table 4-9 Physical and Mechanical Properties of PTFE

Specific gravity	2.13–2.2
Water absorption 24 hr at 73°F (23°C), %	0.01
Tensile strength at 73°F (23°C), psi	2000–6500
Compressive strength, psi	1700
Flexural strength, psi	no break
Flexural modulus, psi $\times$ 10^5	0.7–1.1
Izod impact strength, notched at 73°F (23°C)	3
Coefficient of thermal expansion	
in./in.°F $\times$ 10^{-5}	5.5
in./10°F/100 ft	0.055
Thermal conductivity, Btu/hr/ft^2/°F/in.	1.7
Heat distortion temperature at 66 psi, °F/°C	250/121
Low temperature embrittlement, °F/°C	–450/–268
Resistance to heat at constant drainage, °F/°C	550/288

Table 4-10 Compatibility of PTFE-Lined Pipe with Selected Corrodents
The chemicals listed are in the pure state or in a saturated solution unless otherwise indicated.
Compatibility is shown to the maximum allowable temperature for which data are available.
Incompatibility is shown by an x. A blank space indicates that data are unavailable.

Chemical	Maximum temp.		Chemical	Maximum temp.	
	°F	°C		°F	°C
Acetaldehyde	450	232	Barium carbonate	450	232
Acetamide	450	232	Barium chloride	450	232
Acetic acid 10%	450	232	Barium hydroxide	450	232
Acetic acid 50%	450	232	Barium sulfate	450	232
Acetic acid 80%	450	232	Barium sulfide	450	232
Acetic acid, glacial	450	232	Benzaldehyde	450	232
Acetic anhydride	450	232	Benzene[a]	450	232
Acetone	450	232	Benzene sulfonic acid 10%	450	232
Acetyl chloride	450	232	Benzoic acid	450	232
Acrylic acid			Benzyl alcohol	450	232
Acrylonitrile	450	232	Benzyl chloride	450	232
Adipic acid	450	232	Borax	450	232
Allyl alcohol	450	232	Boric acid	450	232
Allyl chloride	450	232	Bromine gas, dry[a]	450	232
Alum	450	232	Bromine gas, moist		
Aluminum acetate			Bromine liquid[a]	450	232
Aluminum chloride, aqueous	450	232	Butadiene[a]	450	232
Aluminum chloride, dry			Butyl acetate	450	232
Aluminum fluoride	450	232	Butyl alcohol	450	232
Aluminum hydroxide	450	232	n-Butylamine	450	232
Aluminum nitrate	450	232	Butyl phthalate	450	232
Aluminum oxychloride	450	232	Butyric acid	450	232
Aluminum sulfate	450	232	Calcium bisulfide	450	232
Ammonia gas[a]	450	232	Calcium bisulfite	450	232
Ammonium bifluoride	450	232	Calcium carbonate	450	232
Ammonium carbonate	450	232	Calcium chlorate	450	232
Ammonium chloride 10%	450	232	Calcium chloride	450	232
Ammonium chloride 50%	450	232	Calcium hydroxide 10%	450	232
Ammonium chloride, sat.	450	232	Calcium hydroxide, sat.	450	232
Ammonium fluoride 10%	450	232	Calcium hypochlorite	450	232
Ammonium fluoride 25%	450	232	Calcium nitrate	450	232
Ammonium hydroxide 25%	450	232	Calcium oxide	450	232
Ammonium hydroxide, sat.	450	232	Calcium sulfate	450	232
Ammonium nitrate	450	232	Caprylic acid	450	232
Ammonium persulfate	450	232	Carbon bisulfide[a]	450	232
Ammonium phosphate	450	232	Carbon dioxide, dry	450	232
Ammonium sulfate 10–40%	450	232	Carbon dioxide, wet	450	232
Ammonium sulfide	450	232	Carbon disulfide	450	232
Ammonium sulfite			Carbon monoxide	450	232
Amyl acetate	450	232	Carbon tetrachloride[b]	450	232
Amyl alcohol	450	232	Carbonic acid	450	232
Amyl chloride	450	232	Cellosolve		
Aniline	450	232	Chloracetic acid, 50% water	450	232
Antimony trichloride	450	232	Chloracetic acid	450	232
Aqua regia 3:1	450	232	Chlorine gas, dry	x	x

Table 4-10 *Continued*

Chemical	°F	°C	Chemical	°F	°C
Chlorine gas, wet[a]	450	232	Magnesium chloride	450	232
Chlorine, liquid	x	x	Malic acid	450	232
Chlorobenzene[a]	450	232	Manganese chloride		
Chloroform[a]	450	232	Methyl chloride[a]	450	232
Chlorosulfonic acid	450	232	Methyl ethyl ketone[a]	450	232
Chromic acid 10%	450	232	Methyl isobutyl ketone[a]	450	232
Chromic acid 50%	450	232	Muriatic acid[a]	450	232
Chromyl chloride	450	232	Nitric acid 5%[a]	450	232
Citric acid 15%	450	232	Nitric acid 20%[a]	450	232
Citric acid, concentrated	450	232	Nitric acid 70%[a]	450	232
Copper acetate			Nitric acid, anhydrous[a]	450	232
Copper carbonate	450	232	Nitrous acid 10%	450	232
Copper chloride	450	232	Oleum	450	232
Copper cyanide 10%	450	232	Perchloric acid 10%	450	232
Copper sulfate	450	232	Perchloric acid 70%	450	232
Cresol	450	232	Phenol[a]	450	232
Cupric chloride 5%	450	232	Phosphoric acid 50–80%	450	232
Cupric chloride 50%	450	232	Picric acid	450	232
Cyclohexane	450	232	Potassium bromide 30%	450	232
Cyclohexanol	450	232	Salicylic acid	450	232
Dichloroacetic acid	450	232	Silver bromide 10%		
Dichloroethane (ethylene di-chloride)[a]	450	232	Sodium carbonate	450	232
			Sodium chloride	450	232
Ethylene glycol	450	232	Sodium hydroxide 10%	450	232
Ferric chloride	450	232	Sodium hydroxide 50%	450	232
Ferric chloride 50% in water	450	232	Sodium hydroxide, concentrated	450	232
Ferric nitrate 10–50%	450	232	Sodium hypochlorite 20%	450	232
Ferrous chloride	450	232	Sodium hypochlorite, concentrated	450	232
Ferrous nitrate	450	232			
Fluorine gas, dry	x	x	Sodium sulfide to 50%	450	232
Fluorine gas, moist	x	x	Stannic chloride	450	232
Hydrobromic acid, dilute[ab]	450	232	Stannous chloride	450	232
Hydrobromic acid 20%[b]	450	232	Sulfuric acid 10%	450	232
Hydrobromic acid 50%[b]	450	232	Sulfuric acid 50%	450	232
Hydrochloric acid 20%[b]	450	232	Sulfuric acid 70%	450	232
Hydrochloric acid 38%[b]	450	232	Sulfuric acid 90%	450	232
Hydrocyanic acid 10%	450	232	Sulfuric acid 98%	450	232
Hydrofluoric acid 30%[a]	450	232	Sulfuric acid 100%	450	232
Hydrofluoric acid 70%[a]	450	232	Sulfuric acid, fuming[a]	450	232
Hydrofluoric acid 100%[a]	450	232	Sulfurous acid	450	232
Hypochlorous acid	450	232	Thionyl chloride	450	232
Iodine solution 10%[a]	450	232	Toluene[a]	450	232
Ketones, general	450	232	Trichloroacetic acid	450	232
Lactic acid 25%	450	232	White liquor	450	232
Lactic acid, concentrated	450	232	Zinc chloride[c]	450	232

[a]Material will permeate.
[b]Material will cause stress cracking.
[c]Material will be absorbed.
Source: Schweitzer, Philip A. (1991). *Corrosion Resistance Tables*, Marcel Dekker, Inc., New York, Vols. 1 and 2.

Table 4-11 Physical and Mechanical Properties of FEP

Specific gravity	2.15
Water absorption 24 hr at 73°F (23°C), %	<0.01
Tensile strength at 73°F (23°C), psi	2700–3100
Modulus of elasticity in tension at 73°F (23°C) $\times 10^5$ psi	0.9
Compressive strength, psi	16,000
Flexural strength, psi	3000
Izod impact strength, notched at 73°F (23°C)	no break
Coefficient of thermal expansion, in./in.°F $\times 10^{-5}$	8.3–10.5
Thermal conductivity, Btu/hr/ft^2/°F/in.	0.11
Heat distortion temperature, at 66 psi °F/°C	158/70
Resistance to heat at continuous drainage, °F/°C	400/204
Limiting oxygen index, %	95
Flame spread	Nonflammable

Table 4-12 Vapor Permeation of FEP Liners

	Permeation (g/100 in.^{2}/24 hr/mil) at:		
	73°F (23°C)	95°F (35°C)	122°F (50°C)
Gases			
Nitrogen	0.18		
Oxygen	0.39		
Vapors			
Acetic acid		0.42	
Acetone	0.13	0.95	3.29
Acetophenone	0.47		
Benzene	0.15	0.64	
N-Butyl ether	0.08		
Carbon tetrachloride	0.11	0.31	
Decane	0.72		1.03
Ethyl acetate	0.06	0.77	2.9
Ethyl alcohol	0.11	0.69	
Hexane		0.57	
Hydrochloric acid 20%	<0.01		
Methanol			5.61
Sodium hydroxide 50%	4×10^{-5}		
Sulfuric acid 98%	8×10^{-6}		
Toluene	0.37		2.93
Water	0.09	0.45	0.89

Table 4-13 Absorption of Selected Liquids[a] by FEP

Chemical	Temperature (°F/°C)	Range of weight gains (%)
Aniline	365/185	0.3–0.4
Acetophenone	394/201	0.6–0.8
Benzaldehyde	354/179	0.4–0.5
Benzyl alcohol	400/204	0.3–0.4
n-Butylamine	172/78	0.3–0.4
Carbon tetrachloride	172/78	2.3–2.4
Dimethyl sulfoxide	372/190	0.1–0.2
Nitrobenzene	410/210	0.7–0.9
Perchlorethylene	250/121	2.0–2.3
Sulfuryl chloride	154/68	1.7–2.7
Toluene	230/110	0.7–0.8
Tributyl phosphate	392/200[b]	1.8–2.0

[a]168-hour exposure at their boiling points.
[b]Not boiling.

Table 4-14 Compatibility of FEP with Selected Corrodents

The chemicals listed are in the pure state or in a saturated solution unless otherwise indicated. Compatibility is shown to the maximum allowable temperature for which data are available. Incompatibility is shown by an x. A blank space indicates that data are unavailable.

Chemical	Maximum temp. °F	Maximum temp. °C	Chemical	Maximum temp. °F	Maximum temp. °C
Acetaldehyde	200	93	Barium carbonate	400	204
Acetamide	400	204	Barium chloride	400	204
Acetic acid 10%	400	204	Barium hydroxide	400	204
Acetic acid 50%	400	204	Barium sulfate	400	204
Acetic acid 80%	400	204	Barium sulfide	400	204
Acetic acid, glacial	400	204	Benzaldehyde[a]	400	204
Acetic anhydride	400	204	Benzene[a,b]	400	204
Acetone[a]	400	204	Benzene sulfonic acid 10%	400	204
Acetyl chloride	400	204	Benzoic acid	400	204
Acrylic acid	200	93	Benzyl alcohol	400	204
Acrylonitrile	400	204	Benzyl chloride	400	204
Adipic acid	400	204	Borax	400	204
Allyl alcohol	400	204	Boric acid	400	204
Allyl chloride	400	204	Bromine gas, dry[b]	200	93
Alum	400	204	Bromine gas, moist[b]	200	93
Aluminum acetate	400	204	Bromine liquid[a,b]	400	204
Aluminum chloride, aqueous	400	204	Butadiene[b]	400	204
Aluminum chloride, dry	300	149	Butyl acetate	400	204
Aluminum fluoride[b]	400	204	Butyl alcohol	400	204
Aluminum hydroxide	400	204	n-Butylamine[a]	400	204
Aluminum nitrate	400	204	Butyl phthalate	400	204
Aluminum oxychloride	400	204	Butyric acid	400	204
Aluminum sulfate	400	204	Calcium bisulfide	400	204
Ammonia gas[b]	400	204	Calcium bisulfite	400	204
Ammonium bifluoride[b]	400	204	Calcium carbonate	400	204
Ammonium carbonate	400	204	Calcium chlorate	400	204
Ammonium chloride 10%	400	204	Calcium chloride	400	204
Ammonium chloride 50%	400	204	Calcium hydroxide 10%	400	204
Ammonium chloride, sat.	400	204	Calcium hydroxide, sat.	400	204
Ammonium fluoride 10%[b]	400	204	Calcium hypochlorite	400	204
Ammonium fluoride 25%[b]	400	204	Calcium nitrate	400	204
Ammonium hydroxide 25%	400	204	Calcium oxide	400	204
Ammonium hydroxide, sat.	400	204	Calcium sulfate	400	204
Ammonium nitrate	400	204	Caprylic acid	400	204
Ammonium persulfate	400	204	Carbon bisulfide[b]	400	204
Ammonium phosphate	400	204	Carbon dioxide, dry	400	204
Ammonium sulfate 10–40%	400	204	Carbon dioxide, wet	400	204
Ammonium sulfide	400	204	Carbon disulfide	400	204
Ammonium sulfite	400	204	Carbon monoxide	400	204
Amyl acetate	400	204	Carbon tetrachloride[a,b,c]	400	204
Amyl alcohol	400	204	Carbonic acid	400	204
Amyl chloride	400	204	Cellosolve	400	204
Aniline[a]	400	204	Chloracetic acid, 50% water	400	204
Antimony trichloride	250	121	Chloracetic acid	400	204
Aqua regia 3:1	400	204	Chlorine gas, dry	x	x

Table 4-14 *Continued*

Chemical	°F	°C	Chemical	°F	°C
Chlorine gas, wet[b]	400	204	Magnesium chloride	400	204
Chlorine, liquid[a]	400	204	Malic acid	400	204
Chlorobenzene[b]	400	204	Manganese chloride	300	149
Chloroform[b]	400	204	Methyl chloride[b]	400	204
Chlorosulfonic acid[a]	400	204	Methyl ethyl ketone[b]	400	204
Chromic acid 10%	400	204	Methyl isobutyl ketone[b]	400	204
Chromic acid 50%[a]	400	204	Muriatic acid[b]	400	204
Chromyl chloride	400	204	Nitric acid 5%[b]	400	204
Citric acid 15%	400	204	Nitric acid 20%[b]	400	204
Citric acid, concentrated	400	204	Nitric acid 70%[b]	400	204
Copper acetate	400	204	Nitric acid, anhydrous[b]	400	204
Copper carbonate	400	204	Nitrous acid, concentrated	400	204
Copper chloride	400	204	Oleum	400	204
Copper cyanide	400	204	Perchloric acid 10%	400	204
Copper sulfate	400	204	Perchloric acid 70%	400	204
Cresol	400	204	Phenol[b]	400	204
Cupric chloride 5%	400	204	Phosphoric acid 50–80%	400	204
Cupric chloride 50%	400	204	Picric acid	400	204
Cyclohexane	400	204	Potassium bromide 30%	400	204
Cyclohexanol	400	204	Salicylic acid	400	204
Dichloroacetic acid	400	204	Silver bromide 10%	400	204
Dichloroethane (ethylene di-chloride)[b]	400	204	Sodium carbonate	400	204
			Sodium chloride	400	204
Ethylene glycol	400	204	Sodium hydroxide 10%[a]	400	204
Ferric chloride	400	204	Sodium hydroxide 50%	400	204
Ferric chloride 50% in water[a]	260	127	Sodium hydroxide, concen-trated	400	204
Ferric nitrate 10–50%	260	127	Sodium hypochlorite 20%	400	204
Ferrous chloride	400	204	Sodium hypochlorite, con-centrated	400	204
Ferrous nitrate	400	204			
Fluorine gas, dry	200	93	Sodium sulfide to 50%	400	204
Fluorine gas, moist	x	x	Stannic chloride	400	204
Hydrobromic acid, dilute	400	204	Stannous chloride	400	204
Hydrobromic acid 20%[b,c]	400	204	Sulfuric acid 10%	400	204
Hydrobromic acid 50%[b,c]	400	204	Sulfuric acid 50%	400	204
Hydrochloric acid 20%[b,c]	400	204	Sulfuric acid 70%	400	204
Hydrochloric acid 38%[b,c]	400	204	Sulfuric acid 90%	400	204
Hydrocyanic acid 10%	400	204	Sulfuric acid 98%	400	204
Hydrofluoric acid 30%[b]	400	204	Sulfuric acid 100%	400	204
Hydrofluoric acid 70%[b]	400	204	Sulfuric acid, fuming[b]	400	204
Hydrofluoric acid 100%[b]	400	204	Sulfurous acid	400	204
Hypochlorous acid	400	204	Thionyl chloride[b]	400	204
Iodine solution 10%[b]	400	204	Toluene[b]	400	204
Ketones, general	400	204	Trichloroacetic acid	400	204
Lactic acid 25%	400	204	White liquor	400	204
Lactic acid, concentrated	400	204	Zinc chloride[c]	400	204

[a]Material will be absorbed.

[b]Material will permeate.

[c]Material can cause stress cracking.

Source: Schweitzer, Philip A. (1991). *Corrosion Resistance Tables*, Marcel Dekker, Inc., New York, Vols. 1 and 2.

Table 4-15 Physical and Mechanical Properties of PFA

Specific gravity	2.12–2.17
Water absorption 24 hr at 73°F (23°C), %	<0.03
Tensile strength at	
73°F (23°C), psi	4000
482°F (250°C), psi	2000
Modulus of elasticity in tension at	
73°F (23°C), psi	40,000
482°F (250°C), psi	6000
Compressive strength at	
73°F (23°C), psi	3500
–320°F (–196°C), psi	60,000
Fluxural modulus at	
73°F (23°C), psi	90,000
482°F (250°C), psi	10,000
Izod impact strength, notched at 73°F/23°C	No break
Coefficient of linear thermal expansion at	
70–212°F (20–100°C), in./in.°F	7.8×10^{-5}
212–300°F (100–150°C), in./in.°F	9.8×10^{-5}
300–480°F (150–210°C), in./in.°F	12.1×10^{-5}
Heat distortion temperature	
at 66 psi, °F/°C	164/73
at 264 psi, °F/°C	118/48
Resistance to heat at continuous drainage, °F/°C	500/260
Limiting oxygen index, %	<95
Flame spread	10
Underwriters lab rating (Sub. 94)	94V-0

Table 4-16 Permeation of Various Gases in
PFA at 77°F (25°C)

Gas	Permeation (cc mil thickness/ 100 in.^{2} 24 hr atm)
Carbon dioxide	2260
Nitrogen	291
Oxygen	881

Table 4-17 Absorption of Representative Liquids in PFA

Liquid[a]	Temperature (°F/°C)	Range of Weight gains (%)
Aniline	365/185	0.3–0.4
Acetophenone	394/201	0.6–0.8
Benzaldehyde	354/179	0.4–0.5
Benzyl alcohol	400/204	0.3–0.4
n-Butylamine	172/78	0.3–0.4
Carbon tetrachloride	172/78	2.3–2.4
Dimethyl sulfoxide	372/190	0.1–0.2
Freon 113	117/47	1.2
Isooctane	210/99	0.7–0.8
Nitrobenzene	410/210	0.7–0.9
Perchlorethylene	250/121	2.0–2.3
Sulfuryl chloride	154–68	1.7–2.7
Toluene	230/110	0.7–0.8
Tributyl phosphate	392/200[b]	1.8–2.0
Bromine, anhydrous	−5/−22	0.5
Chlorine, anhydrous	248/120	0.5–0.6
Chlorosulfonic acid	302/150	0.7–0.8
Chromic acid 50%	248/120	0.00–0.01
Ferric chloride	212/100	0.00–0.01
Hydrochloric acid 37%	248/120	0.00–0.03
Phosphoric acid, concentrated	212/100	0.00–0.01
Zinc chloride	212/100	0.00–0.03

[a]Liquids were exposed for 168 hours at the boiling point of the solvents. The acidic reagents were exposed for 168 hours.
[b]Not boiling.

Table 4-18 Comparison of Mechanical Properties of PFA at Room Temperature and Cryogenic Temperature

Property	Temperature 73°F (23°C)	−320°F (−196°C)
Yield strength, psi	2100	no yield
Ultimate tensile strength, psi	2600	18,700
Elongation, %	260	8
Flexural modulus	81,000	840,000
Impact strength notched Izod	No break	12
Compressive strength, psi	3500	60,000
Compressive strain, %	20	35
Modulus of elasticity, psi	10,000	680,000

Table 4-19 Compatibility of PFA-Lined Pipe with Selected Corrodents
The chemicals listed are in the pure state or in a saturated solution unless otherwise indicated.
Compatibility is shown to the maximum allowable temperature for which data are available.
Incompatibility is shown by an x. A blank space indicates that data are unavailable.

Chemical	Maximum temp. °F	Maximum temp. °C	Chemical	Maximum temp. °F	Maximum temp. °C
Acetaldehyde	450	232	Barium carbonate	450	232
Acetamide	450	232	Barium chloride	450	232
Acetic acid 10%	450	232	Barium hydroxide	450	232
Acetic acid 50%	450	232	Barium sulfate	450	232
Acetic acid 80%	450	232	Barium sulfide	450	232
Acetic acid, glacial	450	232	Benzaldehyde[b]	450	232
Acetic anhydride	450	232	Benzene[a]	450	232
Acetone	450	232	Benzene sulfonic acid 10%	450	232
Acetyl chloride	450	232	Benzoic acid	450	232
Acrylic acid			Benzyl alcohol[b]	450	232
Acrylonitrile	450	232	Benzyl chloride[a]	450	232
Adipic acid	450	232	Borax	450	232
Allyl alcohol	450	232	Boric acid	450	232
Allyl chloride	450	232	Bromine gas, dry[a]	450	232
Alum	450	232	Bromine gas, moist		
Aluminum acetate			Bromine liquid[a,b]	450	232
Aluminum chloride, aqueous	450	232	Butadiene[a]	450	232
Aluminum chloride, dry			Butyl acetate	450	232
Aluminum fluoride	450	232	Butyl alcohol	450	232
Aluminum hydroxide	450	232	n-Butylamine[b]	450	232
Aluminum nitrate	450	232	Butyl phthalate	450	232
Aluminum oxychloride	450	232	Butyric acid	450	232
Aluminum sulfate	450	232	Calcium bisulfide	450	232
Ammonia gas[a]	450	232	Calcium bisulfite	450	232
Ammonium bifluoride[a]	450	232	Calcium carbonate	450	232
Ammonium carbonate	450	232	Calcium chlorate	450	232
Ammonium chloride 10%	450	232	Calcium chloride	450	232
Ammonium chloride 50%	450	232	Calcium hydroxide 10%	450	232
Ammonium chloride, sat.	450	232	Calcium hydroxide, sat.	450	232
Ammonium fluoride 10%[a]	450	232	Calcium hypochlorite	450	232
Ammonium fluoride 25%[a]	450	232	Calcium nitrate	450	232
Ammonium hydroxide 25%	450	232	Calcium oxide	450	232
Ammonium hydroxide, sat.	450	232	Calcium sulfate	450	232
Ammonium nitrate	450	232	Caprylic acid	450	232
Ammonium persulfate	450	232	Carbon bisulfide[a]	450	232
Ammonium phosphate	450	232	Carbon dioxide, dry	450	232
Ammonium sulfate 10–40%	450	232	Carbon dioxide, wet	450	232
Ammonium sulfide	450	232	Carbon disulfide[a]	450	232
Ammonium sulfite			Carbon monoxide	450	232
Amyl acetate	450	232	Carbon tetrachloride[a,b,c]	450	232
Amyl alcohol	450	232	Carbonic acid	450	232
Amyl chloride	450	232	Cellosolve		
Aniline[b]	450	232	Chloracetic acid, 50% water	450	232
Antimony trichloride	450	232	Chloracetic acid	450	232
Aqua regia 3:1	450	232	Chlorine gas, dry	x	x

Table 4-19 *Continued*

Chemical	°F	°C	Chemical	°F	°C
Chlorine gas, wet[a]	450	232	Magnesium chloride	450	232
Chlorine, liquid[b]	x	x	Malic acid	450	232
Chlorobenzene[a]	450	232	Manganese chloride		
Chloroform[a]	450	232	Methyl chloride[a]	450	232
Chlorosulfonic acid[b]	450	232	Methyl ethyl ketone[a]	450	232
Chromic acid 10%	450	232	Methyl isobutyl ketone[a]	450	232
Chromic acid 50%[b]	450	232	Muriatic acid[a]	450	232
Chromyl chloride	450	232	Nitric acid 5%[a]	450	232
Citric acid 15%	450	232	Nitric acid 20%[a]	450	232
Citric acid, concentrated	450	232	Nitric acid 70%[a]	450	232
Copper acetate			Nitric acid, anhydrous[a]	450	232
Copper carbonate	450	232	Nitrous acid 10%	450	232
Copper chloride	450	232	Oleum	450	232
Copper cyanide	450	232	Perchloric acid 10%	450	232
Copper sulfate	450	232	Perchloric acid 70%	450	232
Cresol	450	232	Phenol[a]	450	232
Cupric chloride 5%	450	232	Phosphoric acid 50–80%[b]	450	232
Cupric chloride 50%	450	232	Picric acid	450	232
Cyclohexane	450	232	Potassium bromide 30%	450	232
Cyclohexanol	450	232	Salicylic acid	450	232
Dichloroacetic acid	450	232	Silver bromide 10%		
Dichloroethane (ethylene di-chloride)[a]	450	232	Sodium carbonate	450	232
			Sodium chloride	450	232
Ethylene glycol	450	232	Sodium hydroxide 10%	450	232
Ferric chloride	450	232	Sodium hydroxide 50%	450	232
Ferric chloride 50% in water[b]	450	232	Sodium hydroxide, concen-trated	450	232
Ferric nitrate 10–50%	450	232	Sodium hypochlorite 20%	450	232
Ferrous chloride	450	232	Sodium hypochlorite, con-centrated	450	232
Ferrous nitrate	450	232			
Fluorine gas, dry	x	x	Sodium sulfide to 50%	450	232
Fluorine gas, moist	x	x	Stannic chloride	450	232
Hydrobromic acid, dilute[a,c]	450	232	Stannous chloride	450	232
Hydrobromic acid 20%[a,c]	450	232	Sulfuric acid 10%	450	232
Hydrobromic acid 50%[a,c]	450	232	Sulfuric acid 50%	450	232
Hydrochloric acid 20%[a,c]	450	232	Sulfuric acid 70%	450	232
Hydrochloric acid 38%[a,c]	450	232	Sulfuric acid 90%	450	232
Hydrocyanic acid 10%	450	232	Sulfuric acid 98%	450	232
Hydrofluoric acid 30%[a]	450	232	Sulfuric acid 100%	450	232
Hydrofluoric acid 70%[a]	450	232	Sulfuric acid, fuming[a]	450	232
Hydrofluoric acid 100%[a]	450	232	Sulfurous acid	450	232
Hypochlorous acid	450	232	Thionyl chloride[a]	450	232
Iodine solution 10%[a]	450	232	Toluene[a]	450	232
Ketones, general	450	232	Trichloroacetic acid	450	232
Lactic acid 25%	450	232	White liquor	450	232
Lactic acid, concentrated	450	232	Zinc chloride[b]	450	232

[a]Material will permeate.
[b]Material will be absorbed.
[c]Material will cause stress cracking.
Source: Schweitzer, Philip A. (1991). *Corrosion Resistance Tables*, Marcel Dekker, Inc., New York, Vols. 1 and 2.

Table 4-20 Physical and Mechanical Properties of PVDF-Lined Pipe

Specific gravity of liner	1.75–1.78
Tensile strength of liner at yield 73°F (23°C), psi	5000–6000
Compressive strength of liner at yield, psi	10,000–16,000
Coefficient of thermal expansion of liner	
unswaged in./in./°F × 10⁻⁵	6.6
in./10°F/100 ft	.066
Thermal conductivity of liner, Btu/hr/ft²/°F/in.	0.9
Heat distortion temperature of liner at 66 psi, °F/°C	270–300/132–149
Resistance to heat at continuous drainage, °F/°C	275/135

Table 4-21 Vacuum Rating of PVDF-Lined Pipe

Type of liner	Temperature (°F/°C)	Vacuum rating (Hg) at pipe size (in.):							
		1	1½	2	3	4	6	8	10
Swaged	70/21	FV	FV	FV	FV	FV	FV	FV	
	180/82	FV	FV	FV	FV	FV	FV	FV	
	275/135	FV	FV	FV	FV	FV	FV	FV	
Interference fit	70/21	FV	FV	FV	FV	FV	FV	FV	
	180/82	FV	FV	FV	FV	FV	FV	FV	
	275/135	FV	FV	FV	FV	FV	FV	FV	
Loose	70/21	FV	FV	FV	FV	FV	FV	FV	FV
	180/82	FV	FV	FV	FV	FV	FV	FV	16
	275/135	FV	FV	FV	FV	FV	19	19	13

Table 4-22 Typical PVDF Liner Thickness to Prevent Permeation

Nominal pipe size (in.)	Liner thickness (in.)
1	0.150
1½	0.160
2	0.172
3	0.175
4	0.207
6	0.218
8	0.218

Table 4-23 Compatibility of Lined PVDF Pipe with Selected Corrodents
The chemicals listed are in the pure state or in a saturated solution unless otherwise indicated.
Compatibility is shown to the maximum allowable temperature for which data are available.
Incompatibility is shown by an x. A blank space indicates that data are unavailable.

Chemical	Maximum temp. °F	Maximum temp. °C	Chemical	Maximum temp. °F	Maximum temp. °C
Acetaldehyde	x	x	Barium carbonate	275	135
Acetamide	75	24	Barium chloride	275	135
Acetic acid 10%	225	107	Barium hydroxide	275	135
Acetic acid 50%	200	93	Barium sulfate	275	135
Acetic acid 80%	175	79	Barium sulfide	275	135
Acetic acid, glacial	125	52	Benzaldehyde	75	24
Acetic anhydride	x	x	Benzene	150	66
Acetone	x	x	Benzene sulfonic acid 10%	125	52
Acetyl chloride	125	52	Benzoic acid	225	107
Acrylic acid			Benzyl alcohol	250	121
Acrylonitrile	75	24	Benzyl chloride	275	135
Adipic acid	150	66	Borax	275	135
Allyl alcohol	125	52	Boric acid	275	135
Allyl chloride	175	79	Bromine gas, dry	150	66
Alum	275	135	Bromine gas, moist		
Aluminum acetate			Bromine liquid	150	66
Aluminum chloride, aqueous	275	135	Butadiene	250	121
Aluminum chloride, dry			Butyl acetate	75	24
Aluminum fluoride	275	135	Butyl alcohol		
Aluminum hydroxide	275	135	*n*-Butylamine		
Aluminum nitrate	275	135	Butyl phthalate	x	x
Aluminum oxychloride	275	135	Butyric acid	225	107
Aluminum sulfate	275	135	Calcium bisulfide	275	135
Ammonia gas	x	x	Calcium bisulfite	275	135
Ammonium bifluoride	150	66	Calcium carbonate	275	135
Ammonium carbonate	275	135	Calcium chlorate	275	135
Ammonium chloride 10%			Calcium chloride	275	135
Ammonium chloride 50%			Calcium hydroxide 10%	275	135
Ammonium chloride, sat.	275	135	Calcium hydroxide, sat.	275	135
Ammonium fluoride 10%	275	135	Calcium hypochlorite	200	93
Ammonium fluoride 25%	275	135	Calcium nitrate	275	135
Ammonium hydroxide 25%	225	107	Calcium oxide	250	121
Ammonium hydroxide, sat.	225	107	Calcium sulfate	275	135
Ammonium nitrate	275	135	Caprylic acid	175	29
Ammonium persulfate	75	24	Carbon bisulfide	75	24
Ammonium phosphate	275	135	Carbon dioxide, dry	275	135
Ammonium sulfate 10–40%	275	135	Carbon dioxide, wet	275	135
Ammonium sulfide	125	52	Carbon disulfide	75	24
Ammonium sulfite			Carbon monoxide	275	135
Amyl acetate	125	52	Carbon tetrachloride	275	135
Amyl alcohol	275	135	Carbonic acid	275	135
Amyl chloride	275	135	Cellosolve		
Aniline	125	52	Chloracetic acid, 50% water	x	x
Antimony trichloride	75	24	Chloracetic acid	x	x
Aqua regia 3:1	75	24	Chlorine gas, dry	x	x

Table 4-23 *Continued*

Chemical	Maximum temp. °F	°C	Chemical	Maximum temp. °F	°C
Chlorine gas, wet	200	93	Magnesium chloride	275	135
Chlorine, liquid	x	x	Malic acid	250	121
Chlorobenzene	175	79	Manganese chloride		
Chloroform	125	52	Methyl chloride	275	135
Chlorosulfonic acid	x	x	Methyl ethyl ketone	x	x
Chromic acid 10%			Methyl isobutyl ketone	x	x
Chromic acid 50%	125	52	Muriatic acid	275	135
Chromyl chloride			Nitric acid 5%	175	79
Citric acid 15%	275	135	Nitric acid 20%	125	52
Citric acid, concentrated	275	135	Nitric acid 70%	x	x
Copper acetate			Nitric acid, anhydrous	x	x
Copper carbonate	275	135	Nitrous acid 10%	200	93
Copper chloride	275	135	Oleum	x	x
Copper cyanide 10%	275	135	Perchloric acid 10%	200	93
Copper sulfate	275	135	Perchloric acid 70%	125	52
Cresol	150	66	Phenol	125	52
Cupric chloride 5%	275	135	Phosphoric acid 50–80%	225	107
Cupric chloride 50%	275	135	Picric acid	75	24
Cyclohexane	275	135	Potassium bromide 30%	275	135
Cyclohexanol	150	66	Salicylic acid	200	93
Dichloroacetic acid	125	52	Silver bromide 10%		
Dichloroethane (ethylene di-chloride)	175	79	Sodium carbonate	275	135
			Sodium chloride	275	135
Ethylene glycol	275	135	Sodium hydroxide 10%	125	52
Ferric chloride	275	135	Sodium hydroxide 50%	x	x
Ferric chloride 50% in water	275	135	Sodium hydroxide, con-centrated	x	x
Ferric nitrate 10–50%	275	135	Sodium hypochlorite 20%	125	52
Ferrous chloride	275	135	Sodium hypochlorite, con-centrated	125	52
Ferrous nitrate	275	135			
Fluorine gas, dry	75	24	Sodium sulfide to 50%	275	135
Fluorine gas, moist			Stannic chloride	275	135
Hydrobromic acid, dilute	275	135	Stannous chloride 50%	275	135
Hydrobromic acid 20%	275	135	Sulfuric acid 10%	250	121
Hydrobromic acid 50%	275	135	Sulfuric acid 50%	250	121
Hydrochloric acid 20%	275	135	Sulfuric acid 70%	200	93
Hydrochloric acid 38%	275	135	Sulfuric acid 90%	200	93
Hydrocyanic acid 10%	275	135	Sulfuric acid 98%	150	66
Hydrofluoric acid 30%	250	121	Sulfuric acid 100%	x	x
Hydrofluoric acid 70%	200	93	Sulfuric acid, fuming	x	x
Hydrofluoric acid 100%	200	93	Sulfurous acid	200	93
Hypochlorous acid	75	24	Thionyl chloride	x	x
Iodine solution 10%	150	66	Toluene	175	79
Ketones, general			Trichloroacetic acid	125	52
Lactic acid 25%	125	52	White liquor		
Lactic acid, concentrated	125	52	Zinc chloride	275	135

Source: Schweitzer, Philip A. (1991). *Corrosion Resistance Tables*, Marcel Dekker, Inc., New York, Vols. 1 and 2.

Table 4-24 Typical Polypropylene-Lined Pipe Dimensions

Nominal pipe size (in.)	Lined Pipe O.D. (in.)	Liner thickness (in.)	Lined pipe I.D. (in.)
1	1.325	0.150	0.74
1½	1.900	0.160	1.27
2	2.375	0.172	1.72
2½	2.875	0.175	2.12
3	3.500	0.175	2.70
4	4.500	0.207	3.64
6	6.625	0.218	5.62
8	8.625	0.218	7.62
10	10.75	0.290	9.56
12	12.75	0.340	11.41

Table 4-25 Compatibility of PP-Lined Pipe with Selected Corrodents
The chemicals listed are in the pure state or in a saturated solution unless otherwise indicated.
Compatibility is shown to the maximum allowable temperature for which data are available.
Incompatibility is shown by an x. A blank space indicates that data are unavailable.

Chemical	Maximum temp.		Chemical	Maximum temp.	
	°F	°C		°F	°C
Acetaldehyde	75	24	Barium carbonate	200	93
Acetamide	150	66	Barium chloride	200	93
Acetic acid 10%	200	93	Barium hydroxide	200	93
Acetic acid 50%	200	93	Barium sulfate	200	93
Acetic acid 80%	125	52	Barium sulfide	200	93
Acetic acid, glacial	125	52	Benzaldehyde	75	24
Acetic anhydride	75	24	Benzene	x	x
Acetone	125	52	Benzene sulfonic acid 10%	75	24
Acetyl chloride	x	x	Benzoic acid	150	66
Acrylic acid	x	x	Benzyl alcohol	125	52
Acrylonitrile	125	52	Benzyl chloride	75	24
Adipic acid	150	66	Borax	175	79
Allyl alcohol	150	66	Boric acid	225	107
Allyl chloride	75	24	Bromine gas, dry	x	x
Alum	225	107	Bromine gas, moist		
Aluminum acetate			Bromine liquid	x	x
Aluminum chloride, aqueous	225	107	Butadiene	x	x
Aluminum chloride, dry			Butyl acetate	x	x
Aluminum fluoride	225	107	Butyl alcohol		
Aluminum hydroxide	200	93	n-Butylamine		
Aluminum nitrate	200	93	Butyl phthalate	125	52
Aluminum oxychloride	125	52	Butyric acid	175	79
Aluminum sulfate	225	107	Calcium bisulfide	200	93
Ammonia gas	150	66	Calcium bisulfite	200	93
Ammonium bifluoride	200	93	Calcium carbonate	225	107
Ammonium carbonate	225	107	Calcium chlorate	200	93
Ammonium chloride 10%			Calcium chloride	225	107
Ammonium chloride 50%			Calcium hydroxide 10%		
Ammonium chloride, sat.	225	107	Calcium hydroxide, sat.	225	107
Ammonium fluoride 10%	200	93	Calcium hypochlorite	175	79
Ammonium fluoride 25%	200	93	Calcium nitrate	200	93
Ammonium hydroxide 25%	225	107	Calcium oxide	225	107
Ammonium hydroxide, sat.	225	107	Calcium sulfate	225	107
Ammonium nitrate	150	66	Caprylic acid	125	52
Ammonium persulfate	150	66	Carbon bisulfide	x	x
Ammonium phosphate	225	107	Carbon dioxide, dry	225	107
Ammonium sulfate 10–40%	225	107	Carbon dioxide, wet	225	107
Ammonium sulfide	150	66	Carbon disulfide	x	x
Ammonium sulfite			Carbon monoxide	225	107
Amyl acetate	75	24	Carbon tetrachloride	x	x
Amyl alcohol	75	24	Carbonic acid	225	107
Amyl chloride	x	x	Cellosolve		
Aniline	125	52	Chloracetic acid, 50% water	125	52
Antimony trichloride	150	66	Chloracetic acid	125	52
Aqua regia 3 : 1	75	24	Chlorine gas, dry	x	x

Table 4-25 *Continued*

Chemical	Maximum temp.		Chemical	Maximum temp.	
	°F	°C		°F	°C
Chlorine gas, wet	x	x	Magnesium chloride	225	107
Chlorine, liquid	x	x	Malic acid	125	52
Chlorobenzene	x	x	Manganese chloride		
Chloroform	x	x	Methyl chloride	x	x
Chlorosulfonic acid	x	x	Methyl ethyl ketone	125	52
Chromic acid 10%			Methyl isobutyl ketone	75	24
Chromic acid 50%[a]	125	52	Muriatic acid		
Chromyl chloride	125	52	Nitric acid 5%	175	79
Citric acid 15%	225	107	Nitric acid 20%	150	66
Citric acid, concentrated	225	107	Nitric acid 70%	x	x
Copper acetate			Nitric acid, anhydrous	x	x
Copper carbonate	200	93	Nitrous acid, concentrated	x	x
Copper chloride	200	93	Oleum	x	x
Copper cyanide 10%	200	93	Perchloric acid 10%	150	66
Copper sulfate	200	93	Perchloric acid 70%	75	24
Cresol	x	x	Phenol	150	66
Cupric chloride 5%	200	93	Phosphoric acid 50–80%	225	107
Cupric chloride 50%	200	93	Picric acid	75	24
Cyclohexane	x	x	Potassium bromide 30%	225	107
Cyclohexanol	75	24	Salicylic acid	125	52
Dichloroacetic acid	125	52	Silver bromide 10%		
Dichloroethane (ethylene di-chloride)	75	24	Sodium carbonate	225	107
			Sodium chloride	225	107
Ethylene glycol	125	52	Sodium hydroxide 10%	200	93
Ferric chloride	200	93	Sodium hydroxide 50%	200	93
Ferric chloride 50% in water			Sodium hydroxide, con-centrated	150	66
Ferric nitrate 10–50%	200	93	Sodium hypochlorite 20%	150	66
Ferrous chloride[a]	200	93	Sodium hypochlorite, con-centrated	150	66
Ferrous nitrate	200	93			
Fluorine gas, dry	x	x	Sodium sulfide to 50%	150	66
Fluorine gas, moist	x	x	Stannic chloride	225	107
Hydrobromic acid, dilute	225	107	Stannous chloride 50%	175	79
Hydrobromic acid 20%	175	79	Sulfuric acid 10%	225	107
Hydrobromic acid 50%	175	79	Sulfuric acid 50%	200	93
Hydrochloric acid 20%			Sulfuric acid 70%	175	79
Hydrochloric acid 38%			Sulfuric acid 90%		
Hydrocyanic acid 10%	150	66	Sulfuric acid 98%		
Hydrofluoric acid 30%	200	93	Sulfuric acid 100%	x	x
Hydrofluoric acid 70%	200	93	Sulfuric acid, fuming	x	x
Hydrofluoric acid 100%	x	x	Sulfurous acid	175	79
Hypochlorous acid	150	66	Thionyl chloride	x	x
Iodine solution 10%	75	24	Toluene	x	x
Ketones, general			Trichloroacetic acid	125	52
Lactic acid 25%	150	66	White liquor		
Lactic acid 80%	150	66	Zinc chloride	175	79

Source: Schweitzer, Philip A. (1991). *Corrosion Resistance Tables*, Marcel Dekker, Inc., New York, Vols. 1 and 2.
[a]Material is subject to stress cracking.

Table 4-26 Dimensions of Saran-Lined Pipe

Nominal pipe size (in.)	Lined Pipe O.D. (in.)	Lined pipe I.D. (in.)	Liner thickness (in.)
1	1.325	0.74	0.150
1½	1.900	1.27	0.160
2	2.375	1.72	0.172
2½	2.875	2.12	0.175
3	3.500	2.70	0.175
4	4.500	3.64	0.207
6	6.625	5.62	0.218
8	8.625	7.62	0.218

Source: Courtesy of Dow Chemical.

Table 4-27 Maximum Allowable Operating Pressure of Polyethylene-Lined Pipe

Type of joint	Pressure (psi) at pipe size (in.):				
	2	3	4	6	8
Bell and spigot (gasket), pipe anchored	275	275	250	200	175
Bell and spigot (epoxied)	200	190	150	150	125
Flare-flange	225	225	200	175	150
Grooved	325	275	225	175	175

Table 4-28 Compatibility of Polyethylene-Lined Pipe with Selected Corrodents
The chemicals listed are in the pure state or in a saturated solution unless otherwise indicated.
Compatibility is shown to the maximum allowable temperature for which data are available.
Incompatibility is shown by an x. A blank space indicates that data are unavailable.

Chemical	Maximum temp.		Chemical	Maximum temp.	
	°F	°C		°F	°C
Acetaldehyde	x	x	Barium carbonate	140	60
Acetamide	140	60	Barium chloride	140	60
Acetic acid 10%	140	60	Barium hydroxide	140	60
Acetic acid 50%	140	60	Barium sulfate	140	60
Acetic acid 80%	80	27	Barium sulfide	140	60
Acetic acid, glacial	80	27	Benzaldehyde	x	x
Acetic anhydride	x	x	Benzene	x	x
Acetone	80	27	Benzene sulfonic acid 10%	90	32
Acetyl chloride	x	x	Benzoic acid	140	60
Acrylic acid			Benzyl alcohol	x	x
Acrylonitrile	150	66	Benzyl chloride		
Adipic acid	140	60	Borax	140	60
Allyl alcohol	140	60	Boric acid	140	60
Allyl chloride	110	43	Bromine gas, dry	x	x
Alum	140	60	Bromine gas, moist	x	x
Aluminum acetate			Bromine liquid	x	x
Aluminum chloride, aqueous	140	60	Butadiene	x	x
Aluminum chloride, dry	140	60	Butyl acetate	90	32
Aluminum fluoride	140	60	Butyl alcohol	140	60
Aluminum hydroxide	140	60	*n*-Butylamine	x	x
Aluminum nitrate	140	60	Butyl phthalate		
Aluminum oxychloride			Butyric acid	x	x
Aluminum sulfate	140	60	Calcium bisulfide	140	60
Ammonia gas	140	60	Calcium bisulfite	140	60
Ammonium bifluoride	140	60	Calcium carbonate	140	60
Ammonium carbonate	140	60	Calcium chlorate	140	60
Ammonium chloride 10%	140	60	Calcium chloride	140	60
Ammonium chloride 50%	140	60	Calcium hydroxide 10%	140	60
Ammonium chloride, sat.	140	60	Calcium hydroxide, sat.	140	60
Ammonium fluoride 10%	140	60	Calcium hypochlorite	140	60
Ammonium fluoride 25%	140	60	Calcium nitrate	140	60
Ammonium hydroxide 25%	140	60	Calcium oxide	140	60
Ammonium hydroxide, sat.	140	60	Calcium sulfate	140	60
Ammonium nitrate	140	60	Caprylic acid		
Ammonium persulfate 5%	150	66	Carbon bisulfide	x	x
Ammonium phosphate	80	27	Carbon dioxide, dry	140	60
Ammonium sulfate 10–40%	140	60	Carbon dioxide, wet	140	60
Ammonium sulfide	140	60	Carbon disulfide	x	x
Ammonium sulfite	140	60	Carbon monoxide	140	60
Amyl acetate	80	27	Carbon tetrachloride	x	x
Amyl alcohol	140	60	Carbonic acid	140	60
Amyl chloride	x	x	Cellosolve	80	27
Aniline	130	54	Chloracetic acid, 50% water	x	x
Antimony trichloride	140	60	Chloracetic acid	x	x
Aqua regia 3:1	130	54	Chlorine gas, dry 10%	80	27

Table 4-28 *Continued*

Chemical	Maximum temp. °F	Maximum temp. °C	Chemical	Maximum temp. °F	Maximum temp. °C
Chlorine gas, wet	x	x	Magnesium chloride	140	60
Chlorine, liquid	x	x	Malic acid	140	60
Chlorobenzene	x	x	Manganese chloride	80	27
Chloroform	x	x	Methyl chloride	x	x
Chlorosulfonic acid	x	x	Methyl ethyl ketone	x	x
Chromic acid 10%	140	60	Methyl isobutyl ketone	80	27
Chromic acid 50%	90	32	Muriatic acid	140	60
Chromyl chloride			Nitric acid 5%	140	60
Citric acid 15%	140	60	Nitric acid 20%	140	60
Citric acid, concentrated	140	60	Nitric acid 70%	x	x
Copper acetate			Nitric acid, anhydrous	x	x
Copper carbonate			Nitrous acid, concentrated	120	49
Copper chloride	140	60	Oleum		
Copper cyanide	140	60	Perchloric acid 10%	140	60
Copper sulfate	140	60	Perchloric acid 70%	x	x
Cresol	x	x	Phenol	100	38
Cupric chloride 5%	140	60	Phosphoric acid 50–80%	100	38
Cupric chloride 50%	140	60	Picric acid	x	x
Cyclohexane	80	27	Potassium bromide 30%	140	60
Cyclohexanol	80	27	Salicylic acid	140	60
Dichloroacetic acid			Silver bromide 10%		
Dichloroethane (ethylene di-chloride)	80	27	Sodium carbonate	140	60
			Sodium chloride	140	60
Ethylene glycol	140	60	Sodium hydroxide 10%	140	60
Ferric chloride	140	60	Sodium hydroxide 50%	140	60
Ferric chloride 50% in water	140	60	Sodium hydroxide, con-centrated		
Ferric nitrate 10–50%	140	60	Sodium hypochlorite 20%	140	60
Ferrous chloride	140	60	Sodium hypochlorite, con-centrated	140	60
Ferrous nitrate	140	60			
Fluorine gas, dry	80	27	Sodium sulfide to 50%	140	60
Fluorine gas, moist	80	27	Stannic chloride	140	60
Hydrobromic acid, dilute	140	60	Stannous chloride	140	60
Hydrobromic acid 20%	140	60	Sulfuric acid 10%	140	60
Hydrobromic acid 50%	140	60	Sulfuric acid 50%	140	60
Hydrochloric acid 20%	140	60	Sulfuric acid 70%	80	27
Hydrochloric acid 38%	140	60	Sulfuric acid 90%	x	x
Hydrocyanic acid 10%	140	60	Sulfuric acid 98%	x	x
Hydrofluoric acid 30%	140	60	Sulfuric acid 100%	x	x
Hydrofluoric acid 70%	x	x	Sulfuric acid, fuming	x	x
Hydrofluoric acid 100%	x	x	Sulfurous acid	140	60
Hypochlorous acid	140	60	Thionyl chloride	x	x
Iodine solution 10%	80	27	Toluene	x	x
Ketones, general	80	27	Trichloroacetic acid	80	27
Lactic acid 25%	140	60	White liquor		
Lactic acid, concentrated			Zinc chloride	140	60

Source: Schweitzer, Philip A. (1991). *Corrosion Resistance Tables*, Marcel Dekker, Inc., New York, Vols. 1 and 2.

5

Miscellaneous Lined Piping Systems

In addition to the thermoplastic-lined piping systems, there are a variety of other lined piping systems. Some have been designed for specific applications, while others are designed for general process use. In some instances the temperature ranges of these linings exceed those of the thermoplastic-lined systems.

Several of these lined piping systems predate the introduction of the plastic linings, being in existence before the development of thermoplastics. They still find application in areas where the thermoplastic materials are not suitable.

5.1 RUBBER-LINED PIPE SYSTEMS

Rubber-lined piping systems are perhaps the oldest of any lined piping system and were in service before the advent of thermoplastics and thermoplastic-lined piping systems. There are many applications where rubber-lined pipe is still used.

It must be kept in mind that the terminology "rubber-lined pipe" by itself is misleading. Rubber is a very general term. There are many rubber compositions and formulations from natural rubber to a wide range of synthetic rubbers, all of which are available for lined piping systems.

Several factors must be taken into account when selecting a specific rubber lining compound. Among these potential considerations are:

1. Corrosion resistance
2. Need to prevent contamination of the material being conveyed
3. Must lining meet Food and Drug Administration requirements?
4. Maximum operating temperature
5. Abrasion resistance

The type of rubber that will provide the most suitable lining for the specific service should be selected. This may be natural rubber or one of the synthetic rubbers. In general, soft rubber is used for abrasion resistance, semi-hard for general service, and hard for the more severe service conditions. Multiple ply lining combinations of hard and soft rubber are also available. Lining thicknesses range from ⅛ to ¼ inch depending upon the type of service, the type of rubber, and the method of lining.

Rubber-lined pipe is most commonly joined by means of flanging. The rubber lining is extended over the face of the flanges. With hard rubber lining, a gasket is required. With soft rubber lining, a coating or a polyethylene sheet is required in place of a gasket to avoid bonding of the lining on one flange face to the lining on the adjacent flange face and to permit disassembly of the joint.

When pressures exceed 125 psi there is a tendency for soft rubber linings to extrude out between the flanges. To prevent this the lining should be terminated inside of the bolt holes and the balance of the space between the flanges filled with a masonite spacer of the proper thickness.

Standard factory pipe lengths are 20 feet maximum. Since this piping cannot be field-fabricated, it is necessary to lay out the pipe line accurately and have each straight section fabricated at the factory. Pressure rating of the pipe is dependent upon the rating of the outer metallic shell, which is generally 150 or 300 psi.

Pipe is available from 1½ through 10 inches as standard, with larger diameters available on special order. Refer to Table 5-1 for dimensions.

5.1.1 Natural Rubber-Lined Pipe

Chemically, natural rubber is a polymer of methyl-butadiene (isoprene), with the following structure:

$$CH_3$$
$$|$$
$$CH_2 = C - CH = CH_2$$

It is prepared by coagulating the latex of the *Hevea brasiliensis* tree, which is cultivated primarily in the Far East. When polymerized, the units link together forming long chains, each containing over 1000 units. Simple butadiene does not yield a good grade of rubber apparently because the chains are too smooth and do not form a strong enough interlock. Synthetic rubbers are produced by introducing side groups into the chain either by modifying butadiene or by producing a copolymer of butadiene and some other compound.

Purified raw rubber becomes sticky in hot weather and brittle in cold weather. Its valuable properties become apparent after vulcanization. Depending upon the degree of curing, natural rubber is classified as soft, semi-hard, or hard rubber. The corrosion-resistant properties vary between the different grades of rubber, as do allowable operating temperatures. Pipe linings are available in soft rubber, semi-hard rubber, hard rubber, and multiple ply linings of soft rubber/hard rubber/soft rubber. Table 5-2 lists the maximum allowable operating temperatures of the various types of natural rubber linings.

Soft rubber–lined pipe has the advantages of:

1. Low cost
2. Unaffected by mechanical stresses or rapid temperature changes, making it ideal for outside pipe lines
3. Good general chemical resistance
4. Excellent to superior physical properties—maximum tensile strength, elongation, abrasion and tear resistance

Among the limitations of soft rubber–lined pipe are:

1. Not oil resistant
2. Not flame resistant
3. Not for use with dilute hydrochloric acid (5–10%) or spent acids
4. Relatively low allowable operating temperature

Semi-hard rubber–lined pipe has the following advantages:

1. Moderate cost
2. Excellent chemical and permeation resistance

Limitations of the semi-hard rubber–lined pipe are:

1. Not oil resistant
2. Not flame resistant
3. Not abrasion resistant

Hard rubber–lined pipe has the advantages of:

1. Moderate cost
2. Better chemical, heat, and permeation resistance than soft rubber
3. Higher allowable operating temperature

The limitations of hard rubber–lined pipe are:

1. Not oil resistant
2. Not flame resistant
3. Subject to damage by cold weather exposure, sudden extreme temperature changes, or mechanical stresses
4. Not abrasion resistant

The multiply lining of soft natural/hard natural/and soft natural rubber has basically the same advantages as soft linings but with a higher allowable operating temperature and the same general limitations.

Natural rubber provides excellent resistance to most inorganic salt solutions, alkalies, and nonoxidizing acids. Dilute and moderate concentrations of hydrochloric acid will react with soft rubber to form rubber hydrochloride, and therefore it is not recommended that soft rubber lining be used when handling hydrochloric acid except for concentrated acid (37%).

Strong oxidizing agents such as nitric acid, concentrated sulfuric acid, permanganates, dichromates, chlorine dioxide, and sodium hypochlorite will severely attack natural rubber. Mineral and vegetable oils, gasoline, benzene, toluene, and chlorinated hydrocarbons also affect natural rubber.

Compatibilities of the various grades of natural rubber linings with selected corrodents are given in Tables 5-3, 5-4, 5-5, and 5-6.

5.1.2 Neoprene-Lined Pipe

Neoprene is one of the oldest of the synthetic rubbers. It is a chlorinated butadiene chemically known as a polychloroprene with the basic formula:

$$CH_2 - \underset{\underset{Cl}{|}}{C} - Ch = CH_2$$

The raw material for the manufacture of neoprene is acetylene. Neoprene was introduced commercially by DuPont in 1932 as an oil-resistant substitute for natural rubber, but its range of chemical resistance overcomes many of the shortcomings of natural rubber.

Neoprene can be ignited by an open flame but will stop burning when the flame is removed. Because of its chlorine content neoprene is more resistant to burning than exclusively hydrocarbon rubbers. Natural rubber and many of the synthetic rubbers will continue to burn once ignited even after the flame is removed. In an actual fire situation neoprene will continue to burn.

Neoprene-lined pipe has an operating temperature range of 0 to 200°F (–18 to 93°C). It is unaffected by cold weather, rapid temperature changes, or mechanical stresses. It also offers excellent abrasion resistance.

One of the outstanding properties of neoprene is its resistance to attack from solvent, waxes, fats, oils, greases, and many other petroleum-based products. Excellent service is also experienced when it is in contact with aliphatic compounds (methyl and ethyl alcohols, ethylene glycols, and etc.), aliphatic hydrocarbons, and most freon refrigerants.

Dilute mineral acids, inorganic salt solutions, or alkalies cause little if any change in the appearance or properties of neoprene. Chlorinated and aromatic hydrocarbons, organic esters, aromatic hydroxy compounds, certain ketones, highly oxidizing acid and salt solutions will have adverse effects on neoprene.

Refer to Table 5–7 for the compatibility of neoprene-lined pipe with selected corrodents.

5.1.3 Butyl Rubber–Lined Pipe

Butyl rubber contains isobutylene as the parent material with small quantities of butadiene or isoprene added. Commercial butyl rubber may contain 5% butadiene as a copolymer. The general formula is:

$$
\begin{array}{c}
CH_3 \\
| \\
C \ - CH_2 \\
| \\
CH_3
\end{array}
$$

Butyl rubber–lined pipe has a maximum operating temperature of 200°F (93°C). It is unaffected by cold water or rapid temperature changes and has excellent sliding abrasion resistance and excellent resistance to aging.

Butyl rubber is very nonpolar. It has exceptional resistance to dilute mineral acids, alkalies, phosphate ester oils, acetone, ethylene, and ethylene glycol. It can be used with hydrochloric acid. Good resistance is also offered to concentrated acids with the exception of nitric and sulfuric. It has poor resistance to petroleum oils, gasoline, and most solvents (except oxygenated solvents). Like natural rubber, it is not flame resistant. Table 5–8 provides the compatibility of butyl rubber–lined pipe with selected corrodents.

5.1.4 Chlorobutyl Rubber–Lined Pipe

Chlorobutyl rubber is chlorinated isobutylene-isoprene, possessing the same general properties as butyl rubber with one important exception: it cannot be used with hydrochloric acid. Table 5-9 lists the compatibility of chlorobutyl rubber with selected corrodents.

5.1.5 Nitrile (Buna-N) Rubber–Lined Pipe

The nitrile rubbers are an outgrowth of the German Buna-N or Perbunan. They are copolymers of butadiene and acrylonitrile having the general formula:

$$CH_2 = CH - C = N$$

The primary advantage of nitrile rubber is its greater resistance to oils, fuels, and solvents as compared to neoprene. Nitrile rubber–lined pipe has a maximum operating temperature of 200°F (93°C). It will support combustion and burn.

The nitrile rubbers exhibit good resistance to solvents, oil, water, and hydraulic fluids. The use of highly polar solvents such as acetone and methyl ethyl ketone, chlorinated hydrocarbons, ozone, nitro compounds, ether, or esters should be avoided since these materials will attack the nitrile rubbers. Refer to Table 5–10 for the compatibility of nitrile rubber–lined pipe with selected corrodents.

5.1.6 EPDM Rubber–Lined Pipe

EPDM is a synthetic hydrocarbon-based rubber made from ethylene-propylenediene monomer. It is combined in such a manner as to produce a rubber with a completely saturated backbone and pendant unsaturation for vulcanization.

EPDM rubber–lined pipe has a maximum operating temperature of 180°F (82°C) and is unaffected by cold or rapid temperature changes. It also exhibits good abrasion resistance.

Ethylene-propylene rubber resists attack from oxygenated solvents such as acetone, methyl ethyl ketone, ethyl acetate, weak acids and alkalies, detergents, phosphate esters, alcohols, and glycols. It is not resistant to hydrocarbon solvents and oils, chlorinated hydrocarbons, or turpentine. Refer to Table 5-11 for the compatibility of EPDM rubber with selected corrodents.

5.1.7 Hypalon Rubber–Lined Pipe

Chlorosulfonated polyethylene is manufactured by DuPont under the trade name of Hypalon. It is similar in many respects to neoprene but has improved chemical resistance. Hypalon-lined pipe has a maximum operating temperature of 200°F (93°C) and is unaffected by cold weather or rapid temperature changes.

Hypalon is highly resistant to attack by hydrocarbon oils and fuels and is also resistant to such oxidizing chemicals as sodium hypochlorite, sodium peroxide, ferric chloride, and sulfuric, chromic, and hydrofluoric acids. Concentrated hydrochloric acid (37%) at temperatures above 158°F (70°C) will attack Hypalon but can be handled without adverse affects at all concentrations below this temperature. Nitric acid at room temperature and up to 60% concentration can be handled without adverse effects. Hypalon is also resistant to salt solutions, alcohols, and both weak and concentrated alkalies.

Hypalon has poor resistance to aliphatic, aromatic, and chlorinated hydrocarbons, aldehydes, and ketones. Refer to Table 5-12 for more complete information on the compatibility of Hypalon-lined pipe with selected corrodents.

5.2 CEMENT MORTAR–LINED DUCTILE IRON PIPE

For centuries gray cast iron pipe has been the standard material used for water and sewer mains. The first official record of a cast iron pipe installation was in 1455 in Sugarland,

Germany. In 1664, King Louis XIV of France ordered the construction of a cast iron pipe main extending 15 miles from a pumping station at Marly-on-Seine to Versailles to supply water to the fountain and town. This pipe is still serving the palace gardens after more than 325 years.

Ductile iron not only retains all of cast iron's attractive qualities, such as machinability and corrosion resistance, but also provides additional strength, toughness, and ductility. Ductile iron differs from cast iron in that its graphite form is spheroidal, or modular, instead of the flake form found in cast iron. Due to its modular graphite form, ductile iron has approximately twice the strength of cast iron as determined by tensile, beam, ring bending, and bursting tests. Its tensile and impact strength and elongation are many times greater than that of cast iron.

Ductile iron pipe is available in nominal diameters of 3–54 inches in standard lengths of 20 feet. Table 5-13 provides the dimensions of ductile iron pipe, while Table 5-14 provides the rated working pressure.

A variety of joints are available for use with ductile iron pipe:

1. *Push-on Joint*—The rubber-gasketed push-on joint is the fastest, easiest to assemble and the most widely used joint for water and waste water service. Because of their tightness, push-on joints can be used in wet trench conditions and underwater applications. Table 5-15 shows the allowable joint deflection.
2. *Mechanical Joint*—Although replaced by the push-on-joint in most applications, the mechanical joint is still used on some underground installations, primarily on fittings. Table 5-16 shows the allowable joint deflection.
3. *Restrained Joint*—A special type of push-on or mechanical joint designed to provide longitudinal restraint, the restrained joint is used in conjunction with or in lieu of thrust blocks to provide restraint against thrust forces due to internal pressures.
4. *Flanged Joint*—A rigid joint primarily used in above-ground installations such as open bays and pipe galleries, the flanged joint employs a gasket inserted between two flanges joined with a series of nuts and bolts to produce a watertight seal.
5. *Ball-and-Socket Joint*—Available in bolted or boltless configurations, ball-and-socket joints provide maximum flexibility (maximum deflection 15° per joint) and restraint against joint separation. These joints are often specified in subaqueous crossings, locations requiring large changes in alignment and grade, and in earthquake-prone areas.

Cement mortar–lined piping systems are used primarily for municipal water supply systems and sewerage systems. Linings are factory applied by either the projection method or the centrifugal process. Either of these methods produce a quality lining which is smooth, uniform, and meets the rigid requirements of ANSI/AWWA Standard C104/ A21.4, "Cement-Mortar Lining for Ductile-Iron Pipe and Fittings for Water." The thicknesses of the linings for pipe and fittings, as stated in this standard, shall not be less than $\frac{1}{16}$ inch for 3- to 12-inch pipe, $\frac{3}{32}$ inch for 14- to 24-inch pipe, and $\frac{1}{8}$ inch for 30- to 54-inch pipe. Increased thicknesses can be furnished when so specified.

When cement linings were first introduced to systems with soft waters, it was found that during the first several months of operation the water hardness increased. This problem was rectified by the application of a thin asphaltic paint coating on the freshly placed cement lining which sealed the surface of the lining. This proved to provide a seasoning benefit since the coating initiated the cure of the lining by minimizing loss of moisture during hydration, which resulted in a controlled cure of the mortar. The use of such seal coats is covered in ANSI/AWWA C104/A21.4.

Cement mortar linings are normally acceptable for service up to the boiling point of water (212°F [100°C]). When the lining has been asphaltic seal coated, the maximum operating temperature is reduced to 150°F (60°C) because of the softening point of the coating.

The protective properties of the linings are the result of two properties of the cement, namely, the chemically alkaline reaction of the cement and the gradual reduction of the amount of water in contact with the iron.

When a cement-lined pipe is initially filled with water, water permeates the pores of the lining, thus freeing a considerable amount of calcium hydrate. The calcium hydrate reacts with the calcium bicarbonate in the water to precipitate calcium carbonate, which tends to clog the pores of the mortar and prevent further passage of water.

The first water through the linings in contact with the iron pipe dissolves some of the iron, but free lime tends to precipitate the iron as iron hydroxide, which also closes the pores of the cement. Sulfates are also precipitated as calcium sulfate. A chemical barrier as well as a physical barrier protect the iron pipe shell from the corrosive action of the water.

Waters carry varying amounts of different ions resulting from dissociation of soluble salts found in soils. Waters with a very low ion content are aggressive to calcium hydroxide contained in hydrated cements due to the waters' low content of carbonates and bicarbonates. Soft waters may also have acidic characteristics due to the presence of free carbon dioxide.

When cement mortar linings are exposed to very soft water, calcium hydroxide is leached out. Although the seal-coat retards this leaching to a great extent, its main function is to assist the curing process of the cement mortar.

Cement mortar–lined ductile iron pipe is generally considered to be suitable for continuous use at a pH between 4 and 12. For service outside this range, the manufacturer should be consulted.

The suitability of cement mortar–lined ductile iron pipe for sewer lines is dependent upon the specific conditions under which the sewer will operate. Mildly corrosive sewage (pH 5–6) flowing through a completely filled pipe line can be handled in unlined pipe or with a lining of ASTM-C150 Type I or Type II Portland cement. With effluents having a pH in the range of 3–4, ASTM-C150 Type V Portland cement (high alumina) linings have given satisfactory service.

When the line flows only partially filled, such as in gravity lines, droplets of sulfuric acid will form at the crown of the pipe. This acid will have a pH considerably less than 3 and will attack and corrode all cements. Since all cement mortars are subject to severe corrosion under septic conditions, unlined or unprotected cement-lined pipes are not recommended for sewer piping systems if they may become septic.

Polyethylene-lined ductile iron pipe or other special organic linings are capable of providing protection against attack resulting from septic conditions. When selecting the proper lining for a sewer pipe, factors such as pH, temperature, abrasiveness of effluent, and septicity must be considered. It is best to select the lining in conjunction with the manufacturer.

Although ductile iron pipe possesses good resistance to corrosion and needs no additional protection in most soils, experience has shown that external corrosion protection is recommended in certain soil environments. Examples include low resistivity soils, anaerobic bacteria, dissimilar metals, differences in soil composition, differential aeration of the soil around the pipe, and external stray direct currents. Soils contaminated by coal mine waters, cinders, refuse, or salts are also generally considered corrosive, as are certain naturally occurring environments such as swamps, peat bogs, expansive clays, and alkali soils.

Ductile iron pipe can be protected in such corrosive areas by a variety of means including cathodic protection, special bonded coatings, or polyethylene encasement. The latter method has proven to be the most effective and economical.

Polyethylene encasement is field applied at the time of installation. The ductile iron pipe is encased with a tube or sheet of 8-ml-thick polyethylene immediately before installing the pipe. The polyethylene acts as an unbonded film, which prevents direct contact of the pipe with the corrosive soil.

Typically some groundwater will seep beneath the wrap. Although the entrapped water initially has the corrosive characteristics of the surrounding soil, the available dissolved oxygen supply beneath the wrap is soon depleted and the oxidation process stops long before any damage can occur.

5.3   GLASS-LINED STEEL PIPE

Glass-lined pipe was formerly available with a choice of gray iron, ductile iron, or steel as a substrate material. However, the gray iron and ductile iron had shortcomings in terms of sensitivity to mechanical shock and thermal shock. An installation was limited by the weakest of the substrate materials employed. Consequently, glass-lined pipe is now available with only a steel substrate in lengths of pipe and fittings.

The steel substrate used in conjunction with code requirements provides the needed structural strength, rigidity, and ruggedness, so that the pipe and fittings can be used for full vacuum or rated pressure service.

Even with the steel substrate, glass-lined pipe is susceptible to thermal shock. This vulnerability need not pose a problem if proper precautions are taken, however. The situations most likely to cause thermal shock damage—pumping a cold charge into a hot glass-lined pipe and vice versa—are often avoidable. Other operating conditions can also introduce temperature differentials, but these are seldom as abrupt or critical.

There are definite temperature limits above which damage can occur. When the glass lining is at a temperature of 250°F (121°C), the maximum safe temperature differential for glass-lined steel pipe is 230°F (128°C).

Glass-lined steel pipe is available in nominal diameters of 1 inch through 12 inches as standard with larger nominal diameters available on special order. The lined pipe has an operating temperature range of from –20 to 450°F (–29 to 232°C). The normal pressure rating for plain glass-lined steel pipe is 300 psi, but this requires optional class 300 split flanges as connectors. The class 150 split flanges normally supplied with the pipe allow a nominal working pressure of 150 psi. Specific ratings for both classes of flanges will be found in Tables 5-17 and 5-18. Jacketed glass-lined steel pipe is also available in standard diameters of 1½–8 inches.

All glass-lined steel pipe is joined by means of split flanges, which require a gasket be inserted between the flanged ends. In order to maintain the corrosion-resistant integrity of the glass-lined piping system, a PTFE envelope gasket is recommended.

It is important to avoid stress at all joints. Keep pipe and fittings free to move to avoid stress when connecting flanges. Use expansion joints between two vessels to permit both axial and lateral adjustment. Use expansion joints in longer lines if they are rigidly connected, and at places where there is a source of vibration, such as pumps.

Never pull or spring pipe into place. Make sure all joints are on the same center line with flanges nearly touching. Use offset arrangements and/or expansion joints to correct misalignment and to adjust length.

Make sure mating flanges are parallel. Tighten bolts evenly around the flanges. Use only the recommended bolt torques found in Table 5–19.

Support lines properly. Place supports at 10-foot intervals to keep joints from sagging. On vertical runs, use riser clamps above or below floors. Where possible, install clamp below a joint. Do not support piping too rigidly. Allow some freedom of movement to relieve thermal stress.

Glass is nonpermeable. Strong oxidizers, gases, and other chemicals cannot permeate the hard glassy layer, so the steel is protected against corrosion from the inside. It provides an extremely broad range of corrosion resistance and resists almost all acid solutions in all concentrations at temperatures up to 250°F (121°C) and resists some as high as 450°F (232°C). The exceptions are aqueous solutions containing fluorides, hot, concentrated phosphoric acid, and hydrofluoric acid. The glass lining is resistant to all concentrations of hydrochloric acid at temperatures up to 300°F (149°C), to dilute concentrations of sulfuric acid at their boiling points, to concentrated solutions of sulfuric acid to about 450°F (232°C), and to all concentrations of nitric acid up to their boiling point. An acid-resistant glass with alkaline resistance is available for use under alkaline conditions up to pH 12 at temperatures of 200°F (93°C). Because the glass lining is inert to so many corrosive chemicals, these piping systems are ideal for preventing contamination in the handling of ultra-pure products.

5.4 NICKEL-LINED PIPE

Nickel-lined pipe is produced by electrodepositing a smooth, ductile, pore-free coating of pure nickel on the inner surfaces of the pipe and fittings in thicknesses of 0.0008–0.015 inch depending upon the needs of the specific application.

Seamless steel pipe, forged steel, and seamless steel welding fittings are used as the substrate. The use of nickel-lined pipe permits the user to take advantage of the durability and corrosion resistance of nickel at a fraction of the cost of a solid nickel piping system.

This piping system was developed to handle uranium hexafluoride in conjunction with the Manhattan Project in 1944. Solid nickel pipe could not be used because of the quantities involved. All of the nickel mined in America for 2 years and the entire free world's production would not have been sufficient to produce the required amount of pipe.

Nickel-lined pipe is available in nominal diameters of 1½–24 inches. The operating pressure of the piping system is dependent upon the substrate material. Operating temperatures are dependent upon the resistivity of the nickel to the corrodents present.

The piping system can be joined by welding or flanging, with welding being the preferred method. There are two welding methods available—the "buttering" method and the EB Weld Insert. The latter method should be used for maximum corrosion resistance and minimum product contamination. Nickel-lined pipe may be heated and bent after the lining has been applied without damaging the liner in any way. Refer to Table 6-75 for the compatibility of nickel-lined pipe with selected corrodents.

Table 5-1 Dimensions of Rubber-Lined Pipe

Nominal pipe size (in.)	Outside diameter (in.)	Inside diameter (in.)
1½	1.900	1.225
2	2.375	1.695
2½	2.875	2.095
3	3.500	2.695
4	4.500	3.665
6	6.625	5.695
8	8.625	7.695
10	10.750	9.815

Table 5-2 Maximum Allowable Operating Temperatures of Natural Rubber–Lined Pipe

Type of Lining	Maximum allowable operating temperature (°F/°C)
Soft	140/60
Semi-hard	180/82
Hard	200/93
Multiple ply	160/71

Table 5-3 Compatibility of Soft Natural Rubber–Lined Pipe with Selected Corrodents
The chemicals listed are in the pure state or in a saturated solution unless otherwise indicated.
Compatibility is shown to the maximum allowable temperature for which data are available.
Incompatibility is shown by an x. A blank space indicates that data are unavailable.

Chemical	Maximum temp.		Chemical	Maximum temp.	
	°F	°C		°F	°C
Acetaldehyde	x	x	Barium carbonate	140	60
Acetamide	x	x	Barium chloride	140	60
Acetic acid 10%	150	66	Barium hydroxide	140	60
Acetic acid 50%	x	x	Barium sulfate	140	60
Acetic acid 80%	x	x	Barium sulfide	140	60
Acetic acid, glacial	x	x	Benzaldehyde	x	x
Acetic anhydride	x	x	Benzene	x	x
Acetone	140	60	Benzene sulfonic acid 10%	x	x
Acetyl chloride	x	x	Benzoic acid	140	60
Acrylic acid			Benzyl alcohol	x	x
Acrylonitrile			Benzyl chloride	x	x
Adipic acid			Borax	140	60
Allyl alcohol			Boric acid	140	60
Allyl chloride			Bromine gas, dry		
Alum	140	60	Bromine gas, moist		
Aluminum acetate			Bromine liquid		
Aluminum chloride, aqueous	140	60	Butadiene		
Aluminum chloride, dry	160	71	Butyl acetate	x	x
Aluminum fluoride	x	x	Butyl alcohol	140	60
Aluminum hydroxide			*n*-Butylamine		
Aluminum nitrate	x	x	Butyl phthalate		
Aluminum oxychloride			Butyric acid	x	x
Aluminum sulfate	140	60	Calcium bisulfide		
Ammonia gas			Calcium bisulfite	140	60
Ammonium bifluoride			Calcium carbonate	140	60
Ammonium carbonate	140	60	Calcium chlorate	140	60
Ammonium chloride 10%	140	60	Calcium chloride	140	60
Ammonium chloride 50%	140	60	Calcium hydroxide 10%	140	60
Ammonium chloride, sat.	140	60	Calcium hydroxide, sat.	140	60
Ammonium fluoride 10%	x	x	Calcium hypochlorite	x	x
Ammonium fluoride 25%	x	x	Calcium nitrate	x	x
Ammonium hydroxide 25%	140	60	Calcium oxide	140	60
Ammonium hydroxide, sat.	140	60	Calcium sulfate	140	60
Ammonium nitrate	140	60	Caprylic acid		
Ammonium persulfate			Carbon bisulfide	x	x
Ammonium phosphate	140	60	Carbon dioxide, dry		
Ammonium sulfate 10–40%	140	60	Carbon dioxide, wet		
Ammonium sulfide	140	60	Carbon disulfide	x	x
Ammonium sulfite			Carbon monoxide	x	x
Amyl acetate	x	x	Carbon tetrachloride	x	x
Amyl alcohol	140	60	Carbonic acid	140	60
Amyl chloride	x	x	Cellosolve	x	x
Aniline	x	x	Chloracetic acid, 50% water	x	x
Antimony trichloride			Chloracetic acid	x	x
Aqua regia 3:1	x	x	Chlorine gas, dry	x	x

Table 5-3 *Continued*

Chemical	Maximum temp. °F	Maximum temp. °C	Chemical	Maximum temp. °F	Maximum temp. °C
Chlorine gas, wet	x	x	Magnesium chloride	140	60
Chlorine, liquid	x	x	Malic acid	x	x
Chlorobenzene	x	x	Manganese chloride		
Chloroform	x	x	Methyl chloride	x	x
Chlorosulfonic acid	x	x	Methyl ethyl ketone	x	x
Chromic acid 10%	x	x	Methyl isobutyl ketone	x	x
Chromic acid 50%	x	x	Muriatic acid	140	60
Chromyl chloride			Nitric acid 5%	x	x
Citric acid 15%	140	60	Nitric acid 20%	x	x
Citric acid, concentrated	x	x	Nitric acid 70%	x	x
Copper acetate			Nitric acid, anhydrous	x	x
Copper carbonate	x	x	Nitrous acid, concentrated	x	x
Copper chloride	x	x	Oleum		
Copper cyanide	140	60	Perchloric acid 10%		
Copper sulfate	140	60	Perchloric acid 70%		
Cresol	x	x	Phenol	x	x
Cupric chloride 5%	x	x	Phosphoric acid 50–80%	140	60
Cupric chloride 50%	x	x	Picric acid		
Cyclohexane	x	x	Potassium bromide 30%	140	60
Cyclohexanol			Salicylic acid		
Dichloroacetic acid			Silver bromide 10%		
Dichloroethane (ethylene di-chloride)	x	x	Sodium carbonate	140	60
			Sodium chloride	140	60
Ethylene glycol	140	60	Sodium hydroxide 10%	140	60
Ferric chloride	140	60	Sodium hydroxide 50%	x	x
Ferric chloride 50% in water	140	60	Sodium hydroxide, con-centrated	x	x
Ferric nitrate 10–50%	x	x	Sodium hypochlorite 20%	x	x
Ferrous chloride	140	60	Sodium hypochlorite, con-centrated	x	x
Ferrous nitrate	x	x			
Fluorine gas, dry	x	x	Sodium sulfide to 50%	140	60
Fluorine gas, moist			Stannic chloride	140	60
Hydrobromic acid, dilute	140	60	Stannous chloride	140	60
Hydrobromic acid 20%	140	60	Sulfuric acid 10%	140	60
Hydrobromic acid 50%	140	60	Sulfuric acid 50%	x	x
Hydrochloric acid 20%	x	x	Sulfuric acid 70%	x	x
Hydrochloric acid 38%	140	60	Sulfuric acid 90%	x	x
Hydrocyanic acid 10%			Sulfuric acid 98%	x	x
Hydrofluoric acid 30%	x	x	Sulfuric acid 100%	x	x
Hydrofluoric acid 70%	x	x	Sulfuric acid, fuming	x	x
Hydrofluoric acid 100%	x	x	Sulfurous acid	x	x
Hypochlorous acid			Thionyl chloride		
Iodine solution 10%			Toluene		
Ketones, general			Trichloroacetic acid		
Lactic acid 25%	x	x	White liquor		
Lactic acid, concentrated	x	x	Zinc chloride	140	60

Source: Schweitzer, Philip A. (1991). *Corrosion Resistance Tables*, Marcel Dekker, Inc., New York, Vols. 1 and 2.

Table 5-4 Compatibility of Semi-Hard Natural Rubber–Lined Pipe with Selected Corrodents
The chemicals listed are in the pure state or in a saturated solution unless otherwise indicated.
Compatibility is shown to the maximum allowable temperature for which data are available.
Incompatibility is shown by an x. A blank space indicates that data are unavailable.

Chemical	Maximum temp.		Chemical	Maximum temp.	
	°F	°C		°F	°C
Acetaldehyde			Barium carbonate	180	82
Acetamide	x	x	Barium chloride	180	82
Acetic acid 10%			Barium hydroxide	180	82
Acetic acid 50%			Barium sulfate	180	82
Acetic acid 80%			Barium sulfide		
Acetic acid, glacial	x	x	Benzaldehyde		
Acetic anhydride	x	x	Benzene	x	x
Acetone	x	x	Benzene sulfonic acid 10%		
Acetyl chloride			Benzoic acid	180	82
Acrylic acid			Benzyl alcohol		
Acrylonitrile			Benzyl chloride		
Adipic acid			Borax	180	82
Allyl alcohol			Boric acid	180	82
Allyl chloride			Bromine gas, dry		
Alum	180	82	Bromine gas, moist		
Aluminum acetate			Bromine liquid		
Aluminum chloride, aqueous	180	82	Butadiene		
Aluminum chloride, dry			Butyl acetate	100	38
Aluminum fluoride	x	x	Butyl alcohol	180	82
Aluminum hydroxide			n-Butylamine		
Aluminum nitrate	100	38	Butyl phthalate		
Aluminum oxychloride			Butyric acid	100	38
Aluminum sulfate	180	82	Calcium bisulfide		
Ammonia gas			Calcium bisulfite	180	82
Ammonium bifluoride			Calcium carbonate	180	82
Ammonium carbonate	180	82	Calcium chlorate		
Ammonium chloride 10%	180	82	Calcium chloride	180	82
Ammonium chloride 50%	180	82	Calcium hydroxide 10%	180	82
Ammonium chloride, sat.	180	82	Calcium hydroxide, sat.	180	82
Ammonium fluoride 10%	x	x	Calcium hypochlorite	x	x
Ammonium fluoride 25%	x	x	Calcium nitrate		
Ammonium hydroxide 25%			Calcium oxide		
Ammonium hydroxide, sat.			Calcium sulfate		
Ammonium nitrate	180	82	Caprylic acid		
Ammonium persulfate			Carbon bisulfide		
Ammonium phosphate	180	82	Carbon dioxide, dry		
Ammonium sulfate 10–40%	180	82	Carbon dioxide, wet		
Ammonium sulfide	180	82	Carbon disulfide		
Ammonium sulfite			Carbon monoxide		
Amyl acetate			Carbon tetrachloride		
Amyl alcohol	180	82	Carbonic acid		
Amyl chloride			Cellosolve		
Aniline	x	x	Chloracetic acid, 50% water	100	38
Antimony trichloride			Chloracetic acid	100	38
Aqua regia 3:1			Chlorine gas, dry		

Table 5-4 *Continued*

Chemical	Maximum temp. °F	Maximum temp. °C	Chemical	Maximum temp. °F	Maximum temp. °C
Chlorine gas, wet			Magnesium chloride	180	82
Chlorine, liquid			Malic acid	100	38
Chlorobenzene			Manganese chloride		
Chloroform			Methyl chloride	100	38
Chlorosulfonic acid			Methyl ethyl ketone		
Chromic acid 10%	x	x	Methyl isobutyl ketone		
Chromic acid 50%	x	x	Muriatic acid		
Chromyl chloride			Nitric acid 5%	100	38
Citric acid 15%	100	38	Nitric acid 20%	x	x
Citric acid, concentrated	100	38	Nitric acid 70%	x	x
Copper acetate			Nitric acid, anhydrous	x	x
Copper carbonate	180	82	Nitrous acid, concentrated	100	38
Copper chloride	x	x	Oleum		
Copper cyanide	180	82	Perchloric acid 10%		
Copper sulfate	180	82	Perchloric acid 70%		
Cresol			Phenol	x	x
Cupric chloride 5%	x	x	Phosphoric acid 50–80%	180	82
Cupric chloride 50%	x	x	Picric acid		
Cyclohexane			Potassium bromide 30%	180	82
Cyclohexanol			Salicylic acid		
Dichloroacetic acid			Silver bromide 10%		
Dichloroethane (ethylene di- chloride)			Sodium carbonate	180	82
			Sodium chloride	180	82
Ethylene glycol	180	82	Sodium hydroxide 10%	180	82
Ferric chloride	180	82	Sodium hydroxide 50%	100	38
Ferric chloride 50% in water	180	82	Sodium hydroxide, con- centrated	100	38
Ferric nitrate 10–50%	100	38	Sodium hypochlorite 20%	x	x
Ferrous chloride	180	82	Sodium hypochlorite, con- centrated	x	x
Ferrous nitrate	100	38			
Fluorine gas, dry			Sodium sulfide to 50%	180	82
Fluorine gas, moist			Stannic chloride	180	82
Hydrobromic acid, dilute	180	82	Stannous chloride	180	82
Hydrobromic acid 20%	180	82	Sulfuric acid 10%	180	82
Hydrobromic acid 50%	180	82	Sulfuric acid 50%	100	38
Hydrochloric acid 20%	180	82	Sulfuric acid 70%	x	x
Hydrochloric acid 38%	180	82	Sulfuric acid 90%	x	x
Hydrocyanic acid 10%			Sulfuric acid 98%	x	x
Hydrofluoric acid 30%			Sulfuric acid 100%	x	x
Hydrofluoric acid 70%	x	x	Sulfuric acid, fuming	x	x
Hydrofluoric acid 100%	x	x	Sulfurous acid	150	66
Hypochlorous acid			Thionyl chloride		
Iodine solution 10%			Toluene		
Ketones, general			Trichloroacetic acid		
Lactic acid 25%	100	38	White liquor		
Lactic acid, concentrated	100	38	Zinc chloride	180	82

Source: Schweitzer, Philip A. (1991). *Corrosion Resistance Tables*, Marcel Dekker, Inc., New York, Vols. 1 and 2.

Table 5-5 Compatibility of Hard Natural Rubber–Lined Pipe with Selected Corrodents
The chemicals listed are in the pure state or in a saturated solution unless otherwise indicated.
Compatibility is shown to the maximum allowable temperature for which data are available.
Incompatibility is shown by an x. A blank space indicates that data are unavailable.

Chemical	Maximum temp.		Chemical	Maximum temp.	
	°F	°C		°F	°C
Acetaldehyde			Barium carbonate	200	93
Acetamide	x	x	Barium chloride	200	93
Acetic acid 10%	200	93	Barium hydroxide	x	x
Acetic acid 50%	200	93	Barium sulfate	200	93
Acetic acid 80%	150	66	Barium sulfide	200	93
Acetic acid, glacial	100	38	Benzaldehyde	x	x
Acetic anhydride	100	38	Benzene	x	x
Acetone	x	x	Benzene sulfonic acid 10%		
Acetyl chloride			Benzoic acid	200	93
Acrylic acid			Benzyl alcohol		
Acrylonitrile	90	32	Benzyl chloride		
Adipic acid	80	27	Borax	200	93
Allyl alcohol	x	x	Boric acid	200	93
Allyl chloride			Bromine gas, dry		
Alum	200	93	Bromine gas, moist		
Aluminum acetate			Bromine liquid		
Aluminum chloride, aqueous	200	93	Butadiene		
Aluminum chloride, dry			Butyl acetate	x	x
Aluminum fluoride	x	x	Butyl alcohol	160	71
Aluminum hydroxide	200	93	*n*-Butylamine		
Aluminum nitrate	190	88	Butyl phthalate		
Aluminum oxychloride			Butyric acid	150	66
Aluminum sulfate	200	93	Calcium bisulfide		
Ammonia gas			Calcium bisulfite	200	93
Ammonium bifluoride	200	93	Calcium carbonate	200	93
Ammonium carbonate	200	93	Calcium chlorate		
Ammonium chloride 10%	200	93	Calcium chloride	200	93
Ammonium chloride 50%	200	93	Calcium hydroxide 10%	200	93
Ammonium chloride, sat.	200	93	Calcium hydroxide, sat.	200	93
Ammonium fluoride 10%	x	x	Calcium hypochlorite	x	x
Ammonium fluoride 25%	x	x	Calcium nitrate	200	93
Ammonium hydroxide 25%	x	x	Calcium oxide	200	93
Ammonium hydroxide, sat.	x	x	Calcium sulfate	200	93
Ammonium nitrate	150	66	Caprylic acid		
Ammonium persulfate	200	93	Carbon bisulfide	x	x
Ammonium phosphate	200	93	Carbon dioxide, dry		
Ammonium sulfate 10–40%	200	93	Carbon dioxide, wet		
Ammonium sulfide	200	93	Carbon disulfide	x	x
Ammonium sulfite			Carbon monoxide	x	x
Amyl acetate			Carbon tetrachloride	x	x
Amyl alcohol	200	93	Carbonic acid	200	93
Amyl chloride			Cellosolve		
Aniline	x	x	Chloracetic acid, 50% water	120	49
Antimony trichloride			Chloracetic acid	120	49
Aqua regia 3:1	x	x	Chlorine gas, dry	x	x

Table 5-5 *Continued*

Chemical	Maximum temp. °F	Maximum temp. °C	Chemical	Maximum temp. °F	Maximum temp. °C
Chlorine gas, wet	190	88			
Chlorine, liquid	x	x	Magnesium chloride	200	93
Chlorobenzene	x	x	Malic acid	150	66
Chloroform	x	x	Manganese chloride		
Chlorosulfonic acid	x	x	Methyl chloride		
Chromic acid 10%	x	x	Methyl ethyl ketone	x	x
Chromic acid 50%	x	x	Methyl isobutyl ketone		
Chromyl chloride			Muriatic acid	200	93
Citric acid 15%	150	66	Nitric acid 5%	150	66
Citric acid, concentrated	150	66	Nitric acid 20%	x	x
Copper acetate			Nitric acid 70%	x	x
Copper carbonate	200	93	Nitric acid, anhydrous	x	x
Copper chloride	100	38	Nitrous acid, concentrated	150	66
Copper cyanide	200	93	Oleum		
Copper sulfate	200	93	Perchloric acid 10%		
Cresol			Perchloric acid 70%		
Cupric chloride 5%			Phenol	x	x
Cupric chloride 50%			Phosphoric acid 50–80%	200	93
Cyclohexane			Picric acid		
Cyclohexanol			Potassium bromide 30%	200	93
Dichloroacetic acid			Salicylic acid		
Dichloroethane (ethylene di-chloride)			Silver bromide 10%		
			Sodium carbonate	200	93
Ethylene glycol	200	93	Sodium chloride	200	93
Ferric chloride	200	93	Sodium hydroxide 10%	200	93
Ferric chloride 50% in water	200	93	Sodium hydroxide 50%	150	66
			Sodium hydroxide, con-centrated	150	66
Ferric nitrate 10–50%	150	66			
Ferrous chloride	200	93	Sodium hypochlorite 20%	x	x
Ferrous nitrate	150	66	Sodium hypochlorite, con-centrated	x	x
Fluorine gas, dry					
Fluorine gas, moist			Sodium sulfide to 50%	200	93
Hydrobromic acid, dilute	200	93	Stannic chloride	200	93
Hydrobromic acid 20%	200	93	Stannous chloride	200	93
Hydrobromic acid 50%	200	93	Sulfuric acid 10%	200	93
Hydrochloric acid 20%	200	93	Sulfuric acid 50%		
Hydrochloric acid 38%	200	93	Sulfuric acid 70%	x	x
Hydrocyanic acid 10%	200	93	Sulfuric acid 90%		
Hydrofluoric acid 30%	x	x	Sulfuric acid 98%		
Hydrofluoric acid 70%	x	x	Sulfuric acid 100%		
Hydrofluoric acid 100%	x	x	Sulfuric acid, fuming		
Hypochlorous acid	150	66	Sulfurous acid	200	93
Iodine solution 10%			Thionyl chloride		
Ketones, general			Toluene		
Lactic acid 25%	150	66	Trichloroacetic acid		
Lactic acid, concentrated	150	66	White liquor		
			Zinc chloride	200	93

Source: Schweitzer, Philip A. (1991). *Corrosion Resistance Tables*, Marcel Dekker, Inc., New York, Vols. 1 and 2.

Table 5-6 Compatibility of Multiple Ply (Soft/Hard/Soft) Natural Rubber with Selected Corrodents

The chemicals listed are in the pure state or in a saturated solution unless otherwise indicated. Compatibility is shown to the maximum allowable temperature for which data are available. Incompatibility is shown by an x. A blank space indicates that data are unavailable.

Chemical	Maximum temp. °F	Maximum temp. °C	Chemical	Maximum temp. °F	Maximum temp. °C
Acetaldehyde	x	x	Barium carbonate	160	71
Acetamide	x	x	Barium chloride	160	71
Acetic acid 10%	x	x	Barium hydroxide	160	71
Acetic acid 50%	x	x	Barium sulfate	160	71
Acetic acid 80%	x	x	Barium sulfide	160	71
Acetic acid, glacial	x	x	Benzaldehyde	x	x
Acetic anhydride	x	x	Benzene	x	x
Acetone	140	60	Benzene sulfonic acid 10%	x	x
Acetyl chloride			Benzoic acid	160	71
Acrylic acid			Benzyl alcohol	x	x
Acrylonitrile			Benzyl chloride	x	x
Adipic acid			Borax	160	71
Allyl alcohol			Boric acid	140	60
Allyl chloride			Bromine gas, dry		
Alum	160	71	Bromine gas, moist		
Aluminum acetate			Bromine liquid		
Aluminum chloride, aqueous	160	71	Butadiene		
Aluminum chloride, dry	160	71	Butyl acetate	x	x
Aluminum fluoride	x	x	Butyl alcohol	160	71
Aluminum hydroxide			n-Butylamine		
Aluminum nitrate	x	x	Butyl phthalate		
Aluminum oxychloride			Butyric acid	x	x
Aluminum sulfate	160	71	Calcium bisulfide		
Ammonia gas			Calcium bisulfite	160	71
Ammonium bifluoride			Calcium carbonate	160	71
Ammonium carbonate	160	71	Calcium chlorate	140	60
Ammonium chloride 10%	160	71	Calcium chloride	140	60
Ammonium chloride 50%	160	71	Calcium hydroxide 10%	160	71
Ammonium chloride, sat.	160	71	Calcium hydroxide, sat.	160	71
Ammonium fluoride 10%	x	x	Calcium hypochlorite	x	x
Ammonium fluoride 25%	x	x	Calcium nitrate	x	x
Ammonium hydroxide 25%	100	38	Calcium oxide	160	71
Ammonium hydroxide, sat.	100	38	Calcium sulfate	160	71
Ammonium nitrate	160	71	Caprylic acid		
Ammonium persulfate			Carbon bisulfide	x	x
Ammonium phosphate	160	71	Carbon dioxide, dry		
Ammonium sulfate 10–40%	160	71	Carbon dioxide, wet		
Ammonium sulfide	160	71	Carbon disulfide	x	x
Ammonium sulfite			Carbon monoxide	x	x
Amyl acetate	x	x	Carbon tetrachloride	x	x
Amyl alcohol	100	38	Carbonic acid	160	71
Amyl chloride	x	x	Cellosolve	x	x
Aniline	x	x	Chloracetic acid, 50% water	x	x
Antimony trichloride			Chloracetic acid	x	x
Aqua regia 3:1	x	x	Chlorine gas, dry	x	x

Table 5-6 *Continued*

Chemical	Maximum temp. °F	Maximum temp. °C	Chemical	Maximum temp. °F	Maximum temp. °C
Chlorine gas, wet	x	x	Magnesium chloride	160	71
Chlorine, liquid	x	x	Malic acid	100	38
Chlorobenzene	x	x	Manganese chloride		
Chloroform	x	x	Methyl chloride	x	x
Chlorosulfonic acid	x	x	Methyl ethyl ketone	x	x
Chromic acid 10%	x	x	Methyl isobutyl ketone	x	x
Chromic acid 50%	x	x	Muriatic acid	140	60
Chromyl chloride			Nitric acid 5%	x	x
Citric acid 15%	x	x	Nitric acid 20%	x	x
Citric acid, concentrated	x	x	Nitric acid 70%	x	x
Copper acetate			Nitric acid, anhydrous	x	x
Copper carbonate	x	x	Nitrous acid, concentrated	x	x
Copper chloride	x	x	Oleum		
Copper cyanide	160	71	Perchloric acid 10%		
Copper sulfate	160	71	Perchloric acid 70%		
Cresol	x	x	Phenol	x	x
Cupric chloride 5%	x	x	Phosphoric acid 50–80%	160	71
Cupric chloride 50%	x	x	Picric acid		
Cyclohexane	x	x	Potassium bromide 30%	160	71
Cyclohexanol			Salicylic acid		
Dichloroacetic acid			Silver bromide 10%		
Dichloroethane (ethylene di-chloride)	x	x	Sodium carbonate	160	71
			Sodium chloride	160	71
Ethylene glycol	160	71	Sodium hydroxide 10%	160	71
Ferric chloride	160	71	Sodium hydroxide 50%	x	x
Ferric chloride 50% in water	160	71	Sodium hydroxide, con-centrated	x	x
Ferric nitrate 10–50%	x	x	Sodium hypochlorite 20%	x	x
Ferrous chloride	140	60	Sodium hypochlorite, con-centrated	x	x
Ferrous nitrate	x	x			
Fluorine gas, dry	x	x	Sodium sulfide to 50%	160	71
Fluorine gas, moist			Stannic chloride	160	71
Hydrobromic acid, dilute	160	71	Stannous chloride	160	71
Hydrobromic acid 20%	160	71	Sulfuric acid 10%	160	71
Hydrobromic acid 50%	160	71	Sulfuric acid 50%	x	x
Hydrochloric acid 20%	x	x	Sulfuric acid 70%	x	x
Hydrochloric acid 38%	160	71	Sulfuric acid 90%	x	x
Hydrocyanic acid 10%			Sulfuric acid 98%	x	x
Hydrofluoric acid 30%	x	x	Sulfuric acid 100%	x	x
Hydrofluoric acid 70%	x	x	Sulfuric acid, fuming	x	x
Hydrofluoric acid 100%	x	x	Sulfurous acid	x	x
Hypochlorous acid			Thionyl chloride		
Iodine solution 10%			Toluene		
Ketones, general			Trichloroacetic acid		
Lactic acid 25%	x	x	White liquor		
Lactic acid, concentrated	x	x	Zinc chloride	160	71

Source: Schweitzer, Philip A. (1991). *Corrosion Resistance Tables*, Marcel Dekker, Inc., New York, Vols. 1 and 2.

Table 5-7 Compatibility of Neoprene-Lined Pipe with Selected Corrodents

The chemicals listed are in the pure state or in a saturated solution unless otherwise indicated. Compatibility is shown to the maximum allowable temperature for which data are available. Incompatibility is shown by an x. A blank space indicates that data are unavailable.

Chemical	Maximum temp.		Chemical	Maximum temp.	
	°F	°C		°F	°C
Acetaldehyde	200	93	Barium carbonate	150	66
Acetamide	200	93	Barium chloride	150	66
Acetic acid 10%	160	71	Barium hydroxide	230	110
Acetic acid 50%	160	71	Barium sulfate	200	93
Acetic acid 80%	160	71	Barium sulfide	200	93
Acetic acid, glacial	x	x	Benzaldehyde	x	x
Acetic anhydride	x	x	Benzene	x	x
Acetone	x	x	Benzene sulfonic acid 10%	100	38
Acetyl chloride	x	x	Benzoic acid	150	66
Acrylic acid	x	x	Benzyl alcohol	x	x
Acrylonitrile	140	60	Benzyl chloride	x	x
Adipic acid	160	71	Borax	200	93
Allyl alcohol	120	49	Boric acid	150	66
Allyl chloride	x	x	Bromine gas, dry	x	x
Alum	200	93	Bromine gas, moist	x	x
Aluminum acetate			Bromine liquid	x	x
Aluminum chloride, aqueous	150	66	Butadiene	140	60
Aluminum chloride, dry			Butyl acetate	60	16
Aluminum fluoride	200	93	Butyl alcohol	200	93
Aluminum hydroxide	180	82	*n*-Butylamine		
Aluminum nitrate	200	93	Butyl phthalate		
Aluminum oxychloride			Butyric acid	x	x
Aluminum sulfate	200	93	Calcium bisulfide		
Ammonia gas	140	60	Calcium bisulfite	x	x
Ammonium bifluoride	x	x	Calcium carbonate	200	93
Ammonium carbonate	200	93	Calcium chlorate	200	93
Ammonium chloride 10%	150	66	Calcium chloride	150	66
Ammonium chloride 50%	150	66	Calcium hydroxide 10%	230	110
Ammonium chloride, sat.	150	66	Calcium hydroxide, sat.	230	110
Ammonium fluoride 10%	200	93	Calcium hypochlorite	x	x
Ammonium fluoride 25%	200	93	Calcium nitrate	150	66
Ammonium hydroxide 25%	200	93	Calcium oxide	200	93
Ammonium hydroxide, sat.	200	93	Calcium sulfate	150	66
Ammonium nitrate	200	93	Caprylic acid		
Ammonium persulfate	200	93	Carbon bisulfide	x	x
Ammonium phosphate	150	66	Carbon dioxide, dry	200	93
Ammonium sulfate 10–40%	150	66	Carbon dioxide, wet	200	93
Ammonium sulfide	160	71	Carbon disulfide	x	x
Ammonium sulfite			Carbon monoxide	x	x
Amyl acetate	x	x	Carbon tetrachloride	x	x
Amyl alcohol	200	93	Carbonic acid	150	66
Amyl chloride	x	x	Cellosolve	x	x
Aniline	x	x	Chloracetic acid, 50% water	x	x
Antimony trichloride	140	60	Chloracetic acid	x	x
Aqua regia 3:1	x	x	Chlorine gas, dry	x	x

Table 5-7 *Continued*

Chemical	Maximum temp. °F	Maximum temp. °C	Chemical	Maximum temp. °F	Maximum temp. °C
Chlorine gas, wet	x	x	Magnesium chloride	200	93
Chlorine, liquid	x	x	Malic acid		
Chlorobenzene	x	x	Manganese chloride	200	93
Chloroform	x	x	Methyl chloride	x	x
Chlorosulfonic acid	x	x	Methyl ethyl ketone	x	x
Chromic acid 10%	140	60	Methyl isobutyl ketone	x	x
Chromic acid 50%	100	38	Muriatic acid	x	x
Chromyl chloride			Nitric acid 5%	x	x
Citric acid 15%	150	66	Nitric acid 20%	x	x
Citric acid, concentrated	150	66	Nitric acid 70%	x	x
Copper acetate	160	71	Nitric acid, anhydrous	x	x
Copper carbonate			Nitrous acid, concentrated	x	x
Copper chloride	200	93	Oleum	x	x
Copper cyanide	160	71	Perchloric acid 10%		
Copper sulfate	200	93	Perchloric acid 70%	x	x
Cresol	x	x	Phenol	x	x
Cupric chloride 5%	200	93	Phosphoric acid 50–80%	150	66
Cupric chloride 50%	160	71	Picric acid	200	93
Cyclohexane	x	x	Potassium bromide 30%	160	71
Cyclohexanol	x	x	Salicylic acid		
Dichloroacetic acid	x	x	Silver bromide 10%		
Dichloroethane (ethylene di-chloride)	x	x	Sodium carbonate	200	93
			Sodium chloride	200	93
Ethylene glycol	100	38	Sodium hydroxide 10%	230	110
Ferric chloride	160	71	Sodium hydroxide 50%	230	110
Ferric chloride 50% in water	160	71	Sodium hydroxide, con-centrated	230	110
Ferric nitrate 10–50%	200	93	Sodium hypochlorite 20%	x	x
Ferrous chloride	90	32	Sodium hypochlorite, con-centrated	x	x
Ferrous nitrate	200	93			
Fluorine gas, dry	x	x	Sodium sulfide to 50%	200	93
Fluorine gas, moist	x	x	Stannic chloride	200	93
Hydrobromic acid, dilute	x	x	Stannous chloride	x	x
Hydrobromic acid 20%	x	x	Sulfuric acid 10%	150	66
Hydrobromic acid 50%	x	x	Sulfuric acid 50%	100	38
Hydrochloric acid 20%	x	x	Sulfuric acid 70%	x	x
Hydrochloric acid 38%	x	x	Sulfuric acid 90%	x	x
Hydrocyanic acid 10%	x	x	Sulfuric acid 98%	x	x
Hydrofluoric acid 30%	x	x	Sulfuric acid 100%	x	x
Hydrofluoric acid 70%	x	x	Sulfuric acid, fuming	x	x
Hydrofluoric acid 100%	x	x	Sulfurous acid	100	38
Hypochlorous acid	x	x	Thionyl chloride	x	x
Iodine solution 10%	80	27	Toluene	x	x
Ketones, general	x	x	Trichloroacetic acid	x	x
Lactic acid 25%	140	60	White liquor	140	60
Lactic acid, concentrated	90	32	Zinc chloride	160	71

Source: Schweitzer, Philip A. (1991). *Corrosion Resistance Tables*, Marcel Dekker, Inc., New York, Vols. 1 and 2.

Table 5-8 Compatibility of Butyl Rubber–Lined Pipe with Selected Corrodents
The chemicals listed are in the pure state or in a saturated solution unless otherwise indicated. Compatibility is shown to the maximum allowable temperature for which data are available. Incompatibility is shown by an x. A blank space indicates that data are unavailable.

Chemical	Maximum temp.		Chemical	Maximum temp.	
	°F	°C		°F	°C
Acetaldehyde	80	27	Barium carbonate		
Acetamide			Barium chloride	150	66
Acetic acid 10%	150	66	Barium hydroxide	190	88
Acetic acid 50%	110	43	Barium sulfate		
Acetic acid 80%	110	43	Barium sulfide	190	88
Acetic acid, glacial	x	x	Benzaldehyde	90	32
Acetic anhydride	x	x	Benzene	x	x
Acetone	100	38	Benzene sulfonic acid 10%	90	32
Acetyl chloride			Benzoic acid	150	66
Acrylic acid			Benzyl alcohol	190	88
Acrylonitrile	x	x	Benzyl chloride	x	x
Adipic acid	x	x	Borax	190	88
Allyl alcohol	190	88	Boric acid	150	66
Allyl chloride	x	x	Bromine gas, dry		
Alum	200	93	Bromine gas, moist		
Aluminum acetate	200	93	Bromine liquid		
Aluminum chloride, aqueous	200	93	Butadiene		
Aluminum chloride, dry	200	93	Butyl acetate	x	x
Aluminum fluoride	180	82	Butyl alcohol	140	60
Aluminum hydroxide	100	38	*n*-Butylamine		
Aluminum nitrate	100	38	Butyl phthalate		
Aluminum oxychloride			Butyric acid	x	x
Aluminum sulfate	200	93	Calcium bisulfide		
Ammonia gas			Calcium bisulfite	120	49
Ammonium bifluoride	x	x	Calcium carbonate	150	66
Ammonium carbonate	190	88	Calcium chlorate	190	88
Ammonium chloride 10%	200	93	Calcium chloride	190	88
Ammonium chloride 50%	200	93	Calcium hydroxide 10%	190	88
Ammonium chloride, sat.	200	93	Calcium hydroxide, sat.	190	88
Ammonium fluoride 10%	150	66	Calcium hypochlorite	x	x
Ammonium fluoride 25%	150	66	Calcium nitrate	190	88
Ammonium hydroxide 25%	190	88	Calcium oxide		
Ammonium hydroxide, sat.	190	88	Calcium sulfate	100	38
Ammonium nitrate	200	93	Caprylic acid		
Ammonium persulfate	190	88	Carbon bisulfide		
Ammonium phosphate	150	66	Carbon dioxide, dry	190	88
Ammonium sulfate 10–40%	150	66	Carbon dioxide, wet	190	88
Ammonium sulfide			Carbon disulfide	190	88
Ammonium sulfite			Carbon monoxide	x	x
Amyl acetate	x	x	Carbon tetrachloride	90	32
Amyl alcohol	150	66	Carbonic acid	150	66
Amyl chloride			Cellosolve	150	66
Aniline	150	66	Chloracetic acid, 50% water	150	66
Antimony trichloride	150	66	Chloracetic acid	100	38
Aqua regia 3:1			Chlorine gas, dry	x	x

Table 5-8 *Continued*

Chemical	Maximum temp. °F	Maximum temp. °C	Chemical	Maximum temp. °F	Maximum temp. °C
Chlorine gas, wet			Magnesium chloride	200	93
Chlorine, liquid	x	x	Malic acid	x	x
Chlorobenzene	x	x	Manganese chloride		
Chloroform	x	x	Methyl chloride	90	32
Chlorosulfonic acid	x	x	Methyl ethyl ketone	100	38
Chromic acid 10%	x	x	Methyl isobutyl ketone	80	27
Chromic acid 50%	x	x	Muriatic acid	x	x
Chromyl chloride			Nitric acid 5%	200	93
Citric acid 15%	190	88	Nitric acid 20%	150	66
Citric acid, concentrated	190	88	Nitric acid 70%	x	x
Copper acetate			Nitric acid, anhydrous	x	x
Copper carbonate			Nitrous acid, concentrated	125	52
Copper chloride	150	66	Oleum	x	x
Copper cyanide			Perchloric acid 10%	150	66
Copper sulfate	190	88	Perchloric acid 70%		
Cresol	x	x	Phenol	150	66
Cupric chloride 5%	150	66	Phosphoric acid 50–80%	150	66
Cupric chloride 50%	150	66	Picric acid		
Cyclohexane	x	x	Potassium bromide 30%		
Cyclohexanol			Salicylic acid	80	27
Dichloroacetic acid			Silver bromide 10%		
Dichloroethane (ethylene di-chloride)	x	x	Sodium carbonate		
			Sodium chloride	200	93
Ethylene glycol	200	93	Sodium hydroxide 10%	150	66
Ferric chloride	175	79	Sodium hydroxide 50%	150	66
Ferric chloride 50% in water	160	71	Sodium hydroxide, concentrated	150	66
Ferric nitrate 10–50%	190	88	Sodium hypochlorite 20%	x	x
Ferrous chloride	175	79	Sodium hypochlorite, concentrated	x	x
Ferrous nitrate	190	88			
Fluorine gas, dry	x	x	Sodium sulfide to 50%	150	66
Fluorine gas, moist			Stannic chloride	150	66
Hydrobromic acid, dilute	125	52	Stannous chloride	150	66
Hydrobromic acid 20%	125	52	Sulfuric acid 10%	200	93
Hydrobromic acid 50%	125	52	Sulfuric acid 50%	150	66
Hydrochloric acid 20%	125	52	Sulfuric acid 70%	x	x
Hydrochloric acid 38%	125	52	Sulfuric acid 90%	x	x
Hydrocyanic acid 10%	140	60	Sulfuric acid 98%	x	x
Hydrofluoric acid 30%	150	66	Sulfuric acid 100%	x	x
Hydrofluoric acid 70%	150	66	Sulfuric acid, fuming	x	x
Hydrofluoric acid 100%	150	66	Sulfurous acid	200	93
Hypochlorous acid	x	x	Thionyl chloride	x	x
Iodine solution 10%			Toluene	x	x
Ketones, general			Trichloroacetic acid	x	x
Lactic acid 25%	125	52	White liquor		
Lactic acid, concentrated	125	52	Zinc chloride	200	93

Source: Schweitzer, Philip A. (1991). *Corrosion Resistance Tables*, Marcel Dekker, Inc., New York, Vols. 1 and 2.

Table 5-9 Compatibility of Chlorobutyl Rubber–Lined Pipe with Selected Corrodents
The chemicals listed are in the pure state or in a saturated solution unless otherwise indicated.
Compatibility is shown to the maximum allowable temperature for which data are available.
Incompatibility is shown by an x. A blank space indicates that data are unavailable.

Chemical	Maximum temp.		Chemical	Maximum temp.	
	°F	°C		°F	°C
Acetaldehyde			Barium carbonate		
Acetamide			Barium chloride	150	66
Acetic acid 10%	150	60	Barium hydroxide		
Acetic acid 50%	150	60	Barium sulfate		
Acetic acid 80%	150	60	Barium sulfide		
Acetic acid, glacial	x	x	Benzaldehyde		
Acetic anhydride	x	x	Benzene		
Acetone	100	38	Benzene sulfonic acid 10%		
Acetyl chloride			Benzoic acid	150	66
Acrylic acid			Benzyl alcohol		
Acrylonitrile			Benzyl chloride		
Adipic acid			Borax		
Allyl alcohol			Boric acid	150	66
Allyl chloride			Bromine gas, dry		
Alum	200	93	Bromine gas, moist		
Aluminum acetate			Bromine liquid		
Aluminum chloride, aqueous	200	93	Butadiene		
Aluminum chloride, dry			Butyl acetate		
Aluminum fluoride			Butyl alcohol		
Aluminum hydroxide			n-Butylamine		
Aluminum nitrate	190	88	Butyl phthalate		
Aluminum oxychloride			Butyric acid		
Aluminum sulfate	200	93	Calcium bisulfide		
Ammonia gas			Calcium bisulfite		
Ammonium bifluoride			Calcium carbonate		
Ammonium carbonate	200	93	Calcium chlorate		
Ammonium chloride 10%	200	93	Calcium chloride	160	71
Ammonium chloride 50%	200	93	Calcium hydroxide 10%		
Ammonium chloride, sat.	200	93	Calcium hydroxide, sat.		
Ammonium fluoride 10%			Calcium hypochlorite		
Ammonium fluoride 25%			Calcium nitrate	160	71
Ammonium hydroxide 25%			Calcium oxide		
Ammonium hydroxide, sat.			Calcium sulfate	160	71
Ammonium nitrate	200	93	Caprylic acid		
Ammonium persulfate			Carbon bisulfide		
Ammonium phosphate	150	66	Carbon dioxide, dry		
Ammonium sulfate 10–40%	150	66	Carbon dioxide, wet		
Ammonium sulfide			Carbon disulfide		
Ammonium sulfite			Carbon monoxide	100	38
Amyl acetate			Carbon tetrachloride		
Amyl alcohol	150	66	Carbonic acid	150	66
Amyl chloride			Cellosolve		
Aniline	150	66	Chloracetic acid, 50% water		
Antimony trichloride	150	66	Chloracetic acid	100	38
Aqua regia 3:1			Chlorine gas, dry		

Table 5-9 *Continued*

Chemical	Maximum temp. °F	Maximum temp. °C	Chemical	Maximum temp. °F	Maximum temp. °C
Chlorine gas, wet			Magnesium chloride	200	93
Chlorine, liquid			Malic acid		
Chlorobenzene			Manganese chloride		
Chloroform			Methyl chloride		
Chlorosulfonic acid			Methyl ethyl ketone		
Chromic acid 10%	x	x	Methyl isobutyl ketone		
Chromic acid 50%	x	x	Muriatic acid		
Chromyl chloride			Nitric acid 5%	200	93
Citric acid 15%	90	32	Nitric acid 20%	150	66
Citric acid, concentrated			Nitric acid 70%	x	x
Copper acetate			Nitric acid, anhydrous	x	x
Copper carbonate			Nitrous acid, concentrated	125	52
Copper chloride	150	66	Oleum		
Copper cyanide	160	71	Perchloric acid 10%		
Copper sulfate	160	71	Perchloric acid 70%		
Cresol			Phenol	150	66
Cupric chloride 5%	150	66	Phosphoric acid 50–80%	150	66
Cupric chloride 50%	150	66	Picric acid		
Cyclohexane			Potassium bromide 30%		
Cyclohexanol			Salicylic acid		
Dichloroacetic acid			Silver bromide 10%		
Dichloroethane (ethylene di-			Sodium carbonate		
chloride)			Sodium chloride	200	93
Ethylene glycol	200	93	Sodium hydroxide 10%	150	66
Ferric chloride	175	79	Sodium hydroxide 50%		
Ferric chloride 50% in	100	38	Sodium hydroxide, con-		
water			centrated		
Ferric nitrate 10–50%	160	71	Sodium hypochlorite 20%		
Ferrous chloride	175	79	Sodium hypochlorite, con-		
Ferrous nitrate			centrated		
Fluorine gas, dry			Sodium sulfide to 50%	150	66
Fluorine gas, moist			Stannic chloride		
Hydrobromic acid, dilute	125	52	Stannous chloride		
Hydrobromic acid 20%	125	52	Sulfuric acid 10%	200	93
Hydrobromic acid 50%	125	52	Sulfuric acid 50%		
Hydrochloric acid 20%	x	x	Sulfuric acid 70%	x	x
Hydrochloric acid 38%	x	x	Sulfuric acid 90%	x	x
Hydrocyanic acid 10%			Sulfuric acid 98%	x	x
Hydrofluoric acid 30%			Sulfuric acid 100%	x	x
Hydrofluoric acid 70%	x	x	Sulfuric acid, fuming	x	x
Hydrofluoric acid 100%	x	x	Sulfurous acid	200	93
Hypochlorous acid			Thionyl chloride		
Iodine solution 10%			Toluene		
Ketones, general			Trichloroacetic acid		
Lactic acid 25%	125	52	White liquor		
Lactic acid, concentrated	125	52	Zinc chloride	200	93

Source: Schweitzer, Philip A. (1991). *Corrosion Resistance Tables*, Marcel Dekker, Inc., New York, Vols. 1 and 2.

Table 5-10 Compatibility of Hard Nitrile Rubber–Lined Pipe with Selected Corrodents
The chemicals listed are in the pure state or in a saturated solution unless otherwise indicated.
Compatibility is shown to the maximum allowable temperature for which data are available.
Incompatibility is shown by an x. A blank space indicates that data are unavailable.

Chemical	Maximum temp. °F	Maximum temp. °C	Chemical	Maximum temp. °F	Maximum temp. °C
Acetaldehyde	x	x	Barium carbonate		
Acetamide	180	82	Barium chloride	125	52
Acetic acid 10%	x	x	Barium hydroxide		
Acetic acid 50%	x	x	Barium sulfate		
Acetic acid 80%	x	x	Barium sulfide		
Acetic acid, glacial	x	x	Benzaldehyde		
Acetic anhydride	x	x	Benzene	150	66
Acetone	x	x	Benzene sulfonic acid 10%		
Acetyl chloride	x	x	Benzoic acid	150	66
Acrylic acid	x	x	Benzyl alcohol		
Acrylonitrile	x	x	Benzyl chloride		
Adipic acid	180	82	Borax		
Allyl alcohol	180	82	Boric acid	150	66
Allyl chloride	x	x	Bromine gas, dry		
Alum	150	66	Bromine gas, moist		
Aluminum acetate			Bromine liquid		
Aluminum chloride, aqueous	150	66	Butadiene		
Aluminum chloride, dry			Butyl acetate		
Aluminum fluoride			Butyl alcohol		
Aluminum hydroxide	180	82	*n*-Butylamine		
Aluminum nitrate	190	88	Butyl phthalate		
Aluminum oxychloride			Butyric acid		
Aluminum sulfate	200	93	Calcium bisulfide		
Ammonia gas	190	88	Calcium bisulfite		
Ammonium bifluoride			Calcium carbonate		
Ammonium carbonate	x	x	Calcium chlorate		
Ammonium chloride 10%			Calcium chloride		
Ammonium chloride 50%			Calcium hydroxide 10%		
Ammonium chloride, sat.			Calcium hydroxide, sat.		
Ammonium fluoride 10%			Calcium hypochlorite	x	x
Ammonium fluoride 25%			Calcium nitrate		
Ammonium hydroxide 25%			Calcium oxide		
Ammonium hydroxide, sat.			Calcium sulfate		
Ammonium nitrate	150	66	Caprylic acid		
Ammonium persulfate			Carbon bisulfide		
Ammonium phosphate	150	66	Carbon dioxide, dry		
Ammonium sulfate 10–40%			Carbon dioxide, wet		
Ammonium sulfide			Carbon disulfide		
Ammonium sulfite			Carbon monoxide		
Amyl acetate			Carbon tetrachloride		
Amyl alcohol	150	66	Carbonic acid	100	38
Amyl chloride			Cellosolve		
Aniline	x	x	Chloracetic acid, 50% water		
Antimony trichloride			Chloracetic acid		
Aqua regia 3:1			Chlorine gas, dry		

Table 5-10 *Continued*

Chemical	Maximum temp. °F	Maximum temp. °C	Chemical	Maximum temp. °F	Maximum temp. °C
Chlorine gas, wet			Magnesium chloride		
Chlorine, liquid			Malic acid		
Chlorobenzene			Manganese chloride		
Chloroform			Methyl chloride		
Chlorosulfonic acid			Methyl ethyl ketone		
Chromic acid 10%			Methyl isobutyl ketone		
Chromic acid 50%			Muriatic acid		
Chromyl chloride			Nitric acid 5%		
Citric acid 15%			Nitric acid 20%	x	x
Citric acid, concentrated			Nitric acid 70%	x	x
Copper acetate			Nitric acid, anhydrous	x	x
Copper carbonate			Nitrous acid, concentrated		
Copper chloride			Oleum		
Copper cyanide			Perchloric acid 10%		
Copper sulfate			Perchloric acid 70%		
Cresol			Phenol	x	x
Cupric chloride 5%			Phosphoric acid 50–80%	150	66
Cupric chloride 50%			Picric acid		
Cyclohexane			Potassium bromide 30%		
Cyclohexanol			Salicylic acid		
Dichloroacetic acid			Silver bromide 10%		
Dichloroethane (ethylene di-chloride)			Sodium carbonate	125	52
			Sodium chloride	200	93
Ethylene glycol	100	38	Sodium hydroxide 10%	150	66
Ferric chloride	150	66	Sodium hydroxide 50%		
Ferric chloride 50% in water			Sodium hydroxide, con-centrated		
Ferric nitrate 10–50%	150	66	Sodium hypochlorite 20%	x	x
Ferrous chloride			Sodium hypochlorite, con-centrated	x	x
Ferrous nitrate					
Fluorine gas, dry			Sodium sulfide to 50%		
Fluorine gas, moist			Stannic chloride	150	66
Hydrobromic acid, dilute			Stannous chloride		
Hydrobromic acid 20%			Sulfuric acid 10%	150	66
Hydrobromic acid 50%			Sulfuric acid 50%	150	66
Hydrochloric acid 20%			Sulfuric acid 70%	x	x
Hydrochloric acid 38%			Sulfuric acid 90%	x	x
Hydrocyanic acid 10%			Sulfuric acid 98%	x	x
Hydrofluoric acid 30%			Sulfuric acid 100%	x	x
Hydrofluoric acid 70%	x	x	Sulfuric acid, fuming	x	x
Hydrofluoric acid 100%	x	x	Sulfurous acid		
Hypochlorous acid			Thionyl chloride		
Iodine solution 10%			Toluene		
Ketones, general			Trichloroacetic acid		
Lactic acid 25%			White liquor		
Lactic acid, concentrated			Zinc chloride	150	66

Source: Schweitzer, Philip A. (1991). *Corrosion Resistance Tables*, Marcel Dekker, Inc., New York, Vols. 1 and 2.

Table 5-11 Compatibility of EPDM Rubber–Lined Pipe with Selected Corrodents
The chemicals listed are in the pure state or in a saturated solution unless otherwise indicated.
Compatibility is shown to the maximum allowable temperature for which data are available.
Incompatibility is shown by an x. A blank space indicates that data are unavailable.

Chemical	Maximum temp. °F	Maximum temp. °C	Chemical	Maximum temp. °F	Maximum temp. °C
Acetaldehyde	200	93	Barium carbonate	200	93
Acetamide	200	93	Barium chloride	200	93
Acetic acid 10%	140	60	Barium hydroxide	200	93
Acetic acid 50%	140	60	Barium sulfate	200	93
Acetic acid 80%	140	60	Barium sulfide	140	60
Acetic acid, glacial	140	60	Benzaldehyde	150	66
Acetic anhydride	x	x	Benzene	x	x
Acetone	200	93	Benzene sulfonic acid 10%	x	x
Acetyl chloride	x	x	Benzoic acid	x	x
Acrylic acid			Benzyl alcohol	x	x
Acrylonitrile	140	60	Benzyl chloride	x	x
Adipic acid	200	93	Borax	200	93
Allyl alcohol	200	93	Boric acid	190	88
Allyl chloride	x	x	Bromine gas, dry	x	x
Alum	200	93	Bromine gas, moist	x	x
Aluminum acetate			Bromine liquid	x	x
Aluminum chloride, aqueous			Butadiene	x	x
Aluminum chloride, dry			Butyl acetate	140	60
Aluminum fluoride	190	88	Butyl alcohol	200	93
Aluminum hydroxide	200	93	*n*-Butylamine		
Aluminum nitrate	200	93	Butyl phthalate		
Aluminum oxychloride			Butyric acid	140	60
Aluminum sulfate	190	88	Calcium bisulfide		
Ammonia gas	200	93	Calcium bisulfite	x	x
Ammonium bifluoride	200	93	Calcium carbonate	200	93
Ammonium carbonate	200	93	Calcium chlorate	140	60
Ammonium chloride 10%	200	93	Calcium chloride	200	93
Ammonium chloride 50%	200	93	Calcium hydroxide 10%	200	93
Ammonium chloride, sat.	200	93	Calcium hydroxide, sat.	200	93
Ammonium fluoride 10%	200	93	Calcium hypochlorite	200	93
Ammonium fluoride 25%	200	93	Calcium nitrate	200	93
Ammonium hydroxide 25%	100	38	Calcium oxide	200	93
Ammonium hydroxide, sat.	100	38	Calcium sulfate	200	93
Ammonium nitrate	200	93	Caprylic acid		
Ammonium persulfate	200	93	Carbon bisulfide	x	x
Ammonium phosphate	200	93	Carbon dioxide, dry	200	93
Ammonium sulfate 10–40%	200	93	Carbon dioxide, wet	200	93
Ammonium sulfide	200	93	Carbon disulfide	200	93
Ammonium sulfite			Carbon monoxide	x	x
Amyl acetate	200	93	Carbon tetrachloride	200	93
Amyl alcohol	200	93	Carbonic acid	x	x
Amyl chloride	x	x	Cellosolve	200	93
Aniline	140	60	Chloracetic acid, 50% water		
Antimony trichloride	200	93	Chloracetic acid	160	71
Aqua regia 3:1	x	x	Chlorine gas, dry	x	x

Table 5-11 *Continued*

Chemical	Maximum temp. °F	Maximum temp. °C	Chemical	Maximum temp. °F	Maximum temp. °C
Chlorine gas, wet	x	x	Magnesium chloride	200	93
Chlorine, liquid	x	x	Malic acid	x	x
Chlorobenzene	x	x	Manganese chloride		
Chloroform	x	x	Methyl chloride	x	x
Chlorosulfonic acid	x	x	Methyl ethyl ketone	80	27
Chromic acid 10%			Methyl isobutyl ketone	60	16
Chromic acid 50%	x	x	Muriatic acid		
Chromyl chloride			Nitric acid 5%	60	16
Citric acid 15%	200	93	Nitric acid 20%	60	16
Citric acid, concentrated	200	93	Nitric acid 70%	x	x
Copper acetate	100	38	Nitric acid, anhydrous	x	x
Copper carbonate	200	93	Nitrous acid, concentrated		
Copper chloride	200	93	Oleum	x	x
Copper cyanide	200	93	Perchloric acid 10%	140	60
Copper sulfate	200	93	Perchloric acid 70%		
Cresol	x	x	Phenol		
Cupric chloride 5%	200	93	Phosphoric acid 50–80%	140	60
Cupric chloride 50%	200	93	Picric acid	200	93
Cyclohexane	x	x	Potassium bromide 30%	200	93
Cyclohexanol	x	x	Salicylic acid	200	93
Dichloroacetic acid			Silver bromide 10%		
Dichloroethane (ethylene di-chloride)	x	x	Sodium carbonate	200	93
			Sodium chloride	140	60
Ethylene glycol	200	93	Sodium hydroxide 10%	200	93
Ferric chloride	200	93	Sodium hydroxide 50%	180	82
Ferric chloride 50% in water	200	93	Sodium hydroxide, con-centrated	180	82
Ferric nitrate 10–50%	200	93	Sodium hypochlorite 20%	200	93
Ferrous chloride	200	93	Sodium hypochlorite, con-centrated	200	93
Ferrous nitrate	200	93	Sodium sulfide to 50%	200	93
Fluorine gas, dry			Stannic chloride	200	93
Fluorine gas, moist	60	16	Stannous chloride	200	93
Hydrobromic acid, dilute	90	32	Sulfuric acid 10%	150	66
Hydrobromic acid 20%	140	60	Sulfuric acid 50%	150	66
Hydrobromic acid 50%	140	60	Sulfuric acid 70%	140	60
Hydrochloric acid 20%	100	38	Sulfuric acid 90%	x	x
Hydrochloric acid 38%	90	32	Sulfuric acid 98%	x	x
Hydrocyanic acid 10%	200	93	Sulfuric acid 100%	x	x
Hydrofluoric acid 30%	60	16	Sulfuric acid, fuming	x	x
Hydrofluoric acid 70%	x	x	Sulfurous acid		
Hydrofluoric acid 100%	x	x	Thionyl chloride		
Hypochlorous acid	200	93	Toluene	x	x
Iodine solution 10%	140	60	Trichloroacetic acid	80	27
Ketones, general	x	x	White liquor	200	93
Lactic acid 25%	140	60	Zinc chloride	200	93
Lactic acid, concentrated					

Source: Schweitzer, Philip A. (1991). *Corrosion Resistance Tables*, Marcel Dekker, Inc., New York, Vols. 1 and 2.

Table 5-12 Compatibility of Hypalon–Lined Pipe with Selected Corrodents
The chemicals listed are in the pure state or in a saturated solution unless otherwise indicated.
Compatibility is shown to the maximum allowable temperature for which data are available.
Incompatibility is shown by an x. A blank space indicates that data are unavailable.

Chemical	Maximum temp.		Chemical	Maximum temp.	
	°F	°C		°F	°C
Acetaldehyde	60	16	Barium carbonate	200	93
Acetamide	x	x	Barium chloride	200	93
Acetic acid 10%	200	93	Barium hydroxide	200	93
Acetic acid 50%	200	93	Barium sulfate	200	93
Acetic acid 80%	200	93	Barium sulfide	200	93
Acetic acid, glacial	x	x	Benzaldehyde	x	x
Acetic anhydride	200	93	Benzene	x	x
Acetone	x	x	Benzene sulfonic acid 10%	x	x
Acetyl chloride	x	x	Benzoic acid	200	93
Acrylic acid			Benzyl alcohol	140	60
Acrylonitrile	140	60	Benzyl chloride	x	x
Adipic acid	140	60	Borax	200	93
Allyl alcohol	200	93	Boric acid	200	93
Allyl chloride			Bromine gas, dry	60	16
Alum	200	93	Bromine gas, moist	60	16
Aluminum acetate			Bromine liquid	60	16
Aluminum chloride, aqueous			Butadiene	x	x
Aluminum chloride, dry			Butyl acetate	60	16
Aluminum fluoride	200	93	Butyl alcohol	200	93
Aluminum hydroxide	200	93	*n*-Butylamine		
Aluminum nitrate	200	93	Butyl phthalate		
Aluminum oxychloride			Butyric acid	x	x
Aluminum sulfate	180	82	Calcium bisulfide		
Ammonia gas	90	32	Calcium bisulfite	200	93
Ammonium bifluoride			Calcium carbonate	90	32
Ammonium carbonate	140	60	Calcium chlorate	90	32
Ammonium chloride 10%	190	88	Calcium chloride	200	93
Ammonium chloride 50%	190	88	Calcium hydroxide 10%	200	93
Ammonium chloride, sat.	190	88	Calcium hydroxide, sat.	200	93
Ammonium fluoride 10%	200	93	Calcium hypochlorite	200	93
Ammonium fluoride 25%			Calcium nitrate	100	38
Ammonium hydroxide 25%	200	93	Calcium oxide	200	93
Ammonium hydroxide, sat.	200	93	Calcium sulfate	200	93
Ammonium nitrate	200	93	Caprylic acid	x	x
Ammonium persulfate	80	27	Carbon bisulfide		
Ammonium phosphate	140	60	Carbon dioxide, dry	200	93
Ammonium sulfate 10–40%	200	93	Carbon dioxide, wet	200	93
Ammonium sulfide	200	93	Carbon disulfide	200	93
Ammonium sulfite			Carbon monoxide	x	x
Amyl acetate	60	16	Carbon tetrachloride	200	93
Amyl alcohol	200	93	Carbonic acid	x	x
Amyl chloride	x	x	Cellosolve		
Aniline	140	60	Chloracetic acid, 50% water		
Antimony trichloride	140	60	Chloracetic acid	x	x
Aqua regia 3:1			Chlorine gas, dry	x	x

Table 5-12 *Continued*

Chemical	°F	°C	Chemical	°F	°C
Chlorine gas, wet	90	32	Magnesium chloride	200	93
Chlorine, liquid			Malic acid		
Chlorobenzene	x	x	Manganese chloride	180	82
Chloroform	x	x	Methyl chloride	x	x
Chlorosulfonic acid	x	x	Methyl ethyl ketone	x	x
Chromic acid 10%	150	66	Methyl isobutyl ketone	x	x
Chromic acid 50%	150	66	Muriatic acid	140	60
Chromyl chloride			Nitric acid 5%	100	38
Citric acid 15%	200	93	Nitric acid 20%	100	38
Citric acid, concentrated	200	93	Nitric acid 70%	x	x
Copper acetate	x	x	Nitric acid, anhydrous	x	x
Copper carbonate			Nitrous acid, concentrated		
Copper chloride	200	93	Oleum	x	x
Copper cyanide	200	93	Perchloric acid 10%	100	38
Copper sulfate	200	93	Perchloric acid 70%	90	32
Cresol	x	x	Phenol	x	x
Cupric chloride 5%	200	93	Phosphoric acid 50–80%	200	93
Cupric chloride 50%	200	93	Picric acid	80	27
Cyclohexane	x	x	Potassium bromide 30%	200	93
Cyclohexanol	x	x	Salicylic acid		
Dichloroacetic acid			Silver bromide 10%		
Dichloroethane (ethylene di-chloride)	x	x	Sodium carbonate	200	93
			Sodium chloride	200	93
Ethylene glycol	200	93	Sodium hydroxide 10%	200	93
Ferric chloride	200	93	Sodium hydroxide 50%	200	93
Ferric chloride 50% in water	200	93	Sodium hydroxide, con-centrated	200	93
Ferric nitrate 10–50%	200	93	Sodium hypochlorite 20%	200	93
Ferrous chloride	200	93	Sodium hypochlorite, con-centrated		
Ferrous nitrate					
Fluorine gas, dry	140	60	Sodium sulfide to 50%	200	93
Fluorine gas, moist			Stannic chloride	90	32
Hydrobromic acid, dilute	90	32	Stannous chloride	200	93
Hydrobromic acid 20%	100	38	Sulfuric acid 10%	200	93
Hydrobromic acid 50%	100	38	Sulfuric acid 50%	200	93
Hydrochloric acid 20%	160	71	Sulfuric acid 70%	160	71
Hydrochloric acid 38%	140	60	Sulfuric acid 90%	x	x
Hydrocyanic acid 10%	90	32	Sulfuric acid 98%	x	x
Hydrofluoric acid 30%	90	32	Sulfuric acid 100%	x	x
Hydrofluoric acid 70%	90	32	Sulfuric acid, fuming		
Hydrofluoric acid 100%	90	32	Sulfurous acid	160	71
Hypochlorous acid	x	x	Thionyl chloride		
Iodine solution 10%			Toluene	x	x
Ketones, general	x	x	Trichloroacetic acid		
Lactic acid 25%	140	60	White liquor		
Lactic acid, concentrated	80	27	Zinc chloride	200	93

Source: Schweitzer, Philip A. (1991). *Corrosion Resistance Tables*, Marcel Dekker, Inc., New York, Vols. 1 and 2.

Table 5-13 Dimensions of Ductile Iron Pipe

Nominal pipe size (in.)	Outside diameter (in.)	Wall thickness of standard classes (in.)						
		50	51	52	53	54	55	56
3	3.96		0.25	0.28	0.31	0.34	0.37	0.40
4	4.80		0.26	0.29	0.32	0.35	0.38	0.41
6	6.90	0.25	0.28	0.31	0.34	0.37	0.40	0.43
8	9.05	0.27	0.30	0.33	0.36	0.39	0.42	0.45
10	11.10	0.29	0.32	0.35	0.38	0.41	0.44	0.47
12	13.20	0.31	0.34	0.37	0.40	0.43	0.46	0.49
14	15.30	0.33	0.36	0.39	0.42	0.45	0.48	0.51
16	17.40	0.34	0.37	0.40	0.43	0.46	0.49	0.52
18	19.50	0.35	0.38	0.41	0.44	0.47	0.50	0.53
20	21.60	0.36	0.39	0.42	0.45	0.48	0.51	0.54
24	25.80	0.38	0.41	0.44	0.47	0.50	0.53	0.56
30	32.00	0.39	0.43	0.47	0.51	0.55	0.59	0.67
36	38.30	0.43	0.48	0.53	0.58	0.63	0.68	0.73
42	44.50	0.47	0.53	0.59	0.65	0.71	0.77	0.83
48	50.80	0.51	0.58	0.65	0.72	0.79	0.86	0.93
54	57.10	0.57	0.65	0.73	0.81	0.89	0.97	1.05

Table 5-14 Rated Working Pressure of Ductile Iron Pipe

Nominal pipe size (in.)	Pressure for class (psi)			
	50	51	52	53
3		350	350	350
4		350	350	350
6	350	350	350	350
8	350	350	350	350
10	350	350	350	350
12	350	350	350	350
14	350	350	350	350
16	350	350	350	350
18	350	350	350	350
20	300	350	350	350
24	250	300	350	350
30	200	250	300	350
36	200	250	300	350
42	200	250	300	350
48	200	250	300	350
54	200	250	300	350

Table 5-15 Allowable Joint Deflection in Push-On Joint Pipe

Nominal pipe size (in.)	Nominal laying length (ft)	Maximum recommended deflection	
		Offset per length (in.)	Deflection angle (degrees)
3	20	21[a]	5
4	20	21[a]	5
6	20	19[b]	5
8	20	19[b]	5
10	20	19[b]	5
12	20	19[b]	5
14	20	15[b]	4
16	20	15[b]	4
18	20	11[b]	3
20	20	11[b]	3
24	20	11[b]	3
30	20	11[b]	3
36	20	11[b]	3

[a]20-foot length.
[b]18-foot length.

Table 5-16 Allowable Joint Deflection in Mechanical Joint Pipe

Nominal pipe size (in.)	Nominal laying length (ft)	Maximum recommended deflection	
		Offset per length (in.)	Deflection angle
3	20	35	8°-18′
4	20	35	8°-18′
6	18	27	7°-07′
8	18	20	5°-21′
10	18	20	5°-21′
12	18	20	5°-21′
14	18	13.5	3°-35′
16	18	13.5	3°-35′
18	18	11	3°-00′
20	18	11	3°-00′
24	18	9	2°-23′

Table 5-17 Pressure Temperature Ratings for Glass-Lined Steel Pipe with Standard Class 150 Split Flanges

Temp. (°F/°C)	Pressure (psi) at pipe diameters (in.):		
	1,1½,2,4,6,8	3	10 & 12
−20 to 100/−29 to 38	235	210	150
150/66	225	210	150
200/93	215	210	150
250/121	212	210	150
300/149	210	210	150
350/177	205	205	150
400/204	200	200	150
450/232	185	185	150

Table 5-18 Pressure Temperature Ratings for Glass-Lined Steel Pipe with Class 300 Split Flanges

Nominal pipe size (in.)	Temperature (°F/°C)	Pressure (psi)
1½	−20 to 450/−29 to 232	350
2	−20 to 450/−29 to 232	250
3	−20 to 450/−29 to 232	210
4	−20 to 450/−29 to 232	350
6	−20 to 450/−29 to 232	240
8	−20 to 450/−29 to 232	375
10	−20 to 450/−29 to 232	150
12	−20 to 450/−29 to 232	200

Table 5-19 Recommended Bolt Torques for Glass-Lined Steel Pipe

Nominal pipe size (in.)	Torque (ft-lb)							
	ANSI class 150				ANSI class 300			
	Bolts		Crt-AF gaskets		Bolts		Crt-AF gaskets	
	No.	Size (in.)	Lub.	Dry	No.	Size (in.)	Lub.	Dry
1	4	½	20	30	4	⅝	40	60
1½	4	½	20	30	4	¾	70	100
2	4	⅝	40	60	8	⅝	40	60
3	4	⅝	40	60	8	¾	70	100
4	8	⅝	40	60	8	¾	70	100
5	8	⅝	40	60				
6	8	¾	70	100	12	¾	70	100
8	8	¾	70	100	12	⅞	110	160

Torques are recommended for well-lubricated (Lub.) and unlubricated (Dry) bolts.
ASME Spec. SA193 Grade B7 identified by B7 stamped on J-bolt or bolt head.
ASME Spec. SA 194 Grade 2H nuts, identified by 2H stamped on tip surface.
Lubrication is strongly recommended to reduce variable binding between the bolt and nut. The nut should also ride freely over the complete length of the bolt threads before assembly.

6

Metallic Piping Systems

A wide variety of metallic piping systems are available, starting with a family of ferrous alloys and alloys of nonferrous materials. In addition, there are systems produced from "pure" metals rather than alloys. Over the years these materials have been produced as needs arose in industry for materials to handle specific corrodents and/or for materials to be able to operate at elevated temperatures in the presence of corrodents.

As may be expected, no one material is completely corrosion resistant, although some come close. However, as the overall corrosion resistance is improved, the cost increases. Because of this, the most corrosion-resistant material is not always selected for use. Compromises must be made. This can be done by making judicious decisions. Although other forms of attack must be considered in special circumstances, uniform attack is the one form most commonly confronting the user of metallic piping systems. The rate of uniform attack is reported in various units. In the United States it is generally reported as *inches penetration per year* (ipy) or *mils penetration per year* (mpy) and *milligrams per square decimeter per day* (mdd). To convert from ipy to mpy, multiply the ipy value by 1000. Conversion of ipy to mdd or vice versa requires knowledge of the metal density. Conversion factors are given in Table 6-1. The subject of uniform corrosion will be discussed later in this chapter.

6.1 CORROSION OF METALS

Corrosion is the destructive attack of a metal by a chemical or electrochemical reaction. Deterioration by physical causes is not called corrosion, but is described as erosion, galling, or wear. In some instances corrosion may accompany physical deterioration and is described in such terms as corrosion-erosion, corrosive wear, or fretting corrosion.

Direct chemical corrosion is limited to unusual conditions involving highly aggressive environments or high temperature or both. Examples are metals in contact with strong acids or alkalies.

Electrochemical reaction is the result of electrical energy passing from a negative area to a positive area through an electrolyte medium. With iron or steel in aerated water, the negative electrodes are portions of the iron surface itself covered by porous rust (iron oxides), and positive electrodes are areas exposed to oxygen, the positive and negative electrode areas interchanging and shifting from place to place as the corrosion reaction proceeds. "Rusting" applies to the corrosion of iron-based alloys with the formation of corrosion products consisting largely of hydrous ferric oxides. Nonferrous metals and alloys therefore corrode, but do not rust.

All structural metals corrode to some extent in natural environments. Bronzes, brasses, stainless steels, zinc, and aluminum corrode so slowly under the service conditions in which they are placed that they are expected to survive for long periods without protection. When these same metals are placed into contact with more aggressive corrodents, they suffer attack and are degraded. Corrosion of structural grades of iron and steel, however, proceeds rapidly unless the metal is amply protected.

Ordinarily iron and steel corrode in the presence of both oxygen and water. If either of these ingredients are absent, corrosion will not take place. Rapid corrosion may take place in water, the rate of corrosion being accelerated by the velocity or the acidity of the water, by the motion of the metal, by an increase in the temperature, by the presence of certain bacteria, or by other factors. Conversely, corrosion is retarded by protective layers (films) consisting of corrosion products or absorbed oxygen. High alkalinity of the water also retards the rate of corrosion on steel surfaces.

There are nine basic forms of corrosion that can be encountered in a piping system:

1. Uniform corrosion
2. Intergranular corrosion
3. Galvanic corrosion
4. Crevice corrosion
5. Pitting
6. Erosion corrosion
7. Stress corrosion cracking
8. Biological corrosion
9. Dezincification

Prevention or control of this corrosion can usually be achieved by use of a suitable material of construction, use of proper design and installation techniques, and if specific in plant procedures are followed, or a combination of these factors.

6.1.1 Uniform Corrosion

Metals resist corrosion by forming a "passive" film on the surface. This film in a sense is corrosion, but once formed further degradation of the base metal is prevented. Most metals will form this film after a period of time of exposure to air. Chemical treatment can help form this film more rapidly. For example, nitric acid when applied to austenitic stainless steel will form this protective film.

As long as this film is intact, whether formed naturally or by chemical treatment, the base metal will be protected. Examples are the patina formed on copper, the rusting of iron, or the tarnishing of silver. "Fogging" of nickel and high-temperature oxidation of metals are also examples.

These films do not provide an overall resistance to chemical attack. They may be destroyed by the presence of various chemicals. The immunity of the film to attack is a function of the film composition, the temperature, and the agressiveness of the corrosive material.

Pure metals and their alloys tend to enter into chemical union with the elements of a corrosive medium to form stable compounds similar to those found in nature. When metal loss occurs in this manner, the compound formed is referred to as the corrosion product and the metal surface is referred to as being corroded. For example, halogens, particularly chlorides, will attack and penetrate this film on stainless steel. This will result in a general corrosion of the pipe. Such interaction between a chemical and a metal can be determined from the corrosion tables. This is known as uniform attack.

Ratings of metals, whenever attack is uniform in the presence of a specific corrodent, are classified into three groups, the classifications being based on the metal loss in inches per year (ipy) or mils per year (mpy).

These classifications are as follows:

1. <5 mpy (0.005 ipy)—Metals in this category are considered as having good corrosion resistance and are recommended for use as a piping material.

2. 5–50 mpy—Metals in this group are considered satisfactory if a higher rate of corrosion can be tolerated.

3. >50 mpy—Usually not a satisfactory choice.

Most corrosion rates are given as <5 mpy, <20 mpy, or >50 mpy. A rating of <5 mpy indicates that the metal has excellent resistance in that specific media and would be an excellent choice to use. The category of <20 mpy would also be considered a satisfactory choice for a piping material, but not as good as a rating of <5 mpy.

6.1.2   Intergranular Corrosion

Intergranular corrosion is a localized attack taking place at and adjacent to grain boundaries. The grains themselves suffer little corrosion. This attack is the result of carbide precipitation during welding operations. Carbide precipitation can be prevented by using alloys with extra low carbon content, by using stabilized grades of alloys, or by specifying solution heat treatment followed by a rapid quench, which will keep the carbides in solution.

Austenitic stainless steels that are not stabilized or of the extra low carbon types are subject to carbide precipitation. When stabilized with columbium or titanium to decrease the carbide formation, or when they contain less than 0.03% carbon, the austenitic stainless steels are not normally subject to carbide precipitation.

It is more practical, in so far as piping systems are concerned, to use either a stabilized or extra low carbon content austenitic stainless steel than to consider a solution heat treatment.

6.1.3   Galvanic Corrosion

When two dissimilar metals or alloys are immersed in a corrosive or conductive solution, corrosion takes place as the result of the difference in potential existing between the materials. Attack can also occur with metals having the same analysis but having different surface conditions with an electrolyte present. Under these conditions an electrolytic cell is formed. Each metal or alloy has "built-in" properties that cause it to act as an anode or cathode. The flow of current is from the anode to the cathode, the anode being the more active metal and the cathode being the less active metal. This results in the anode being corroded.

The particular way in which a material will react can be predicted from the relative position of the material in the galvanic series (see Table 6-2). When it is necessary to use dissimilar materials, two materials should be selected that are relatively close in the galvanic series. The further apart the metals are in the galvanic series, the greater the rate of corrosion.

Relative areas between the anode and cathode affect the rate of corrosion. Since the flow of current is from the anode to the cathode, the combination of a large cathode area and small anode area is unfavorable. Corrosion of the anode can be 100–1000 times greater than if the two areas were equal. Galvanic corrosion can be prevented by insulating the two metals from each other. For example, when bolting flanges of dissimilar metals together, plastic washers can be used to separate the two metals.

6.1.4 Crevice Corrosion

Crevice corrosion occurs within or adjacent to a crevice formed by contact with another piece of the same or another metal, or with a nonmetallic material, such as a gasket. When this occurs, attack is usually more severe than on surrounding areas of the same surface. This form of corrosion can occur as the result of a deficiency of oxygen in the crevice, acidity changes in the crevice, buildup of ions in the crevice, or depletion of an inhibitor.

Prevention can be accomplished by proper design and operating procedures. Nonabsorbent gasketing material should be used at flanged joints, while fully penetrated butt-welded joints are preferable to threaded joints.

6.1.5 Pitting

Pitting is a form of localized attack, the rate of corrosion being greater at some areas than at others. If appreciable attack is confined to a relatively small area of metal acting as an anode, the resulting pits are described as deep. If the area of attack is relatively large and not so deep, the pits are called shallow. Depth of pitting is sometimes referred to as pitting factor, which is the ratio of the deepest metal penetration to average metal penetration as determined by weight loss in the specimin. A pitting factor of one represents uniform attack.

The first step in prevention lies in the proper material selection, followed by a design that prevents stagnation of material and alternate wetting and drying of the surface. Increase in stream velocity will also help.

6.1.6 Erosion Corrosion

Erosion corrosion, sometimes referred to as impingement attack, is the result of the relative movement between a corrodent and the metal surface. High-velocity streams in the piping system can cause this, particularly in elbows and tees. This phenomenon can be present with or without solids in the stream. Prevention can be accomplished by reducing the velocity, selecting a harder metal, and properly designing the piping system.

6.1.7 Stress Corrosion Cracking

All metals can be subject to stress corrosion cracking (SCC). Three conditions are necessary for stress corrosion cracking to occur: a suitable environment, a tensile stress, and a sensitive metal. With copper-containing alloys, an ammonia-containing environment leads to stress corrosion cracking. With the low alloy austenitic stainless steels, a chloride-containing environment is necessary. Concentrated solutions are not necessary to induce SCC. Quite often a solution containing just a few parts per million of the critical ion is all that is required. Temperature and pH are also critical. Usually there is a threshold temperature below which cracking will not occur and a plus or minus pH value to be met before cracking will start.

The failure mechanism is triggered by a tensile stress. SCC will not generally occur if the part is in compression. The tensile stress must approach the yield strength of the metal. Often pits, which act as stress concentration sites, will initiate SCC. Residual stresses from welding, straightening, installation, bending, or accidental denting of the component can lead to cracking.

Alloy content is very important, with nickel being the key. Neither the nickel-free ferritic stainless steels nor the high-nickel alloys are sensitive to stress corrosion cracking. When the nickel content exceeds 30% the alloy is immune to SCC. The most common grades of stainless steel (304, 304L, 316, 316L, 321, 347, 303, 302, and 301), which have a nickel content in the range of 7–10%, are the most susceptible to SCC. Table 6-3 provides a partial listing of alloy systems subject to stress corrosion cracking.

It is important that proper installation techniques be followed so that no stresses are induced into the piping. Piping should not be "sprung" into place, since this will induce a stress. Any stresses induced during fabrication should be removed by an appropriate stress-relieving operation. Avoiding stagnant areas in the design and reducing operating temperatures, if possible, will also help to eliminate SCC.

Examples of stress corrosion cracking include the cracking of cold-formed brass in ammonia environments, cracking of austenitic stainless steels in the presence of chlorides, cracking of monel in hydrofluosilicic acid, and the caustic embrittlement cracking of steel in caustic solutions.

6.1.8 Biological Corrosion

The metabolic activity of microorganisms can directly or indirectly cause deterioration of a metal by the corrosion process. This activity can

1. Produce a corrosive environment
2. Create electrolytic concentration cells on the metal surface
3. Alter the resistance of surface films
4. Have an influence on the rate of anodic or cathodic reaction
5. Alter the environment composition

Such activity can be prevented by selecting a more resistant material of construction or by using biocides. For some species of bacteria, a change in pH will provide control.

6.1.9 Dezincification

When one element in a solid alloy is removed by corrosion, the process is known as dezincification. The most common example is the removal of zinc from brass alloys. Selection of the proper material of construction and the reduction of the amount of oxygen present will help alleviate the situation.

6.2 OPERATING PRESSURES AND OPERATING TEMPERATURES

Corrosion-resistant piping is usually available in four wall thicknesses and is designated as schedule 5S, schedule 10S, schedule 40S, and schedule 80S. Other wall thicknesses are available, but these four are the most commonly used. The allowable operating pressures and temperatures of seamless or welded corrosion-resistant pipe are calculated by the following formula:

$$P = \frac{2SEt}{D_o - 2yt}$$

where:
- P = internal design pressure (psig)
- S = applicable allowable stress (psi)
- D_o = outside diameter (in.)
- E = longitudinal joint factor (= 1.0 for seamless pipe; see Table 6-4 for welded joints)
- t = $t_m - C$ (design thickness in inches)
- t_m = minimum required thickness, satisfying requirements for pressure and mechanical corrosion and erosion allowances (12½% less than the nominal thickness given in Tables 6-5 and 6-6)
- C = mechanical, corrosion, and erosion allowances (in.)
- y = a coefficient having values as shown in Table 6-7 for ductile ferrous materials (For ductile nonferrous materials, it has a value of 0.4, and zero for brittle materials such as cast iron.)

The applicable allowable stress is a function of temperature and decreases as the operating temperature increases. These allowable stress values are given for each specific piping material in the section dealing with that specific piping system.

The minimum pipe wall thickness (t_m) required for a specific application of pressure and temperature can be calculated from the following equation:

$$t_m = \frac{PD_o}{2(SE+Py)} + C$$

where:
- P = design pressure (psi)
- D_o = outside diameter of pipe (in.)
- C = mechanical, corrosion, and erosion allowances, and any thread or groove depth (see Table 6–8)
- S = allowable stress (psi)
- E = longitudinal weld joint factor (=1.0 for seamless pipe; see Table 6-4 for welded joints)
- y = coefficient having values as shown in Table 6-7 for ductile ferrous materials (For ductile nonferrous materials it has a value of 0.4 and zero for brittle materials such as cast iron.)
- t_m = minimum required thickness

The schedule number for corrosion-resistant piping is defined in the piping code as:

$$SN = \frac{1000\, P_i}{S}$$

where

P_i = internal pipe pressure (psig)
S = allowable stress on the pipe

Selection of proper pipe schedule for a particular application can be made using this correlation. Assume a 1-inch 304ss pipe is to operate at 750°F (399°C) with an internal pressure of 100 psig. The allowable stress for 304ss at 750°F (399°C) is 10,400 psi. Substituting in the formula,

$$SN = \frac{1000\ (100)}{10,400} = 9.62$$

Under these conditions, a schedule 10 pipe could be tentatively selected. The minimum required wall thickness must be checked.

Assume the pipe is seamless and C has a value of 0.05:

$$t_m = \frac{PD_o}{2(SE + Py)} + C$$

$$t_m = \frac{100\ (1.315)}{2\ [(10400)(1) + 100\ (0.4)]} + 0.05$$

$$t_m = 0.06$$

The wall thickness of 1-inch schedule 10 pipe is 0.109, therefore schedule 10 can be used.

6.3   JOINING OF METALLIC PIPE

Metallic piping systems may be joined by welding, expanding, flanging, and threading (depending upon the schedule number). Since most of the corrosion-resistant piping systems are produced from relatively expensive materials, large quantities of schedule 5S and schedule 10S pipe are used. These schedules have relatively thin walls, which conserve metal and therefore reduce the cost, and at the same time have allowable operating pressures and temperatures that are compatible with most processing applications.

These particular schedule numbers are joined only by welding or expanding. The wall thicknesses are too thin to permit threading. They may have flanges or unions installed by means of an expanding operation. Fittings in these systems are manufactured with longer legs than conventional welding fittings, which permits the installation of a flange on the end. Flanges and unions are supplied with a serrated insert. The insert is of the same alloy as the piping system, while the flange is of carbon steel or ductile iron. This reduces the cost of the flange.

To secure the flange to the pipe or fitting, the pipe or fitting is expanded into the serrated inside diameter of the insert. The steps involved in the operation are simple:

1.   Expand the pipe or fitting into the insert using the expanders supplied by the manufacturer.
2.   Loosen the flange from the insert after the pipe or fitting has been expanded into it.
3.   Rotate the flange to line up bolt holes with mating flange.
4.   Bolt up the joint.

The pressure ratings of these flanges are a function of the operating temperature and the material of construction of the flange (carbon steel or ductile iron). Table 6-9 lists the allowable operating pressures based on operating temperatures.

Welding of corrosion-resistant piping systems is the overall preferred method of joining. Since the welding techniques vary with each type of material, this technique is discussed under the heading of each material.

Flanging, using solid flanges made of the corrosion-resistant material of the piping system, is not normally done because of the expense.

Threading of corrosion-resistant piping can only be done on schedule 40S and schedule 80S pipe. Schedule 5S and schedule 10S cannot be threaded. One of the problems with threaded alloy pipe is that of galling. Many of the alloys are austenitic stainless steels or other alloys that have as high or higher nickel contents. It is this nickel content that makes these alloys prone to galling.

Although several standards have been established covering pipe threads for various applications, the oldest and most commonly used is the Americal Standard for Pipe Threads for ANSI B2.1. The specified number of threads per inch are as follows:

Pipe size (in.)	Threads per inch
⅛	28
¼, ⅜	19
½, ¾	14
1–6	11
8, 10	10
>12	8

When threading alloy pipe, it is important that clean sharp dies be used. The threads of both parts to be joined must be thoroughly cleaned prior to making up the joint. Any threads that have become burred or bent should be straightened or removed. When joining the threaded pipe, a proper thread lubricant-sealant should be used, such as TFE pipe-thread tape sealant. The lubricant reduces the friction, which permits the two parts to be pulled up further, resulting in a better joint. The joint should not be made up too quickly in order to avoid raising the temperature of the two parts.

In order to make up a tight joint, it is necessary to have a specified amount of thread engagement. The normal thread engagement required to produce a tight joint is as follows:

Pipe size (in.)	Engagement depth (in.)	Pipe size (in.)	Engagement depth (in.)
⅛	0.25	2 ½	0.94
¼	0.38	3	1.00
⅜	0.38	3 ½	1.06
½	0.50	4	1.12
¾	0.56	5	1.25
1	0.69	6	1.31
1 ¼	0.69	8	1.44
1 ½	0.69	10	1.62
2	0.75	12	1.75

Faulty threading or an improper lubricant are the causes of most leaky joints. Threads that appear wavy, shaved, rough, or chewed are the result of using a dull or improperly adjusted threading tool. Wavy threads are apparent both to the eye and touch, due to circumferential waves or longitudinal flats, or a slightly helical form rather than true circular form.

Shaved threads appear to have been cut with two dies, giving the appearance of a double thread at the start of the thread. Rough or chewed threads are visibly rough and torn. Any threads having these defects can result in a leaky joint.

Lightweight schedule 5S and schedule 10S can also be joined by brazing, soldering, or epoxy cementing. Fittings for this purpose are furnished with a belled end. It is important that the material used for joining these fittings be compatible with the corrodent being handled. This type of joint is not recommended for use in the handling of aggressive corrodents.

6.4 THERMAL EXPANSION

When fluctuating or elevated temperatures are anticipated, provision must be made for the piping system to compensate for the thermal expansion and contraction. The same applies when extremely low operating temperatures are anticipated.

Do not overlook the fact that most piping systems are installed at a temperature other than that at which the system will operate. Also, in certain sections of the country wide temperature fluctuations exist between the summer and winter months.

Compensation for this thermal expansion/contraction can be accomplished by the use of properly designed expansion loops and/or properly designed expansion joints. Expansion loops should not be used when handling corrodents that are susceptible to inducing stress corrosion, for example, chlorides. In these cases expansion joints should be considered.

For temperatures not exceeding 450°F (232°C) PTFE expansion joints may be used. These will have the same corrosion resistance as the metallic piping system in most cases. One disadvantage is the bellows construction when installed in a horizontal line. Under these conditions corrodent will collect in the bellows and can pose a safety hazard when dismantling. If possible, install in the vertical position, as they will drain. See Table 4-10 for the compatibility of PTFE with selected corrodents. Additional information on compensation for thermal expansion/contraction can be found in sec. 1.3.3.

6.5 PIPING SUPPORT

Properly designed piping support is essential for the satisfactory performance of the piping system. Inadequate support, resulting from excessive spans between hangers, will induce stress in the pipe. Under normal circumstances this stress is not of a sufficient magnitude to cause mechanical failure, but it can make the pipe susceptible to stress corrosion from the material being handled when otherwise the pipe would not be attacked. Details for calculating the proper support spacing can be found in sec. 1.3.2. In addition, the recommended maximum support spacing for each metallic system is given in the section dealing with that system.

6.6 STAINLESS STEEL PIPE

Stainless steels are alloys of iron to which a minimum of approximately 11% chromium has been added to provide a passive film to resist "rusting" when the material is exposed to the weather. This film is self-forming and self-healing in environments where the stainless steel is resistant. As more chromium is added to the alloy, improved corrosion resistance results, consequently there are stainless steels with chromium contents of 15, 17, and 20% (and even higher). Chromium provides resistance to oxidizing environments such as nitric acid and also provides resistance to pitting and crevice attack.

To provide even greater corrosion resistance and improve mechanical strength, other alloying ingredients are added. Molybdenum is extremely effective in improving pitting and crevice corrosion resistance.

Resistance to reducing environment and stress corrosion cracking is obtained by the addition of nickel. Nitrogen can also be added to improve corrosion resistance to pitting and crevice attack and to improve the strength.

Copper is added to improve resistance to general corrosion in sulfuric acid and strengthen some precipitation hardenable grades. However, in sufficient amounts copper will reduce the pitting resistance of some alloys.

To stabilize carbon, columbium and titanium are added. They form carbides and reduce the amount of carbon available to form chromium carbides, which can be deleterious to corrosion resistance.

As a result of these varying alloying possibilities more than 70 stainless steels are available, not all of which are produced for piping systems. Only the major alloys will be covered in this book.

The stainless steels can be divided into four major groups, depending upon their microstructure, and are classified as:

ferritic
austenitic
martensitic
duplex

6.6.1 Ferritic Stainless Steels

The ferritic stainless steels contain 15–30% chromium with a low carbon content (0.1%). Table 6-10 gives the chemical composition of the members of this group, and Table 6-11 gives the physical and mechanical properties of the ferritic stainless steels used for piping.

The ferritic stainless steels are limited in operating temperatures to 650°F (343°C) since they tend to be embrittled around 850°F (475°C). Maximum allowable stress values for the ferritic stainless steels are given in Table 6-12.

Corrosion resistance of these alloys is rated good, although they do not resist reducing media such as hydrochloric acid. Mildly corrosive solutions and oxidizing media are handled satisfactorily. These alloys exhibit a high level of resistance to pitting and crevice attack in many chloride-containing environments. They are also essentially immune to stress corrosion cracking in chloride environments.

Ferritic stainless steels are especially resistant to organic acids. The resistance to chloride attack makes these alloys ideal candidates for food-processing applications and a wide range of chemical processing uses.

Type 430 exhibits resistance to chloride stress corrosion cracking and to elevated temperature sulfide attack. It is subject to embrittlement at 885°F (475°C) and loss of ductility at subzero temperatures.

Type 444, like all ferritic stainless steels, relies on a passive film to resist corrosion, but exhibits rather high corrosion rates when activated. This characteristic explains the abrupt transition in corrosion rates that occurs at particular acid concentrations. For example, it is resistant to very dilute solutions of sulfuric acid at boiling temperature, but corrodes rapidly at higher concentrations. The corrosion rates of type 444 in strongly concentrated sodium hydroxide solutions are also higher than those for austenitic stainless steels.

Type XM-27 provides excellent resistance to pitting and crevice corrosion in chloride-containing environments, excellent resistance to chloride stress corrosion cracking, resistance to intergranular corrosion, and resistance to a wide variety of other corrosive environments. The alloy is also suitable for use in direct contact with foods.

Table 6-13 provides additional information on the compatibility of ferritic stainless steels in contact with selected corrodents.

6.6.2   Austenitic Stainless Steels

Austenitic stainless steels contain both nickel and chromium. The addition of substantial quantities of nickel to high-chromium alloys stabilizes the austenite at room temperatures. This group of steels contain 16–26% chromium and 6–22% nickel. The carbon content is kept low (0.08%) to minimize carbide precipitation. This series of stainless steels is the most widely used for corrosion resistance of any of the stainless steels.

When dealing with corrosive environments, the piping system designer and specifier should take careful note of the fact that there are significant differences in corrosion resistance and strength among the various stainless alloys. The types of stainless steels comprising the austenitic group are as follows:

201	304L*	314S	330
202	304N	316*	347*
205	305	316F	348
301	308	316L*	20Cb3*
302	309	316N	904L*
302B	309S	317*	20Mo-6*
303	310	317L*	AL-6XN*
304*	310S	321*	

Those types marked with an asterisk are the most commonly used for piping and will be discussed in detail.

Austenitic stainless steels exhibit good ductility and toughness at the most severe of cryogenic temperatures –423°F (–253°C). Impact tests show that type 304 is very stable over long periods of exposure and does not exhibit any marked degradation of toughness. Properly made welds also have excellent low-temperature properties. Austenitic grades cold-worked to high strength levels are also suitable for low-temperature service. Type 310 can be cold-worked as much as 85% and still exhibit a good notched-to-unnotched tensile ratio down to –423°F (–253°C). Table 6-14 provides typical mechanical properties of stainless steels at cryogenic temperatures.

Types 304 and 304L

These grades of stainless, although possessing a wide range of corrosion resistance, have the least overall resistance of any of the austenitic grades we are considering. Type 304 is subject to intergranular corrosion as a result of carbide precipitation. Welding of the pipe can cause this phenomenon, but competent welders, using good welding techniques, can control the problem. Depending upon the particular corrodent being handled, the effect of the carbide precipitation may or may not present a problem. If the corrodent being handled will attack through intergranular corrosion, another alloy should be used or the pipe line should not be welded.

If the carbon content in the alloy is not allowed to exceed 0.03%, carbide precipitation can be controlled. Type 304L is such an alloy. This alloy can be used for welded lines without the danger of carbide precipitation. The physical and mechanical properties of these alloys will be found in Table 6-15.

Allowable operating pressures and temperatures for type 304 and 304L are calculated by the formulas given on pages 230–231. The allowable design stresses for types 304 and 304L are given in Table 6-16.

The allowable operating pressures for different schedules of type 304 stainless steel have been calculated and are given in the following tables: table 6-17 for seamless schedule 5S, table 6-18 for seamless schedule 10S, table 6-19 for seamless schedule 40 plain end, table 6-20 for seamless schedule 80 plain end, and table 6-21 for seamless schedule 40 and schedule 80 threaded ends.

No corrosion allowance has been made in any of the above calculations. In Tables 6-17 through 6-20 for plain end pipe, C=0, E=1.0, and y=0.4. In Table 6-21, C = depth of thread except for pipe diameters of ⅛, ¼, and ⅜ inch, where C = 0.05. In this table, E = 1.0 and y = 0.4.

Operating pressures of pipe produced from other austenitic alloys may be determined by making use of the conversion factors found in Table 6-22. The following example illustrates the use of these factors.

What would be the allowable operating pressure of a 2-inch schedule 10 type 316 stainless steel pipe operating at 200°F (93°C)? From Table 6-18, it is established that a 2-inch schedule 10 type 304 pipe at 200°F (93°C) has a maximum allowable operating pressure of 1110 psi. From Table 6-22 we find the factor for type 316 at 200°F (93°C) is 1.13, therefore the maximum allowable operating pressure of a 2-inch type 316 stainless pipe operating at 200°F/93°C is (1110)(1.13) = 1254 psi.

Type 304 and 304L stainless exhibit good overall corrosion resistance. They are used extensively in the handling of nitric acid and most organic acids. Table 6-23 lists the compatibility of type 304 and 304L stainless steel with selected corrodents.

Types 316 and 316L

These are molybdenum-bearing chromium nickel stainless steels having greatly increased resistance to pitting and crevice corrosion as compared to type 304 stainless. In addition, these grades offer superior creep, stress-to-rupture and tensile strengths at elevated temperatures. They may be susceptible to chloride stress corrosion cracking. Type 316L is a low-carbon variation of type 316 and offers the additional feature of preventing excessive intergranular precipitation of chromium carbides during welding and stress relieving. The mechanical and physical properties are given in Table 6-24.

Maximum allowable operating pressures and temperatures of pipe produced from these alloys can be determined by using the correction factors found in Table 6-22 in conjunction with Tables 6-17 through 6-21 for various schedules of piping. An explanation of the use of these tables will be found on page 236. Table 6-25 lists the maximum allowable design stresses for types 316 and 316L pipe at various temperatures. Type 316 pipe has a maximum operating temperature of 1500°F (816°C), while type 316L has a maximum operating temperature of 800°F (427°C).

In general, these stainless steels are more corrosion resistant than the type 304 stainless steels. With the exception of oxidizing acids such as nitric acid, the type 316 alloys will provide satisfactory resistance to corrodents handled by types 304 with the ability to handle some corrodents that the type 304 alloys cannot handle. Table 6-26 provides a listing of the compatibility of type 316 and 316L stainless steel in contact with selected corrodents.

Type 317 and 317L

These are also molybdenum- and chromium-bearing stainless steels with somewhat higher alloy contents than the 316 and 316L alloys. This provides an improved chloride pitting and crevice corrosion resistance over types 316 and 316L. However, they may still be susceptible to chloride stress corrosion cracking. Type 317L, like type 316L, is a low-carbon variation of the basic alloy, which offers the additional feature of preventing excessive intergranular precipitation of chromium carbide during welding and stress relieving. The mechanical and physical properties are shown in Table 6-27.

The maximum allowable operating pressure and temperature of pipe produced from these alloys can be determined by using the correction factors found in Table 6-22 in conjunction with Tables 6-17 through 6-21 for various schedules of piping. An explanation of the use of these tables may be found on page 236. Table 6-28 gives the maximum design stresses for types 317 and 317L pipe. Type 317 pipe has a maximum operating temperature of 1500°F (816°C).

These alloys have an overall improved corrosion resistance over the 316 series of stainless steels, particularly in the areas of pitting and crevice corrosion resistance. They can be used wherever the 316 series are suitable and in many instances with improved performance. Refer to Table 6-29 for the compatibility of these alloys with selected corrodents.

Type 321

This is a titanium-stabilized stainless steel which has improved intergranular corrosion resistance. It is particularly useful for high-temperature service in the carbide precipitation range. Even though it has improved resistance to carbide precipitation, it may be susceptible to chloride stress corrosion cracking. It has a maximum operating temperature of 1000°F (538°C). The mechanical and physical properties are given in Table 6-30.

The maximum allowable operating pressure of pipe produced from this alloy can be determined by using the correction factors found in Table 6-22 in conjunction with Tables 6-17 through 6-21 for various schedules of piping. An explanation of the use of these tables can be found on page 236. Table 6-31 gives the maximum allowable design stress for type 321 pipe.

Type 321 stainless steel can be used wherever type 316 is suitable, with improved corrosion resistance, particularly in the presence of nitric acid. It also has an improved resistance to carbide precipitation. Refer to Table 6-32 for the compatibility of type 321 stainless steel with selected corrodents.

Type 347

This is a columbium (niobium)-stabilized stainless steel. The addition of columbium reduces the tendency for carbide precipitation, thereby improving the intergranular corrosion resistance. Basically this alloy is equivalent to type 304 stainless steel with the added protection from carbide precipitation. Type 304L also provides this protection but has a maximum operating temperature of 800°F (427°C), while type 347 can be operated to 1000°F (538°C).

Table 6-33 gives the mechanical and physical properties of type 347 stainless steel. Type 347 stainless has higher allowable stress values than type 304 or type 304L, therefore it can be operated at a higher pressure.

The maximum allowable operating pressure of pipe produced from type 347 stainless steel can be determined by using the correction factors found in Table 6-22 in conjunction with Tables 6-17 through 6-21 for various schedules of piping. An explanation of the use of these tables will be found on page 236. Table 6-34 provides the maximum allowable design stress for pipe produced from this alloy.

Type 347 stainless steel can be used wherever type 304 stainless steel can be used. Refer to Table 6-23 for the compatibility of type 347 stainless steel.

20Cb3 Stainless Steel

This alloy is stabilized with columbium and tantalum and has a high nickel content. It exhibits excellent resistance to chloride stress corrosion cracking and general corrosion resistance to sulfuric acid. There is minimal carbide precipitation due to welding. Table 6-35 provides the mechanical and physical properties of 20Cb3. Table 6-36 lists the elevated temperature tensile properties of the alloy.

The maximum allowable operating pressure of pipe produced from 20Cb3 stainless steel can be determined by using the correction factors found in Table 6-22 in conjunction with Tables 6-17 through 6-21 for various schedules of piping. An explanation of the use of these tables may be found on page 236. Table 6-37 provides the maximum allowable design stress for 20Cb3 pipe. The maximum allowable operating temperature is 800°F (427°C).

This alloy is particularly useful in the handling of sulfuric acid. It is resistant to stress corrosion cracking in sulfuric acid at a variety of temperatures and concentrations. The resistance of 20Cb3 to chloride stress corrosion cracking is also increased over type 304 and type 316 stainless steel. For example, 20Cb3 did not crack in a boiling acidified 25% sodium chloride (pH 1.5) solution, whereas type 304 and type 316 stainless steels did crack.

20Cb3 stainless also exhibits excellent resistance to sulfide stress cracking and consequently finds many applications in the oil industry. Table 6-38 lists the compatibility of 20Cb3 stainless with selected corrodents.

904L

This is a fully austenitic low-carbon nickel chromium stainless steel with additions of molybdenum and copper. It offers excellent corrosion resistance in a variety of difficult applications. Mechanical and physical properties of 904L stainless steel are given in Table 6-39.

Its high nickel and chromium contents make alloy 904L resistant to corrosion in a wide variety of both oxidizing and reducing environments. Molybdenum and copper are included in the alloy for increased resistance to pitting and crevice corrosion and to general corrosion in reducing acids. Other advantages of the alloy's composition are sufficient nickel for resistance to chloride-ion stress corrosion cracking and a low carbon content for resistance to intergranular corrosion.

The alloy's outstanding attributes are resistance to nonoxidizing acids along with resistance to pitting, crevice corrosion, and stress corrosion cracking in such media as stack gas condensate and brackish water.

Alloy 904L is especially suited for handling sulfuric acid. Hot solutions at moderate concentrations represent the most corrosive conditions. It also has excellent resistance to phosphoric acid.

20Mo-6

This austenitic stainless steel is resistant to corrosion in hot chloride environments with low pH. It has good resistance to pitting, crevice corrosion, and stress corrosion cracking in chloride environments and is also resistant to oxidizing media. This alloy is designed for applications where better pitting and crevice corrosion resistance than 20Cb3 offers are required. Mechanical and physical properties of 20Mo-6 stainless will be found in Table 6-40.

The maximum allowable operating pressure of pipe produced from 20Mo-6 stainless steel can be determined by using the correction factors found in Table 6-22 in conjunction with Tables 6-17 through 6-21 for various schedules of piping. An explanation of the use of these tables may be found on page 236. Table 6-41 provides the maximum allowable design stress values for 20Mo-6 pipe. This pipe has a maximum operating temperature of 800°F (427°C).

20Mo-6 stainless is melted with low carbon to provide a high level of resistance to intergranular corrosion. It also possesses excellent resistance to chloride stress corrosion cracking. When in contact with sulfuric acid, excellent resistance is shown at 176°F (80°C) with the exception of concentrations in the range of approximately 75–97 wt%. In boiling sulfuric acid, 20Mo-6 stainless has good resistance to general corrosion only in relatively dilute concentrations. At approximately 10% concentration of boiling sulfuric acid, the corrosion rate becomes excessive.

20Mo-6 stainless is also highly resistant to phosphoric acid, both wet process plant acid and reagent grade concentrated phosphoric acid.

Alloy 25-6Mo

This is an austenitic alloy produced by Inco International containing nickel, chromium, molybdenum, copper, nitrogen, carbon, manganese, phosphorus, sulfur, and silicon as alloying ingredients.

The mechanical properties of alloy 25-6Mo are superior to the other austenitic stainless steels such as 316L. Typical mechanical properties are:

Tensile strength: 1000,000 psi
Yield strength (0.2% offset): 48,000 psi
Elongation: 42%

Physical properties of the annealed material at room temperature are:

Density: 0.296 lb/in.^{3}
Specific heat: 0.12 Btu/lb°F
Thermal conductivity: 116 Btu-in./ft^2 hr°F
Coefficient of expansion between 68 and 212°F (20 and 100°C): 8.4 in./in.°F $\times$ 10^{-6}
Young's modulus: 28.1 $\times$ 10^6 psi

One of the outstanding attributes of alloy 25-6Mo is its resistance to environments containing chlorides or other halides. It is especially suited for application in high-chloride environments such as brackish water, sea water, caustic chlorides, and pulp mill bleach systems.

The alloy offers excellent resistance to pitting and crevice corrosion. Performance in these areas is often measured using critical pitting temperatures (CPT), critical crevice temperatures (CCT) and pitting resistance equivalent number (PREN).

As a general rule, the higher the PREN, the better the resistance. Alloys having similar values may differ considerably in actual service. Those alloys with values greater than 38 on the PREN scale offer more corrosion resistance than the austenitic stainless steels. Alloy 25-6Mo has a PREN of 47. The pitting resistance number is determined by the chromium, molybdenum, and nitrogen contents.

$$PREN = \%Cr + 3.3 \times \%Mo + 30 \times \%N$$

For 25-6Mo:

$$PREN = 20 + 3.3(6.5) + 30(0.2) = 47.45$$

The critical pitting temperature for alloy 25-6Mo is 140°F (60°C) or higher, while the critical crevice temperature is 90°F (32.5°C).

In brackish and wastewater systems, microbially influenced corrosion can occur, especially in systems where equipment has been idle for extended periods. A 6% molybdenum alloy offers protection from manganese-bearing, sulfur-bearing, and generally reducing types of bacteria. Because of its resistance to microbially influenced corrosion, alloy 25-6Mo is being used in the wastewater piping systems of power plants.

In saturated sodium chloride environments and pH values of 6–8, alloy 25-6Mo exhibits a corrosion rate of less than 1 mpy. Even under more aggressive oxidizing conditions involving sodium chlorate, alloy 25-6Mo maintains a corrosion rate of less than 1 mpy and shows no pitting even at temperatures up to boiling.

Piping produced from alloy 25-6Mo is manufactured in accordance with USASI standard dimensions, as shown in Tables 6-5 and 6-6.

AL-6XN

This alloy is a low-carbon, high-purity superaustenitic stainless steel containing chromium, nickel, molybdenum, and nitrogen. It has excellent corrosion resistance, strength, and ductility and is readily weldable. The presence of nitrogen provides the alloy with improved pitting and crevice corrosion resistance, greater resistance to localized corrosion in oxidizing chlorides and reducing solutions, plus higher strength. The mechanical and physical properties will be found in Table 6-42.

The maximum allowable operating pressure of pipe produced from AL-6XN stainless can be determined by using the correction factors found in Table 6-22 in conjunction with Tables 6-17 through 6-21 for various schedules of piping. An explanation of the use of these tables may be found on page 236. The maximum allowable design stress values for AL-6XN stainless steel pipe are given in Table 6-43. This pipe has a maximum allowable operating temperature of 800°F (427°C).

The corrosion-resistant properties of AL-6XN show exceptional resistance to pitting, crevice attack, and stress cracking in high chlorides and general resistance in various acid, alkaline, and salt solutions found in chemical processing and other industrial environments. AL-6XN stainless exhibits excellent resistance to oxidizing chlorides, reducing solutions and sea water corrosion. Sulfuric, nitric, phosphoric, acetic, and formic acids can be handled in various concentrations and at a variety of temperatures. The compatibility of AL-6XN stainless steel with selected corrodents is given in Table 6-44.

6.6.3 Martensitic Stainless Steels

Martensitic stainless steels contain relatively lower levels of chromium with sufficient carbon to permit martensite formation with rapid cooling. Martensite is a body-centered tetragonal structure which provides increased strength and hardness versus the annealed stainlesses with other lattice structures. Other elements, such as nickel and molybdenum, may be added for improved corrosion resistance and mechanical properties.

Precipitation-hardening stainless steels contain chromium and nickel and are strengthened by aging due to the presence of elements such as copper, columbium, titanium, or aluminum. Molybdenum may be added for corrosion resistance and columbium may be used to stabilize carbon.

The corrosion resistance of the martensitic stainless steels is inferior to that of the austenitic stainless steels. These alloys are generally used in mildly corrosive services such as atmospheric, fresh water, and organic exposures.

Alloy 350

This is an austenitic/martensitic alloy. It may be subject to intergranular attack unless cooled to subzero temperature before aging. The corrosion resistance of alloy 350 is similar to that of type 304 stainless steel. Refer to Table 6-23 for the compatibility of type 350 stainless steel with selected corrodents. This alloy is used where high strength and corrosion resistance at room temperature are required. The mechanical and physical properties will be found in Table 6-45. To prevent overaging, operating temperatures should be kept below 1000°F (540°C).

410 Stainless Steel

This alloy contains 11.5–13.5% chromium with 1% manganese, 1% silicon, and a maximum of 0.03% sulfur. Mechanical and physical properties are shown in Table 6-46.

In general, corrosion resistance is somewhat less than that for type 304 stainless steel. Refer to Table 6-47 for the compatibility of type 410 stainless steel with selected corrodents.

Custom 450* Stainless Steel

This is a martensitic age-hardenable stainless steel with good corrosion resistance and moderate strength. Its corrosion resistance is similar to that of type 304 stainless steel. This alloy finds applications where type 304 stainless steel is not strong enough or where type 410 is insufficiently corrosion resistant. Refer to Table 6-48 for the mechanical and physical properties of custom 450 stainless steel.

Custom 455* Stainless Steel

This is a martensitic age-hardenable stainless steel having high strength with corrosion resistance better than type 410 stainless steel. It may be subject to hydrogen embrittlement under some conditions. This alloy finds application when high strength and corrosion resistance is required. It resists chloride cracking as well as alkalies and nitric acid. Refer to Table 6-49 for the mechanical and physical properties of custom alloy 455.

Alloy 718

This alloy is an age-hardenable martensitic alloy having high strength, good tensile fatigue, creep, and rupture strength. It has an operating temperature range of –423 to

---

*Custom 450 and custom 455 are a registered trademark of Carpenter Technology Corp.

1300°F (−253 to 704°C). Alloy 718 possesses good corrosion resistance, being compatible with alkalies, organic acids, and sulfuric acid. Excellent oxidation resistance is demonstrated throughout its operating temperature range. Refer to Table 6-50 for the mechanical and physical properties.

Alloy 17-7PH

This is a semi-austenitic stainless steel. In the annealed condition it is austenitic, but in the aged condition it is martensitic. It possesses high strength and has a corrosion resistance equivalent to type 304 stainless steel. Refer to Table 6-51 for the mechanical and physical properties of alloy 17-7PH.

6.6.4   Duplex Stainless Steels

The duplex stainless steels are those alloys whose microstructures are a mixture of austenite and ferrite. These alloys were developed to improve the corrosion resistance of the austenitic stainlesses, particularly in the areas of chloride stress-corrosion cracking, and in maintaining corrosion resistance after welding. The original duplex stainlesses developed did not meet all of the criteria desired. Consequently, additional research was undertaken.

The original duplex stainless steels did not have nitrogen added specifically as an alloying ingredient. By adding 0.15–0.25% nitrogen, the chromium partitioning between the two phases is reduced, resulting in the pitting and crevice corrosion resistance of the austenite being improved. This nitrogen addition also improves the weldability of the stainless steel without losing any of its corrosion resistance. It is not necessary to heat-treat these duplexes after welding.

Because the stainless steels are a mixture of austenite and ferrite, it is only logical that their physical properties would lie between the comparable properties of these microstructures. The duplexes have better toughness than ferritic grades and higher yield strengths than the austenitics.

Since the duplexes contain a large amount of ferrite, they are magnetic. However, unlike the ferritics, they have a high degree of toughness along with their high strength.

Because the duplexes have a higher yield strength than the austenitics, they can provide certain economic advantages. Money can be saved using thinner-walled sections for piping without sacrificing operating pressures. Conversely, piping manufactured from these stainless steels using conventional wall thicknesses can be operated at higher pressures.

The high chromium and molybdenum contents of the duplex stainless steels are particularly important in providing resistance in oxidizing environments and are also responsible for the exceptionally good pitting and crevice corrosion resistance, especially in chloride environments. In general these stainless steels have greater pitting resistance than type 316, and several have an even greater resistance than alloy 904L.

The resistance to crevice corrosion of the duplexes is superior to the resistance of the 300 series austenitics. They also provide an appreciably greater resistance to stress-corrosion cracking. Like 20Cb3, the duplexes are resistant to chloride stress-corrosion cracking in chloride-containing process streams and cooling water. However, under very severe conditions, such as boiling magnesium chloride, the duplexes will crack, as will 20Cb3.

To achieve the desired microstructure, the nickel content of the duplexes is below that of the austenitics. Because the nickel content is a factor for providing corrosion resistance in reducing environments, the duplexes show less resistance in these environments than do the austenitics. However, the high chromium and molybdenum contents partially offset this loss, and consequently, they can be used in some reducing environments, particularly dilute and cooler solutions.

7Mo Plus*

This is a duplex stainless alloy with approximately 40% austenite distributed within a ferrite matrix. The yield strength of annealed 7Mo Plus is greater than twice that of typical austenitic stainless steels. Refer to Table 6-52 for the mechanical and physical properties of this alloy.

The general corrosion resistance of 7Mo Plus is superior to those of stainless steels such as type 304 and type 316 in many environments. Because of its high chromium content, it has good corrosion resistance in strongly oxidizing media such as nitric acid. Molybdenum extends the corrosion resistance of the alloy into the less oxidizing environments. Chromium and molybdenum impact a high level of resistance to pitting and crevice corrosion.

This alloy provides excellent chloride stress-corrosion cracking resistance in applications where type 304 or type 316 stainless steels would be used for general corrosion resistance.

Alloy 2205

Good strength, toughness, and corrosion resistance is exhibited by this alloy. It resists oxidizing mineral acids and most organic acids, in addition to reducing acids and chloride environments. Refer to Table 6-53 for the mechanical and physical properties.

6.6.5   Joining Stainless Steel Piping

There are three basic methods by which stainless steel piping is joined—welding, threading, and expanding. (The two latter systems, threading and expanding, were covered in sec. 6.3.) The preferred method of joining for corrosion resistant service is welding. Flanging is also available, but because of the cost of stainless steel flanges, this method is not normally used for joining of the piping runs. The use of flanges is limited to connecting to equipment and spool pieces, which must be frequently removed for cleaning. These flanges are welded to the pipe or fitting using the same techniques used for joining pipe. The allowable working pressures of stainless steel flanges for various primary service ratings may be found in Tables 6-54 through 6-58. Table 6-59 provides a representative pressure rating of types 304 and 316L Van Stone flanges.

It is important that proper welding techniques be employed to ensure a strong corrosion resistant joint. The details of welding procedures will not be covered in this text: that information may be found in sources devoted to the subject of welding. What will be discussed is weld rod selection. Proper weld or filler rod selection is important to achieve a weld metal with the desired corrosion resistance and strength characteristics. A pipe joint can fail in the weld zone if the weld rod selected results in the weld zone having a lower alloy content than that of the parent metal. The characteristics of the weld metal are primarily dependent on the alloy content of the filler rod and, to a lesser, extent on the degree to which the molten weld metal is protected from the environment. This protection

---

*Registered trademark of Carpenter Technology Corp.

is provided by the shielding gases used in certain welding processes or by the action of chemical fluxes applied to the welding rods.

The first criterion for the weld rod selection is alloy content. A brief discussion will help in understanding which filler metal to use.

For austenitic stainless steels, the general rule is to use the alloy most similar to the bare metal being welded. The greater amount of chromium and nickel in certain alloys, type 308 for example, is useful for welding types 302 and 304 base metals and hence is standard for all the lower chromium-nickel base metals. While the same principle applies to type 316 in that the minimum chromium content is higher in the weld metal than the base metal, the designation of the filler metal is the same.

In selecting welding materials, there is a misconception that the higher the AISI number, the higher the alloy content. This is not always true, as in the case of type 347, which is a stabilized grade for preventing carbide precipitation in high-temperature service. Type 347 should not be used as a "general purpose" filler metal for welding other alloys, because type 347 can be crack sensitive.

The only standard martensitic stainless steels available as either covered electrodes or bare welding wire are types 410 and 420. This sometimes presents a problem when attempting to secure similar properties in the weld metal as in the base metal. Except for 410NiMo, martensitic stainless steels weld metals in the as-deposited condition are low in toughness.

Austenitic stainless steel weld deposits are often used to weld the martensitic grades. These electrodes provide an as-welded deposit of somewhat lower strength but of great toughness. For as-welded applications where thermal compatibility is desired, the 410NiMo filler wire is a good choice.

The weld metal of ferritic stainless steels is usually lower in toughness, ductility, and corrosion resistance than the heat-affected zone of the base metal. Austenitic stainless steel filler metal is used frequently to join ferritic base metal to secure a ductile weld. For example, type 308 filler metal is frequently used to weld type 430. For the low-carbon or stabilized ferritic grades, the use of austenitic filler metal can provide a weld of good mechanical properties. The austenitic weld should be selected as a low grade carbon (e.g., 316L). The filler metal should always be selected so that the chromium and molybdenum content of the filler metal will be at least equal to that of the base metal. This ensures that the weld will have adequate corrosion resistance in severe environments.

The selection of a filler metal to weld precipitation-hardening stainless steels will depend upon the properties required of the weld. If high strength is not needed at the weld, then the filler may be a tough austenitic stainless steel. When mechanical properties comparable to those of the hardened base metal are desired in the weld, the weld metal must also be a precipitation-hardening compound. The weld analysis may be the same as the base metal, or it may be modified slightly to gain optimum weld metal properties.

Refer to Table 6-60 for suggested filler metals for various types of stainless steel pipe.

6.6.6   Installing Stainless Steel Piping

It is important that good piping practices be followed when installing stainless steel piping. Hangers and supports should be properly located and installed so that loads will not be induced into the piping system. Accessory equipment installed in the pipe line should be properly supported independently of the pipe. Pipe should not be "sprung" into place since this places a stress in the pipe. Any external loads induced into the pipe could result in stress-corrosion cracking of the pipe.

Table 6-61 provides the maximum support spacing for stainless steel pipe. This spacing does not take into account concentrated loads which may be present such as valves, meters, etc. which should be independently supported, or the pipe spans reduced. If the lines are to be insulated, the spans should be reduced by 30%. Although longer spans may be permitted, it is usually desirable to provide at least one support for each length of pipe.

6.7   NICKEL PIPE

There are two basic compositions used for the production of nickel pipe—alloy 200 and alloy 201. Alloy 200 is commercially pure wrought nickel, while alloy 201 is the low-carbon version of alloy 200 and is used in applications involving temperatures above 600°F (315°C). The low carbon prevents formation of the grain-boundary graphitic phase, which reduces ductility tremendously. Because of this, nickel alloy 200 is limited to a maximum operating temperature of 600°F (315°C). For applications above this temperature, nickel alloy 201 should be used. Refer to Table 6-62 for the mechanical and physical properties of nickel 200 and nickel 201.

Both alloys of nickel are available in pipe diameters of ½ inch through 12 inches in schedules 5S, 10S, 40S and 80S. They are produced in accordance with USASI standard dimensions as shown in Tables 6-5 and 6-6. Random lengths of 10 and 20 feet are standard. Fittings are available in welding, threaded, or flanged types. Fittings with long tangents are also available for expanding. The preferred method for joining nickel pipe is either by welding or expanding. Only schedules 40S and 80S can be threaded. When welding nickel pipe, alloy 200 or alloy 201, nickel 61 filler material, and 131 or 141 electrodes should be used. The maximum allowable design stress for nickel pipe is shown in Table 6-63. Maximum allowable operating pressure of nickel 200 pipe may be found in Tables 6-64 and 6-65. Maximum operating pressures of nickel alloy 201 pipe will be found in Tables 6-66 through 6-70. In these tables no allowance has been made for corrosion. Therefore, in Tables 6-64, 6-66, 6-67, 6-68, and 6-69, $C = 0$, $y = 0.4$, and E = 1. In Tables 6-65 and 6-70, C = depth of thread, $y = 0.4$, and E = 1.

Nickel pipe should be installed following good piping practices. No external stresses should be transmitted to the pipe. The recommended maximum distances between supports should be followed. Distances between supports are given in Table 6-71. These distances are based on uninsulated pipe carrying a liquid having a maximum specific gravity of 1.35. All accessory items, such as valves, meters, etc., should be supported independently of the pipe, or the support spacing should be reduced.

Flanges are used for joining pipe to equipment such as vessels, pumps, etc. Because of cost, it is not recommended that flanges be used for joining pipe sections. Welding or expansion are the preferred methods. Table 6-72 supplies the maximum nonshock rating of nickel flanges having primary ratings.

The outstanding characteristic of nickel 200 is its resistance to caustic soda and other alkalies. (Ammonium hydroxide is an exception.) Nickel 200 is not attacked by anhydrous ammonia or ammonium hydroxide in concentrations of 1% or less. Stronger concentrations can cause rapid attack.

Nickel 200 shows excellent resistance to all concentrations of caustic soda at temperatures up to and including the molten state. Below 50%, the corrosion rates are negligible, being usually less than 0.2 mpy even in boiling solutions. As concentrations and temperatures increase, corrosion rates increase very slowly.

Nickel 200 is not subject to stress corrosion cracking in any of the chloride salts, and it has excellent general resistance to nonoxidizing halides. The oxidizing acid chlorides, such as ferric, cupric, and mercuric, are very corrosive and should not be used.

Nickel 201 is extremely useful in the handling of dry chlorine and hydrogen chloride at elevated temperatures.

If aeration is not high, nickel 200 has excellent resistance to most organic acids, particularly fatty acids such as stearic and oleic.

Nickel is also used in handling food and synthetic fibers for maintaining product purity. For food products, the presence of nickel ions is not detrimental to the flavor and is nontoxic. For many organic chemicals, such as phenol and viscose rayon, nickel does not discolor the solutions, unlike iron and copper. Refer to Table 6-73 for the compatibility of nickel with selected corrodents. The corrosion resistances of nickel 200 and nickel 201 are the same.

6.8 MONEL* PIPE

The first nickel alloy, invented in 1905, was approximately two-thirds nickel and one-third copper. The present equivalent of that alloy, Monel alloy 400, remains one of the widely used nickel alloys.

Nickel copper alloys offer somewhat higher strength than unalloyed nickel, with no sacrifice of ductility. The thermal conductivity of Monel, although lower than that of nickel, is significantly higher than that of nickel alloys containing substantial amounts of chromium or iron. Table 6-74 provides the mechanical and physical properties of Monel 400.

Monel pipe is produced in schedules 5S, 10S, 40S, and 80S, in diameters of ½ inch through 12 inches, in accordance with USASI standard dimensions as shown in Tables 6-5 and 6-6. The maximum allowable design stress for Monel 400 pipe at various operating temperatures may be found in Table 6-75.

Operating pressures of Monel 400 pipe have been calculated using the maximum allowable stress values in accordance with the formula on page 230. The results are given in Tables 6-76 through 6-79 for plain end pipe. In these calculations, no allowance had been made for corrosion, therefore C in the formula has been set at 0, y = 0.4, and E (since the tables are based on seamless pipe) is equal to 1. In Tables 6-80 and 6-81 for threaded end pipe, y = 0.4, E = 1.0, and C = depth of thread. Again, no allowance has been made for corrosion.

Monel pipe can be joined by means of welding, threading, or expanding. The preferred method when handling corrosive materials is welding. Threading can be used only on schedule 40 or schedule 80 pipe. Flanging is acceptable, but because of cost the use of flanges is usually limited to making connections to equipment or on spool pieces that must be periodically removed. The allowable operating pressures of forged Monel 400 flanges are shown in Table 6-82.

Installation of Monel 400 pipe should follow good piping practices, which include proper support and a stress-free installation. Maximum support spacing is given in Table

---

*Monel 400 is the trademark of International Nickel.

6-83. All valves and other accessory equipment installed in the pipe line are to be individually supported. It is also recommended that at least one hanger be used on each length of pipe, although allowable support spacing may be greater.

The alloying of 30–33% copper with nickel, producing Monel 400, provides an alloy with many of the characteristics of chemically pure nickel but improves on others. Water handling, including brackish and sea waters, is a major area of application. It gives excellent service under high-velocity conditions. The alloy can pit in stagnant sea water, as does nickel 200, however, the rates of attack are significantly diminished. The absence of chloride stress corrosion cracking is also a factor in the selection of this alloy.

The general corrosion resistance of Monel 400 in the nonoxidizing acids such as sulfuric, hydrochloric, and phosphoric is improved over that of pure nickel. The alloy is not resistant to oxidizing media, such as nitric acid, ferric chloride, chromic acid, wet chlorine, sulfur dioxide, or ammonia.

Monel 400 does have excellent resistance to hydrofluoric acid solutions at various concentrations and temperatures. It is subject to stress corrosion cracking in moist, aerated hydrofluoric or hydrofluosilicic acid vapor. Cracking is unlikely when completely immersed in the acid. The corrosion of Monel 400 is negligible in all types of atmospheres. Indoor exposure produces a very light tarnish that is easily removed by occasional wiping. Outdoor surfaces that are exposed to rain produce a thin gray-green patina. In sulfurous atmospheres a smooth, brown, adherent film forms.

Because of its high nickel content, Monel 400 is practically as resistant as nickel 200 to caustic soda throughout most of the concentration range. It is also resistant to anhydrous ammonia and to ammonium hydroxide solutions of up to 3% concentration. Refer to Table 6-84 for the compatibility of Monel 400 with selected corrodents.

6.9 ALLOY 600 PIPE

Alloy 600 is a nickel-chromium-iron alloy for use in environments requiring resistance to heat and corrosion. The high nickel content of this alloy makes it resistant to corrosion by a number of organic and inorganic compounds and gives it excellent resistance to chloride ion stress corrosion cracking. Its chromium content gives the alloy resistance to sulfur compounds and various oxidizing environments.

In addition, alloy 600 has excellent mechanical properties and a combination of high strength and good workability. It performs well in applications with temperatures from cryogenic to 1200°F (649°C). Refer to Table 6-85 for mechanical and physical properties.

Alloy 600 cold drawn and annealed pipe is furnished in schedules 5S, 10S, 40S, 80S, and 160S. Pipe dimensions are in accordance with USASI standards, as shown in Tables 6-5 and 6-6. Fittings are available for welding or threading. Flanges are also available for threading or welding to the pipe.

The operating pressures of alloy 600 pipe have been calculated using the formula on page 230 and the maximum allowable stress values given in Table 6-86. In the calculations for Tables 6-87 through 6-90 for plain end pipe, no allowance was made for corrosion, therefore C = 0. In Table 6-91 for threaded pipe, C = 0.05 for the ¼-inch and ⅜-inch sizes and C = depth of thread for the remaining sizes. In all cases, y = 0.4 or the appropriate factor based on temperature and, since the pipe is seamless, E = 1.0.

Alloy 600 pipe can be joined by welding or threading. Flanges are also available but are used for connections to equipment or for spool pieces that may have to be removed frequently for cleaning. The preferred method of joining is by welding. Forged alloy 600 flanges conforming to USAS B16.5 steel flange standards have different pressure temperature ratings than the corresponding primary service rating of comparable steel flanges. Tables 6-92 and 6-93 show the service ratings of alloy 600 forged flanges.

Standard piping techniques should be followed when installing alloy 600 pipe. Care should be taken that the pipe is properly lined up to ensure that the completed installation will be free of installation stresses. If installation stresses are present, the corrosion resistance of the pipe will be adversely affected. Adequate support for the piping must also be supplied. The maximum allowable span between supports is shown in Table 6-94. All valves, instruments, or other accessory items installed in the line must be independently supported. Although fairly long spans are permissible, it is usually advantageous to install at least one pipe support per length of pipe.

Alloy 600 is practically free from corrosion by fresh waters, including the most corrosive of natural waters containing free carbon dioxide, iron compounds, and dissolved air. It remains free from stress corrosion cracking even in boiling magnesium chloride.

Because of its chromium content, alloy 600 exhibits greater resistance to sulfuric acid under oxidizing conditions than either nickel 200 or Monel alloy 400. The addition of oxidizing salts to sulfuric acid tends to passivate alloy 600, which makes it suitable for use with acid mine waters or brass pickling solutions, applications where Monel alloy 400 cannot be used.

Alloy 600 is not subject to stress corrosion cracking in any of the chloride salts and has excellent resistance to all of the nonoxidizing halides. Refer to Table 6-95 for the compatibility of alloy 600 with selected corrodents.

6.10 ALLOY 625 PIPE

Alloy 625 is a nickel-chromium-molybdenum-columbium alloy well suited for applications where strength and corrosion resistance are required. The alloy exhibits exceptional fatigue strength and superior strength and toughness at temperatures ranging from cryogenic to 2000°F (1093°C). Mechanical and physical properties are shown in Table 6-96.

Alloy 625 is resistant to oxidation, general corrosion, pitting, and crevice corrosion and is virtually immune to chloride ion stress corrosion cracking. These properties are derived from additions of molybdenum and columbium to the alloy's basic nickel-chromium composition.

Resistance to aqueous solutions is good in a variety of applications including organic acids, sulfuric acid, and hydrochloric acid at temperatures below 150°F (65°C). The alloy has satisfactory resistance to hydrofluoric acid. Although nickel-based alloys are not normally used in nitric acid service because of cost, alloy 625 is resistant to mixtures of nitric-hydrofluoric, where stainless steel loses its resistance.

Alloy 625 has excellent resistance to phosphoric acid solutions, including commercial grades of acids that contain fluorides, sulfates, and chlorides in the production of superphosphoric acid (72% P_2O_s). In general, alloy 625 can be used in applications that alloy 600 is not suitable for. Refer to Table 6-95 for the compatibility of alloy 625 with selected corrodents.

6.11 ALLOY 800 PIPE

Alloy 800 contains about 20% chromium, 32% nickel, and 46% iron as balance. The alloy is used primarily for its oxidation resistance and strength at elevated temperatures. It is particularly useful for high-temperature applications because the alloy does not form the embrittling sigma phase after long exposures at 1200–1600°F (649–871°C). High creep and rupture strengths are other factors that contribute to its performance in many other applications. Refer to Table 6-97 for the mechanical and physical properties of alloy 800.

Cold-drawn seamless alloy 800 pipe is furnished in the annealed and ground condition below 4 inches in diameter and furnished in the annealed and pickled condition above 4 inches in diameter in schedules 5S, 10S, 40S, and 80S. They are produced in accordance with USASI standard dimensions as shown in Tables 6-5 and 6-6. A complete line of welding fittings is available in each of the schedule numbers.

Allowable operating pressures of alloy 800 pipe have been calculated using the formula found on page 230 and the allowable design stresses shown in Table 6-98. The maximum allowable operating pressure for alloy 800 pipe is given in Tables 6-99 through 6-102. No allowance has been made for corrosion, consequently C = 0. Since we are dealing with seamless metallic pipe, E = 1.0 and y = 0.4.

Alloy 800 pipe should be joined by welding. Inconel filler metal 82 should be used for the root pass and to complete the joint.

As with all piping systems, good piping practices should be followed. All piping should be properly aligned, otherwise installation stresses will be induced into the piping system, which will reduce its corrosion resistance and may result in premature failure. Refer to Table 6-103 for the recommended maximum support spacing. No allowance has been made for concentrated loads. Accessory items installed in the line, such as valves, meters, etc., should be individually supported. Although long spans between supports are permitted, it is good piping practice to install a support on each pipe length. The support spacing shown for schedule 40S pipe may also be used for schedule 80S pipe.

At moderate temperatures the general corrosion resistance of alloy 800 is similar to that of the other austenitic nickle-iron-chromium alloys. However, as the temperature increases, alloy 800 continues to exhibit good corrosion resistance, while other austenitic alloys are unsatisfactory for the service.

Alloy 800 has excellent resistance to nitric acid at concentrations up to about 70%. It resists a variety of oxidizing salts, but not halide salts. It also has good resistance to organic acids, such as formic, acetic, and propionic. Alloy 800 is particularly suited for the handling of hot corrosive gases such as hydrogen sulfide.

In aqueous corrosion service alloy 800 has general resistance that falls between 304 and 316 stainless steels. Thus, the alloy is not widely used for aqueous service. The stress corrosion–cracking resistance of alloy 800 is, while not immune, better than that of the 300 series of stainless steels and may be substituted on this basis. Table 6-104 provides the compatibility of alloy 800 with selected corrodents.

6.12   ALLOY 825 PIPE

Alloy 825 is very similar to alloy 800, however, the composition has been modified to provide for improved aqueous corrosion resistance. Table 6-105 provides the mechanical and physical properties of the alloy.

The higher nickel content of alloy 825, as compared to alloy 800, makes it resistant to chloride ion stress corrosion cracking. Additions of molybdenum and copper give resistance to pitting and to corrosion in reducing acid environments such as sulfuric or phosphoric acid solutions. Alloy 825 is resistant to pure sulfuric acid solutions up to 40% by weight at boiling temperatures and at all concentrations at a maximum temperature of 150°F (60°C). In dilute solutions, the presence of oxidizing salts, such as cupric or ferric, actually lowers the corrosion rates. It has limited use in hydrochloric or hydrofluoric acids.

The alloy's chromium content gives it resistance to various oxidizing environments such as nitrates, nitric acid solutions, and oxidizing salts. Alloy 825 is not fully resistant to stress corrosion cracking when tested in boiling magnesium chloride, but it has good resistance in neutral chloride environments. If localized corrosion is a problem with 300 series stainless steel, alloy 825 may be substituted. In addition alloy 825 offers excellent resistance to corrosion by sea water. Refer to Table 6-104 for the compatibility of alloy 825 with selected corrodents.

6.13   ALLOY B-2 PIPE

Within the nickel-molybdenum series, one major alloy—alloy B-2—is a low-carbon and low-silicon (0.02%, 0.08% maximum) version of alloy B. The alloy is uniquely different from other corrosion-resistant alloys because it does not contain chromium. Molybdenum is the primary alloying ingredient and provides a significant corrosion resistance to reducing environments. Table 6-106 provides the mechanical and physical properties of alloy B-2.

Alloy B-2 pipe is available in schedules 5S, 10S, 40S, and 80S. Dimensions are in accordance with USASI standards, as shown in Tables 6-5 and 6-6. Standard pipe lengths are 20 feet. Allowable operating pressures of alloy B-2 pipe have been calculated using the formula found on page 230 along with the allowable design stresses found in Table 6-107. These operating pressures may be found in Tables 6-108 through 6-111 for seamless plain end pipe. Since we are considering seamless metallic pipe, $E = 1.0$ and $y = 0.4$. No allowance has been made for corrosion, therefore $C = 0$. Tables 6-112 and 6-113 give the maximum allowable operation pressure for threaded pipe. In these tables $C = $ depth of thread except for the $\frac{3}{8}$-inch-diameter pipe where $C = 0.5$. Corrosion allowance has not been applied to these tables either.

Alloy B-2 pipe can be joined by welding, threading, or flanging. Because of the expense, the use of flanges is usually restricted to making connections to pumps and/or equipment. The preferred method is welding, although schedules 40S and 80S can be threaded. When welded, a matching alloy B-2 filler should be used. Table 6-114 tabulates the maximum nonshock ratings of forged alloy B-2 flanges.

When installing alloy B-2 piping, care should be taken to ensure that installation stresses are not induced into the piping system. Such stresses can reduce the corrosion resistance of the piping and lead to premature failure. Proper alignment of each section and proper support will produce a stress-free installation. Maximum support spacing is given in Table 6-115. If valves, instruments, or other accessory items are to be installed in the piping system, such items should be independently supported. If the lines are to be insulated, the spans should be reduced by 30%. The support spacing shown in Table 6-115 may also be used for heavier schedule pipe.

Since alloy B-2 was developed to handle hydrochloric acid, it is recommended for the handling of all concentrations in the temperature range of 158–212°F (70–100°C) and for the handling of wet hydrogen chloride gas.

Alloy B-2 has excellent resistance to pure sulfuric acid at all concentrations and temperatures below 60% acid and good resistance to 212°F (100°C) above 60% acid. It is also resistant to a number of nonoxidizing environments, such as hydrofluoric and phosphoric acids, and numerous organic acids, such as acetic, formic, and crysylic acids.

Alloy B-2 is also resistant to many nonoxidizing chloride-bearing salts such as aluminum chloride, magnesium chloride, and antimony chloride.

Because alloy B-2 is nickel rich (approximately 70%), it is resistant to chloride-induced stress corrosion cracking. By virtue of the high molybdenum content it is highly resistant to pitting attack in acid chloride environments.

Alloy B-2 is not recommended for elevated temperature service. Piping is limited to 650°F (343°C) operation. The major factor that limits the use of alloy B-2 is the poor corrosion resistance in oxidizing environments. It has virtually no corrosion resistance to oxidizing acids such as nitric and chromic or to oxidizing salts such as ferric chloride or cupric chloride. The presence of oxidizing salts in reducing acids must be considered. Oxidizing salts such as ferric chloride, ferric sulfate, or cupric chloride, even when present in the part per million range, can accelerate the attack in hydrochloric or sulfuric acids. Even dissolved oxygen has sufficient oxidizing power to affect the corrosion rates in hydrochloric acid. Table 6-116 provides corrosion data of alloy B-2 in the presence of selected corrodents.

6.14 ALLOY C-276 PIPE

Alloy C-276 is a nickel-molybdenum-chromium-tungsten alloy with outstanding corrosion resistance that is maintained even in the welded condition. The high nickel and molybdenum contents provide good corrosion resistance in reducing environments, while the chromium imparts resistance to oxidizing media. The molybdenum also aids resistance to localized corrosion, such as pitting. The low carbon content minimizes carbide precipitation during welding to maintain resistance to corrosion (intergranular attack) in heat-affected zones of welded joints. Alloy C-276 also has high strength at room and elevated temperatures. Refer to Table 6-117 for the mechanical and physical properties of alloy C-276.

Alloy C-276 pipe is available in schedule 5S, 10S, 40S, and 80S. Pipe dimensions are in accordance with USASI standards as shown in Tables 6-5 and 6-6. Fittings are available in welded, threaded, or flanged designs.

The allowable operating pressures of alloy C-276 pipe have been calculated using the formula found on page 230. The design stresses used were taken from Table 6-118. Tables 6-119 through 6-122 list the maximum allowable operating pressure for seamless plain end pipe in various schedules. Since the pipe is seamless, E = 1.0 and y = 0.4. No allowance has been made for corrosion, therefore C = 0. Tables 6-123 and 6-124 list the maximum operating pressures for seamless threaded end pipe in schedules 40S and 80S. In these two tables, C = depth of thread except for ⅜-inch-diameter pipe, where C = 0.05. No corrosion allowance was applied in this case either.

Alloy C-276 pipe can be joined by welding, threading or flanging. Because of expense, the latter method is no longer ordinarily used except for connecting to pumps or other equipment. Ratings of alloy C-276 flanges are given in Table 6-114. The preferred method of joining is welding.

Care must be taken during installation that no stresses be imposed upon the pipe system. All pipes should be properly aligned and supported. Recommended maximum support spacings are given in Table 6-115. These spans should not be exceeded. All accessory equipment, such as valves, meters, etc., should be individually supported. If the line is to be insulated, the support spans given in Table 6-115 should be reduced 30%.

Alloy C-276 is extremely versatile because it possesses good resistance in both oxidizing and reducing media, and this includes conditions with halogen ion contamination. The pitting and crevice corrosion resistance of alloy C-276 makes it an excellent choice when dealing with acid chlorine salts.

Alloy C-276 has exceptional corrosion resistance to many chemical process materials, including highly oxidizing neutral and acid chlorides, solvents, chlorine, formic and acetic acid, and acetic anhydride. It also resists highly corrosive agents such as wet chlorine gas, hypochlorites, and chlorine solutions.

Alloy C-276 has excellent resistance to phosphoric acid. At all temperatures below the boiling point of phosphoric acid, when concentrations are less than 65% by weight, tests have shown corrosion rates of less than 5 mpy. At acid concentrations above 65% by weight and up to 85% alloy C-276 displays similar corrosive behavior, except at temperatures between about 240°F (116°C) and the boiling point, where corrosion rates may be erratic and may reach 25 mpy. Refer to Table 6-116 for the compatibility of alloy C-276 with selected corrodents.

6.15   ALLOY G AND G-3 PIPES

Alloy G and its modification, alloy G-3, are nickel-chromium-iron-molybdenum-copper alloys exhibiting excellent resistance to a variety of corrosive conditions. The alloys are highly resistant to pitting and stress corrosion cracking in both acid and alkaline environments, including hot sulfuric and phosphoric acids, hydrofluoric and contaminated nitric acids, mixed acids, and sulfate compounds.

Alloy G-3 was developed with a lower carbon content to prevent precipitation at the welds. The corrosion resistance of alloy G and alloy G-3 are approximately the same. However, the thermal stability of alloy G-3 is greater.

The mechanical and physical properties of alloy G and alloy G-3 are given in Table 6-125. Piping and fittings are available in both alloy G and alloy G-3 in schedules 5S, 10S, 40S, and 80S.

6.16   ALUMINUM PIPE

Aluminum is, next to iron, the most important metal of commerce in the United States in terms of production and consumption. Aluminum is the third most abundant metal in the crust of the earth, almost twice as abundant as iron.

In addition to low cost, aluminum and its alloys provide a high strength/weight ratio and are readily joined by most common methods used. They have a high resistance to corrosion in most atmospheres and waters and many chemical and other materials. They are nontoxic, permitting applications with foods, beverages, and pharmaceuticals, and they are white or colorless, permitting applications with chemicals and other materials without discoloration being imparted to the materials being handled.

Wrought alloys are of two types: non–heat treatable (of the 1xxx, 2xxx, 3xxx, 4xxx, and 5xxx series) and heat treatable (of the 2xxx, 6xxx and 7xxx series). For piping purposes, we will consider only the 1xxx, 3xxx, 5xxx, and 6xxx series. In the non–heat treatable types, strengthening is produced by strain hardening, which may be augmented by solid solution and dispersion hardening. In the heat treatable types, strengthening is produced by (1) a solution heat treatment at 860–1050°F (460–565°C) to dissolve soluble alloying elements, (2) quenching to retain them in solid solution, and (3) a precipitation or aging treatment, either naturally at ambient temperatures or, more commonly, artificially at 240–380°F (115–195°C) to precipitate those elements in an optimum size and distribution.

Strengthened tempers of non–heat treatable alloys are designated by an "H" following the alloy designation, and those of heat treatable alloys by a "T." Suffix digits designate the specific treatment, e.g., 3003-H112 and 6061-T6. In both, the annealed temper representing a condition of maximum softness is designated by an "O."

All non–heat treatable alloys have a high resistance to general corrosion. Aluminums of the 1xxx series, representing unalloyed aluminum, have a relatively low strength. Alloys of the 3xxx series (aluminum-manganese, aluminum-manganese-magnesium) have the same desirable characteristics as those of the 1xxx series and somewhat higher strength.

Alloys of the 5xxx series (aluminum-magnesium) are the strongest of the non–heat treatable aluminum alloys and not only have the same high resistance to general corrosion of other non–heat treatable alloys in most environments, but in slightly alkaline solutions a better resistance than that of any other aluminum alloys.

Among the heat treatable alloys, those of the 6xxx series, which are moderate-strength alloys, provide a high resistance to general corrosion, equal to or approaching that of the non–heat-treatable alloys.

The aluminum alloys most commonly used for piping systems are alloys 1060, 3003, 5052, 6061, and 6063. Alloy 6063 is the most widely used since it has good mechanical properties at reasonable cost. Refer to Table 6-126 for the mechanical and physical properties of these alloys.

As the temperatures decrease, aluminum shows increased values of tensile and yield strengths, with equal or improved ductility or impact resistance. Consequently, aluminum piping has found many applications in the cryogenic field. The specific alloys most commonly used for cryogenic applications are alloys 3003 and 5052.

Aluminum pipe is available in schedules 5S, 10S, 40S, and 80S, but for corrosion resistance purposes schedules 40S and 80S are most frequently used. Pipe and fittings are produced in accordance with standard USASI dimensions as shown in Tables 6-5 and 6-6.

Operating pressures have been calculated in accordance with the formula found on page 230 using the allowable stress values from Tables 6-127 and 6-128. These stress values are based on welded construction, since this is the most common method of joining, and because of the fact that aluminum alloys, particularly the standard pipe alloys 6061-T and 6063-T6, suffer a reduction in tensile strength caused by the heat of welding. These reductions in tensile strength are approximately 37 and 43%, respectively, in the welding area. All of the ratings in the tables have taken this into account. Tables 6-129 through 6-135 contain the maximum allowable operating pressures for various grades of aluminum pipe. In calculating the tables, y = 0.4 and C = 0. No allowance has been made for corrosion.

Aluminum pipe can be joined by welding, expanding, mechanical coupling, and threading. Although schedule 40 and schedule 80 pipe may be joined by threading, it is not recommended when the piping system is intended to handle corrosives.

The preferred method for joining aluminum pipe used to handle corrosives is by means of welding. When welding alloy 3003 to itself, alloy 110 filler wire should be used. For welding alloy 6061 or alloy 6063 to each other or to themselves, as well as welding these alloys to alloy 3003, filler wire of alloy 4043 should be used.

Forged aluminum flanges are also available. These are not normally used to join pipe sections. Their use is usually restricted to making connections to pumps and/or other equipment. The maximum nonshock pressure ratings of these flanges are given in Table 6-136.

Normal standard pipe-fitting techniques should be employed when installing aluminum pipe. In order to prevent installation stresses, the pipe should be properly aligned and not "sprung into place". Proper support for the piping is also essential. Aluminum is subject to galvanic corrosion in the presence of an electrolyte, as are other metals. When in the presence of copper, brass, nickel, Monel, tin, lead, and carbon steel, aluminum will be corroded. Because of this, care should be taken in supporting the piping system. Conventional steel pipe hangers should be avoided. Use either aluminum or padded hangers. Galvanized hangers may be used but will supply protection only as long as the galvanizing is intact.

The recommended support spacing for aluminum pipe may be found in Tables 6-137 to 6-141. Although the recommended spacing does not specify that at least one hanger be on each pipe length, this is good practice.

Aluminum alloys are compatible with dry salts of most inorganic salts and, within their passive range of pH 4–9, with aqueous solutions of many of them. Corrosion of the pitting type is produced by aqueous solutions, mostly halide salts, under conditions at which the alloys are polarized to their pitting potentials. In most other solutions where conditions are less likely to occur that will polarize the alloys to these potentials, pitting is not a problem. Aluminum alloys are not compatible with most inorganic acids, bases, and salts with a pH outside the passive range of the alloys (pH 4–9).

Aluminum alloys are resistant to a wide variety of organic compounds, including most aldehydes, esters, ethers, hydrocarbons, ketones, mercaptans, other sulfur-containing compounds, and nitro compounds. They are also resistant to most acids, alcohols, and phenols, except when these compounds are nearly dry and near their boiling points. Carbon tetrachloride also exhibits the same behavior.

Aluminum alloys are most resistant to organic compounds halogenated with chlorine, bromine, and iodine. They are also resistant to highly polymerized compounds.

It is also important to recognize that the compatibility of aluminum alloys with mixtures of organic compounds cannot always be predicted from their compatibility with each of the compounds. For example, some aluminum alloys are corroded severely in mixtures of carbon tetrachloride and methyl alcohol, even though they are resistant to each compound alone. Caution should also be exercised in using data for pure organic compounds to predict performances of the alloys with commercial grades that may contain contaminants. Ions of halides and reducible metals, commonly chloride and copper, frequently have been found to be the cause of excessive corrosion of aluminum alloys in commercial grades of organic chemicals that would not have been predicted from their resistance to pure compounds.

Refer to Table 6-142 for the compatibility of aluminum alloys with selected corrodents.

6.17   COPPER PIPE

Since before the dawn of history, when primitive people first discovered the red metal, copper has been serving us. The craftsmen who built the great pyramid for the Egyptian pharaoh Cheops used copper pipe to convey water to the royal bath. A remnant of this pipe was unearthed some years ago, still in usable condition, a testimonial to copper's durability and resistance to corrosion. Today, nearly 5000 years after Cheops, copper is still used to convey water and is a prime material for this purpose.

Copper pipe is normally produced from pure electrolytically refined copper that has been deoxidized with phosphorus. This material is referred to as copper number 122 and is available in annealed, hard-drawn, and light-drawn forms. The mechanical and physical properties of copper may be found in Table 6-143.

Copper pipe is available threaded or plain end in diameters of ⅛ inch through 12 inches. The normal length supplied is 12 feet. Pipe is available in thin-wall, regular, extra strong, and double extra strong schedules. Thin-wall plain end pipe can not be used for threading. It is used with brazing or special flanged fittings and is available in 20-foot lengths, but only one wall thickness. Dimensions of copper pipe are shown in Tables 6-144 and 6-145.

Allowable operating pressures of copper pipe have been calculated using the formula found on page 230 and the allowable design stress values from Table 6-146. In Table 6-147 the maximum allowable operating pressures of threaded pipe are given. In calculating these values using the standard formula, the following values were used: $y = 0.4$, $C = 0.05$ for diameters ⅜ inch and smaller, and the depth of thread for other sizes of threaded pipe. Tables 6-148 and 6-149 provide allowable operating pressures for plain end pipe. For these tables, $C = 0$. In all cases $E = 1.0$, since all of the pipes are seamless. No allowance has been made for corrosion.

Copper pipe can be joined by threading, soldering, or brazing. Fittings are available for each of these methods.

Since copper is softer and more ductile than steel, certain precautions must be taken when threading this material. Since conventional vise jaws will tend to mar or dent the pipe, vises equipped with friction jaws should be used to hold the pipe.

Special dies for threading copper pipe should be used. These dies should be well sharpened and maintained in good condition. A set of iron pipe dies may be used provided they are used only on copper pipe. Strap wrenches should be used to make up the joint.

When copper pipe is joined by either soldering or brazing, the maximum allowable operating pressure will depend upon the solder or brazing material used (see Table 6-150). The 50-50 tin-lead solder is gradually being phased out of use on water or food lines.

Good piping practices should be followed when installing copper pipe. All piping should be properly aligned before joining so that it does not have to be "sprung" into place. Recommended spacings of hanger supports are given in Table 6-151. This table does not make allowance for any concentrated loads in the piping system. If valves, meters, or other accessory items are installed in the line, they must be individually supported.

Because of galvanic action, padded hangers should be used. If steel hangers are employed, they will corrode as a result of galvanic action. Eventually the pipe line would lose its support. Although Table 6-151 may recommend spacing greater than a pipe length, it is generally advisable to provide at least one hanger for each length of pipe.

Copper exhibits good corrosion resistance. The corrosion of metal is usually an electrochemical reaction involving solution of the metal as ions at anodic areas and deposition of hydrogen from the electrolyte at the cathodic areas. Copper, being a noble metal, does not normally displace hydrogen from a solution containing hydrogen ions. With copper, the predominant cathodic reaction is the reduction of oxygen to form hydroxide ions. Therefore, the presence of oxygen or another oxidizing agent is required for corrosion to take place.

Copper is rapidly corroded by oxidizing acids such as nitric and chromic. Organic acids are generally less corrosive than the mineral acids.

Sodium and potassium hydroxide solutions can be handled at room temperature by copper in all concentrations. Copper is not corroded by perfectly dry ammonia but may be rapidly corroded by moist ammonia and ammonium hydroxide solutions. Alkaline salts, such as sodium carbonate, sodium phosphate, or sodium silicate, act like hydroxides but are less corrosive. Table 6-152 gives the compatibility of copper with selected corrodents.

6.18   ALUMINUM BRONZE PIPE

The addition of aluminum to copper was originally done to give strength to the copper while maintaining the corrosion resistance of the base metal. It was found, however, that aluminum bronze was more resistant to direct chemical attack because aluminum oxide plus copper gives the alloy superior corrosion resistance.

Typical mechanical properties for aluminum bronze pipe are:

Tensile strength: 82×10^3 psi
Yield strength, 0.2% offset: 42×10^3 psi
Elongation: 35%
Hardness Brinell: 180

Pipe is available in sizes of ½ inch through 10 inches in diameter in schedules 40 and 80. Schedule 10 and 20 pipes are available on special order. Dimensions are in accordance with USASI standard dimensions in Tables 6-5 and 6-6.

Pipe can be joined by welding or threading. Aluminum bronze pipe is resistant to grain boundary stress corrosion cracking, therefore welding does not affect the pipe's performance.

In the aluminum bronze system there are actually three distinct groups of alloys, each group having a characteristic type of microstructure. These groups are (1) single phase (alpha), (2) duplex (alpha plus beta), and (3) hypereutectoid (beta plus gamma).

Piping is usually produced from the group 1 alloy. These are formulated with an aluminum content in the range of 5–5.5%. These alloys have high strength, good corrosion resistance, and excellent resistance to cavitation and erosion.

Aluminum bronzes are not affected by pitting, crevice corrosion, or stress corrosion. They have excellent resistance to corrodents ranging from sulfuric acid boiling (up to 50%) to hot concentrated solutions and are recommended for such chemicals as phosphoric acid, acetic acid, phthalic anhydride, phenols, and furfural. Refer to Table 6-152 for the compatibility of aluminum bronze with selected corrodents.

6.19   90-10 COPPER-NICKEL ALLOY PIPE (ALLOY 706)

This piping material is specifically designed to handle sea water. Of the several commercial copper-nickel alloys available, alloy 706 offers the best combination of properties for marine application and has the broadest application in sea water service. Mechanical and physical properties are given in Table 6-153.

Alloy 706 has been used aboard ships for sea water distribution and shipboard fire protection. It is also used in many desalting plants. Exposed to sea water, alloy 706 forms a thin but tightly adhering oxide film on its surface. To the extent that this film forms, copper-nickel does in fact "corrode" in marine environments. However, the copper-nickel oxide film is firmly bonded to the underlying metal and is nearly insoluble in sea water. It therefore protects the alloy against further attack once it has formed. Initial corrosion rates may be in the range of from less than 1.0 to about 2.5 mils per year. In the absence of turbulence, as would be the case in a properly designed piping system, the copper-nickel's corrosion rate will decrease with time, eventually dropping to as low as 0.05 mpy after several years of service.

Another important advantage of the oxide film developed on alloy 706 is that it is an extremely poor medium for the adherence and growth of marine life forms. Algae and barnacles, the two most common forms of marine biofouling, simply will not grow on alloy 706. Alloy 706 piping therefore remains clean and smooth, neither corroding appreciably nor becoming encrusted with growth.

Flow velocities play an important role in seawater corrosion. Velocity-dependent erosion-corrosion is often associated with turbulence and entrained particles. Recommended velocities for continuous service are as follows:

Pipe size (in.)	Velocity (fps)
≤3	5
4–8 with short radius bends	6.5
≥4 with long radius bends	11
≥8	11

Alloy 706 pipe is available in standard sizes, as are alloy 706 fittings and flanges. However, smooth long-radius bends should be specified rather than fittings, especially below 6 inches. Alloy 706 pipe, seamless or welded, is readily bent cold. Maximum distance between pipe supports is given in Table 6-154.

Alloy 706 pipe can be joined by welding or brazing.

6.20 RED BRASS PIPE

Red brass pipe is an alloy containing 85% copper and 15% zinc. It has the basic corrosion resistance of copper but with greater tensile strength. Refer to Table 6-155 for the mechanical and physical properties of red brass pipe.

Threaded or plain end red brass pipe is available in diameters of ⅛ inch through 12 inches. It is normally supplied in 12-foot lengths in regular, extra strong, and double extra strong schedules. Dimensions of red brass pipe are the same as those of copper pipe found in Table 6-145.

Allowable operating pressures of red brass pipe have been calculated using the stress values given in Table 6-156 and the formula found on page 230. For Table 6-157, the value of $y = 0.4$ and $C = 0.05$ for diameters ⅜ inch and smaller and the depth of thread for other sizes of threaded pipe. For Table 6-158, the value of $C = 0$. Since seamless pipe was used in both tables, $E = 0$. In neither table was any corrosion allowance used.

Pipe may be joined by threading, soldering, or brazing. Flanges may be installed by any of these methods. Because of the higher copper content, red brass is not subject to dezincification or stress corrosion cracking.

Pipe should be installed using good piping practices so as not to induce stress into the piping system. Table 6-159 gives the recommended support spacing. Even though the recommended spans may be greater, it is advisable to install at least one support per pipe length. All accessory items such as valves, instruments, etc. should be individually supported. To prevent galvanic corrosion, the pipe hangers should be of copper or brass construction. If they are of steel construction, they should be padded to prevent contact with the brass pipe. If the pipe line is to be insulated, the support spacing should be reduced by 30%.

Red brass has basically the same corrosion resistance as copper. Refer to Table 6-152 for the compatibility of red brass with selected corrodents.

6.21   TITANIUM PIPE

Titanium is quite plentiful in the earth's crust, being the ninth most abundant element. It is the fourth most abundant metal and is more plentiful than chromium, nickel, or copper, which are commonly employed in alloys used to resist corrosion. However, in spite of early recognition of its light weight, strength, and corrosion resistance, titanium metal is a relative newcomer to the industrial scene.

Manufacturing processes for producing elemental titanium were developed in the eighteenth century. However, it has only been in the last 50 years through the efforts of the U.S. Bureau of mines scientist W. J. Kroll that the process has been refined sufficiently to produce elemental titanium in commercial quantities.

The industrial utilization of titanium results mainly from the excellent corrosion resistance that this metal has to offer. The strength properties are also utilized, but to a smaller extent.

There are basically four grades of titanium. Unalloyed titanium, represented by grade 2, is most often used for corrosion resistance.

Titanium is a light metal with a density slightly more than half that of iron- or copper-based alloys. The modulus of elasticity of titanium is also about half that of steel. Specific heat and thermal conductivity are similar to those of stainless steel. Titanium has a low expansion coefficient. The mechanical and physical properties of titanium are given in Table 6-160.

Titanium pipe is produced to standard USASI dimensions shown in Tables 6-5 and 6-6. Pipe is available in schedules 5S, 10S, 40S, and 80S. All schedules are available for welding or flanged connections. Schedules 40S and 80S are also available for threaded connections.

Safe operating pressures for seamless, plain end titanium pipe have been calculated using the formula found on page 230. Table 6-161 tabulates the maximum allowable operating pressures up to a maximum temperature of 650°F (343°C). In these calculations a maximum allowable stress of 5450 psi was used with $y = 0.04$, $C = 0$, and $E = 0$. No corrosion allowance was applied.

The maximum allowable stress values for welded and annealed titanium pipe grade 2 is as follows:

Temperature (°F/°C)	Maximum allowable stress (psi)
100/38	10,600
150/66	10,200
200/93	9300
250/121	8400
300/149	7700
350/177	7100
400/204	6500
450/232	6100
500/260	5600
550/288	5300
600/316	4800

A threaded pipe system in schedule 40S is available. The complete system, pipe and fittings, is rated for 150 psi at 650°F (343°C) in all sizes.

Normal pipe-installation techniques should be followed when installing titanium pipe. Table 6-162 provides the recommended support spacing. No allowance has been made for concentrated loads. All valves or accessories installed in the pipe line must be independently supported. If carbon steel hangers are to be used, they should be padded to insulate them from the titanium pipe to prevent galvanic corrosion. In this case the carbon steel hanger would be the one to corrode, which would eventually leave the pipe unsupported.

Titanium's resistance is due to a stable, protective, strongly adherent oxide film that covers its surface. This film forms instantly when a fresh surface is exposed to air and moisture. The oxide film on titanium, although very thin, is very stable and is attacked by only a few substances, most notable of which is hydrofluoric acid. Because of its strong affinity for oxygen, titanium is capable of healing ruptures in this film almost instantly in any environment where a trace of moisture or oxygen is present. Therefore, we find that titanium is impervious to attack in moist chlorine gas. If the moisture content of dry chlorine gas falls below a critical level of 0.5%, rapid or even catastrophic attack can occur.

Anhydrous conditions in the absence of a source of oxygen should be avoided with titanium because the protective film may not be regenerated if it is damaged.

Titanium is immune to all forms of corrosive attack in sea water and chloride salt solutions at ambient temperatures. Under certain conditions in chloride solutions, titanium is subject to pitting and crevice corrosion. The temperature at which the attack occurs is dependent upon the pH of the solution. If the pH exceeds approximately 7.5, pitting and crevice corrosion will take place at about 300°F (149°C).

Titanium offers excellent resistance to oxidizing acids such as nitric and chromic. It has limited resistance to reducing acids such as hydrochloric, sulfuric, and phosphoric.

Table 6-163 provides additional information on the compatibility of titanium with selected corrodents.

6.22   ZIRCONIUM PIPE

Zirconium's primary ore source is zircon, which occurs in several regions of the world in the form of beach sand. It is ranked nineteenth in abundance of the chemical elements occurring in the earth's crust, and it is more abundant than several common metals, such as nickel, chromium, and cobalt.

Hafnium occurs naturally in ores with zirconium. It has chemical and metallurgical properties similar to those of zirconium. Grade 702 (ASTM designation R60702) is an unalloyed zirconium and contains approximately 4.5% hafnium. It is the most widely used zirconium in the chemical processing industry because it has the best overall corrosion resistance but is the lowest in strength.

Alloy 705 (ASTM designation R60705) exhibits similar corrosion resistance to 702 in most environments, but has a significant strength advantage. The strength of alloy 705 is nearly double that of unalloyed 702. Alloy 705 has niobium added to give it increased strength. Refer to Table 6-164 for the mechanical and physical properties of zirconiums 702 and 705.

Zirconium pipe is available in nominal pipe sizes of ⅛ inch through 12 inches in schedules 5, 10, 40, and 80. It is produced in accordance with the dimensions shown in

Table 6-165. The pipe may be joined by welding, flanging, or threading of schedules 40 and 80 only. Of the three methods, welding is preferred. The maximum allowable stress values of zirconium pipe at various temperatures are given in Table 6-166, while Tables 6-167 through 6-170 provide the maximum allowable operating pressures of grade 702 seamless pipe. These pressures were calculated using the equation on page 230 with y = 0.4, C = 0, and E = 0. No allowance has been made for corrosion.

The wide use of zirconium is due to its excellent resistance to many chemical solutions, even at elevated temperatures and pressures. When exposed to an oxygen-containing environment, an adherent protective oxide film forms on its surface. This film is formed spontaneously in air or water at ambient temperature and below, is self-healing, and protects the base metal from chemical and mechanical attack at temperatures up to 572°F (300°C).

As a result, zirconium resists attack by mineral acids, alkaline solutions, most organic and salt solutions, and molten alkalies. It has excellent oxidation resistance up to 752°F (400°C) in air, steam, carbon dioxide, sulfur dioxide, nitrogen, and oxygen. It will not be attacked by oxidizing media unless halides are present.

It is attacked by hydrofluoric acid, wet chlorine, concentrated sulfuric acid, aqua regia, ferric chloride, and cupric chloride. Refer to Table 6-163 for additional information on the compatibility of zirconium with selected corrodents.

6.23 TANTALUM PIPE

Tantalum is not a new product. It came into commercial use at the turn of the century, being used for filaments in light bulbs. When its superior corrosion-resistant properties became apparent, it found applications in the chemical-processing industry.

The general physical properties of tantalum are similar to those of mild steel except that tantalum has a higher melting point. Tantalum used for the manufacture of pipe is produced in two grades:

1. Grade VM, a vacuum-melted product
2. Grade PM, a powder metallurgical product

The mechanical and physical properties of tantalum pipe are given in Table 6-171.

Because of the relatively high cost of tantalum, applications are limited to extremely corrosive conditions. In order to reduce costs, most tantalum piping systems consist of relatively thin-wall tantalum tubing, which is available in either welded or seamless design. Welded tubing is produced with outside diameters of 0.375–9.00 inches, while seamless tubing is available in outside diameters of 0.125–2.00 inches. Wall thicknesses of welded tubing are given in Table 6-172. Standard fittings are available.

If a tantalum piping system is to be considered, it is recommended that the manufacturer design the system to meet the operating conditions specified. At 70°F (21°C), an allowable stress of 10,000 psi can be used in the design of the piping system. Because of the exceptional corrosion resistance of tantalum, most applications can be designed without a corrosion allowance.

A solid tantalum piping system should be joined and installed as per specific instruction from the manufacturer. Joining can be accomplished by inert-gas welding, flare coupling, or flanging.

Since most tantalum piping systems are constructed of relatively thin-wall tubing, it is advisable to provide external protection against mechanical damage. A plastic shield would suffice.

Tantalum is inert to practically all organic and inorganic compounds at temperatures below 302°F (150°C). The only exceptions to this are hydrofluoric acid and fuming sulfuric acid. At temperatures below 302°F (150°C), it is inert to all concentrations of hydrochloric acid, to all concentrations of nitric acid (except fuming), to 98% sulfuric acid, to 85% phosphoric acid, and to aqua regia.

Fuming sulfuric acid attacks tantalum even at room temperature. Similarly, hydrofluoric acid, anhydrous hydrogen fluoride, or any acid medium containing fluoride ion will rapidly attack the metal. Commercial phosphoric acid may attack tantalum if small amounts of fluoride impurity are present. Hot oxalic acid is the only organic acid known to attack tantalum. It is also attacked by concentrated alkaline solutions at room temperature, although it is fairly resistant to dilute solutions. Table 6-163 provides additional information on the corrosion resistance of tantalum in the presence of selected corrodents.

6.24   HIGH-SILICON IRON PIPE

There are two high-silicon iron alloys that are of value in the piping field, both of which are manufactured by the Duriron Company. The first alloy, known as Duriron, contains nominally 14.5% silicon, 1% carbon, with the balance iron. When 4% chromium is added, the product is known as Durichlor. These are cast alloys.

The high silicon content improves corrosion resistance, but it also lowers some of the mechanical properties as compared with gray iron. Silicon irons are hard and brittle and therefore do not stand up well under shock and impact. Because of their hardness, high silicon irons are good for combined corrosion-erosion service. These alloys cannot withstand any substantial stressing or impact, and they cannot be subjected to sudden fluctuations in temperature. Refer to Table 6-173 for the mechanical and physical properties of the high-silicon irons.

Pipe is available in nominal diameters of 1 inch through 12 inches, with larger diameters up to 16 inches available on special order. Pipe and fittings are joined by means of flanging. Pipe lengths and fittings are supplied with "beaded ends." Malleable iron split flanges are supplied with 125-pound USASI drillings, which are used to make up the joint. It is necessary to use a gasket to seal the joint. Care must be taken that the gasket is compatible with the material being handled. Table 6-174 lists the maximum pipe lengths available by diameter. Note that pipe lengths are short.

Because of such factors as thermal shock and pulsating pressures, the pressure ratings of Duriron and Durichlor pipe are set at 50 psig. In destruction tests conducted by the manufacturer with water at ambient temperature, rupture does not occur below 400 psig, even in the larger sizes. On this basis it is possible to use the pipe at higher pressure with the approval of the manufacturer for the specific application. The upper temperature limit is 1400°F (760°C).

Since the high-silicon iron pipes have such low tensile strength, care should be taken during installation that the pipe not be subjected to mechanical or thermal shock. In addition, it is imperative that proper pipe-fitting techniques be followed during the installation. All pipe sections must be properly aligned before flange bolts are tightened.

Duriron and Durichlor pipe cannot be field fabricated. It must be ordered to exact length. This requires the preparation of a detailed piping drawing with the length of each straight piece calculated. Small gaps in the line can be closed by the use of spacers, which range in length from ¼ inch to 3 inches. These are installed between two factory ends and tie-rodded between the flanges on each beaded end.

It is very important that high-silicon iron pipe be properly supported. Horizontal lines should be supported at every joint. Vertical lines should be supported every 10 feet, with an additional support at the bottom of the riser.

High-silicon irons have excellent corrosion resistance to a wide range of chemicals. One of the major applications is in handling sulfuric acid. It is resistant to all concentrations of sulfuric acid, up to and including the normal boiling point. This alloy will handle nitric acid above 30% to the boiling point. Below 30% the temperature is limited to about 180°F (82°C).

Durichlor is resistant to environments such as severe chloride-containing solutions and other strongly oxidizing services. It is capable of handling hydrochloric acid up to a temperature of 80°F (27°C). Refer to Table 6-175 for the compatibility of high-silicon iron with selected corrodents.

Table 6-1 Conversion Factors from ipy to mdd[a]

Metal	Density (g/cc)	0.00144 / Density × 10^{-3}	696 × Density
Aluminum	2.72	0.529	1890
Brass (red)	8.75	0.164	6100
Brass (yellow)	8.47	0.170	5880
Cadmium	8.65	0.167	6020
Columbium	8.4	0.171	5850
Copper	8.92	0.161	6210
Copper-nickel (70–30)	8.95	0.161	6210
Iron	7.87	0.183	5480
Duriron	7.0	0.205	4870
Lead (chemical)	11.35	0.127	7900
Magnesium	1.74	0.826	1210
Nickel	8.89	0.162	6180
Monel	8.84	0.163	6140
Silver	10.50	0.137	7300
Tantalum	16.6	0.0868	11550
Titanium	4.54	0.317	3160
Tin	7.29	0.198	5070
Zinc	7.14	0.202	4970
Zirconium	6.45	0.223	4490

[a]Multiply ipy by (696 × density) to obtain mdd. Multiply mdd by (0.00144/density) to obtain ipy.

Table 6-2 Galvanic Series of Metals and Alloys

Corroded end (anodic)	
Magnesium	Muntz metal
Magnesium alloys	Naval bronze
Zinc	Nickel (active)
Galvanized steel	Inconel (active)
Aluminum 6053	Hastelloy C (active)
Aluminum 3003	Yellow brass
Aluminum 2024	Admiralty brass
Aluminum	Aluminum bronze
Alclad	Red brass
Cadmium	Copper
Mild steel	Silicon bronze
Wrought iron	70–30 Cupro-nickel
Cast iron	Nickel (passive)
Ni-resist	Inconel (passive)
13% Chromium stainless steel (active)	Monel
50-50 Lead tin solder	18-8 Stainless steel type 304 (passive)
Ferretic stainless steel 400 series	18-8-3 Stainless steel type 316 (passive)
18-8 Stainless steel type 304 (active)	Silver
18-8-3 Stainless steel type 316 (active)	Graphite
	Gold
Lead	
	Platinum
Tin	
	Protected end (cathodic)

Table 6-3 Alloy Systems Subject to Stress Corrosion Cracking

Alloy	Environment
Aluminum base	Air, sea water, salt and chemical combination
Magnesium base	Nitric acid, caustic, HF solutions, salts, coastal atmospheres
Copper base	Primarily ammonia and ammonium hydroxide, amines, mercury
Martensitic and precipitation hardening stainless steels	Sea water, chlorides, H_2S solutions
Austenitic stainless steels	Chlorides (inorganic and organic), caustic solutions, sulfurous and polythionic acids
Nickel base	Caustic above 600°F (315°C), fused caustic, hydrofluoric acid
Titanium	Sea water, salt atmospheres, fused salt
Carbon steel	Caustic, anhydrous ammonia, nitrate solutions

Table 6-4 Longitudinal Joint Factor E

Type of joint	E
Arc or gas weld:	
Double-welded butt	0.85
Double-welded butt[a]	0.90
Double-welded butt with 100% radiography in accordance with 336.4.2C and conforming with requirements of 327.4.3	1.00
Single-welded butt	0.80
Single-welded butt[a]	0.90
Single-welded butt with 100% radiography in accordance with 336.4.2C and conforming with requirements of 327.4.3	1.00
Spiral-welded	0.75
Electric resistance weld and electric flash weld	0.85
Furnace weld:	
Lap weld	0.75
Butt weld	0.60

[a]Welds with 0.90 joint factor shall be finished, random-radiographed by the technique, and evaluated in accordance with UW 51 of Section VIII of the A.S.M.E. Boiler and Pressure Vessel Code. Radiographing (random) shall consist of not less than 12 inches of radiogaphy per 100 feet of weld with reexamination and repair in accordance with UW 51 of Section VIII of the A.S.M.E. Boiler and Pressure Vessel Code.

Table 6-5 USASI Dimensions of Corrosion-Resistant Pipe

Nominal pipe size (in.)	Outside diameter (in.)	Schedule 5S		Schedule 10S	
		Inside diameter (in.)	Wall thickness (in.)	Inside diameter (in.)	Wall thickness (in.)
⅛	0.405			0.307	0.049
¼	0.540			0.410	0.065
⅜	0.675			0.545	0.065
½	0.840	0.710	0.065	0.674	0.083
¾	1.050	0.920	0.065	0.884	0.083
1	1.315	1.185	0.065	1.097	0.109
1½	1.900	1.770	0.065	1.682	0.109
2	2.375	2.245	0.065	2.157	0.109
3	3.500	3.334	0.083	3.260	0.120
4	4.500	4.334	0.083	4.260	0.120
5	5.563	5.345	0.109	5.295	0.134
6	6.625	6.407	0.109	6.357	0.134
8	8.625	8.407	0.109	8.329	0.148
10	10.750	10.482	0.134	10.420	0.165
12	12.750	12.438	0.156	12.390	0.180
14	14.000	13.688	0.156	13.624	0.188
16	16.000	15.670	0.165	15.624	0.188
18	18.000	17.670	0.165	17.624	0.188
20	20.000	19.624	0.188	19.564	0.218
24	24.000	23.565	0.218	23.500	0.250
30	30.000	29.500	0.250	29.376	0.312

Table 6-6 USASI Dimensions of Corrosion-Resistant Pipe

Nominal pipe size (in.)	Outside diameter (in.)	Schedule 40S		Schedule 80S	
		Inside diameter (in.)	Wall thickness (in.)	Inside diameter (in.)	Wall thickness (in.)
1/8	0.405	0.269	0.068	0.215	0.095
1/4	0.540	0.364	0.088	0.215	0.119
3/8	0.675	0.493	0.091	0.423	0.126
1/2	0.840	0.622	0.109	0.546	0.147
3/4	1.050	0.824	0.113	0.742	0.154
1	1.315	1.049	0.133	0.957	0.179
1 1/2	1.900	1.610	0.145	1.500	0.200
2	2.375	2.067	0.154	1.959	0.218
3	3.500	3.068	0.216	2.900	0.300
4	4.500	4.026	0.237	3.826	0.337
5	5.563	5.047	0.258	4.813	0.375
6	6.625	6.065	0.280	5.761	0.432
8	8.625	7.981	0.322	7.625	0.500
10	10.750	10.020	0.365	9.750	0.500
12	12.750	12.000	0.375	11.750	0.500
14	14.000	13.250	0.375	13.000	0.500
16	16.000	15.250	0.375	15.000	0.500
18	18.000	17.250	0.375	17.000	0.500
20	20.000	19.250	0.375	19.000	0.500
24	24.000	23.250	0.375	23.000	0.500
30	30.000	29.250	0.375	29.000	0.500
36	36.000	35.250	0.375	35.000	0.500

Table 6-7 Values of y

	y at °F/°C					
	900/482 and below	950/510	1000/538	1050/560	1100/593	1150/621 and above
Ferritic steels	0.4	0.5	0.7	0.7	0.7	0.7
Austenitic steels	0.4	0.4	0.4	0.4	0.5	0.7

Table 6-8 Values of C

Type of pipe	C (in.)
Cast iron, centrifugally cast	0.14
Cast iron, pit cast	0.18
Threaded steel, wrought iron, or nonferrous pipe:	
$\frac{3}{8}$ in. and smaller	0.05
$\frac{1}{2}$ in. and larger	depth of thread
Grooved steel, wrought iron, or nonferrous pipe	depth of groove
Plain end steel, wrought iron, or tube:	
1 in. and smaller	0.05
$1\frac{1}{4}$ in. and larger	0.065
Plain end nonferrous pipe or tube	0.000

Table 6-9 Maximum Allowable Nonshock Pressure Ratings of Insert Flanges

Operating temperature (°F/°C)	Maximum allowable operating presure (psi)	
	Carbon steel	Ductile iron
–20 to 100/–29 to 38	220	176
150/66	205	164
200/93	190	152
250/121	180	144
300/149	170	136
350/177	155	124
400/204	145	116
450/232	130	104
500/260	120	96

Table 6-10 Chemical Composition of Ferritic Stainless Steels

AISI type	Nominal composition (%)				
	C max	Mn max	Si max	Cr	Other[a]
405	0.08	1.00	1.00	11.50–14.50	0.10–0.30 Al
403	0.12	1.00	1.00	14.00–18.00	
430F	0.12	1.25	1.00	14.00–18.00	0.15 S min
430(Se)	0.12	1.25	1.00	14.00–18.00	0.15 Se min
444	0.025	1.00	1.00 max	17.5 –19.5	1.75–2.50 Mo
446	0.20	1.50	1.00	23.00–17.00	0.25 max N
XM-27[b]	0.002	0.10	0.20	26.00	

[a]Elements in addition to those shown are as follows: phosphorus—0.06% max in type 430F and 430(Se), .015% in XM-27; sulfur—0.03% max in types 405, 430, 444, and 446; 0.15% min type 430F, 0.01% in XM-27; nickel—1.00% max in type 444, 0.15% in XM-27; titanium + niobium—0.80% max in type 444; copper—0.02% in XM-27; nitrogen—0.010% in XM-27.
[b]E-Brite 26-1 trademark of Allegheny Ludlum Industries Inc.

Table 6-11 Physical and Mechanical Properties of Ferritic Stainless Steels

Property	Type of alloy		
	430	444	XM-27
Modulus of elasticity $\times$ 10^6	29	29	
Tensile strength $\times$ 10^3, psi	60	60	70
Yield strength 0.2% offset $\times$ 10^3, psi	35	40	56
Elongation in 2 in., %	20	20	30
Hardness, Brinell	B-165	217	Rock. B-83
Density, lb/in.^{3}	0.278	0.28	0.28
Specific gravity	7.75	7.75	7.66
Specific heat (32–212°F), Btu/lb°F	0.11	0.102	0.102
Thermal conductivity, Btu/lb°F			
at 70°F (20°C)	15.1	15.5	
at 1500°F (815°C)	15.2		
Thermal expansion coefficient			
(32–212°F) $\times$ 10^{-6} in./in.°F	6.0	6.1	5.9

Table 6-12 Allowable Stress Values of Ferritic Stainless Steels Used for Piping

SS type	Allowable stress value (ksi) at temperature not exceeding (°F/°C)						
	100/38	200/93	300/149	400/204	500/260	600/316	650/343
444	12.8	12.2	11.8	11.3	10.9	10.5	10.4
XM-27	12.75	12.75	12.5	12.5	12.5	12.5	12.5

Table 6-13 Compatibility of Ferritic Stainless Steels with Selected Corrodents

The chemicals listed are in the pure state or in a saturated solution unless otherwise indicated. Compatibility is shown to the maximum allowable temperature for which data are available. Incompatibility is shown by an x. A blank space indicates that data are unavailable. When compatible, the corrosion rate is <20 MPY.

Chemical	430 (°F/°C)	444 (°F/°C)	XM-27 (°F/°C)
Acetaldehyde			
Acetamide			
Acetic acid 10%	70/21	200/93	200/93
Acetic acid 50%	x	200/93	200/93
Acetic acid 80%	70/21	200/93	130/54
Acetic acid, glacial	70/21		140/60
Acetic anhydride 90%	150/66		300/149
Acetone			
Acetyl chloride			
Acrylic acid			
Acrylonitrile			
Adipic acid			
Allyl alcohol			
Allyl chloride			
Alum			
Aluminum acetate			
Aluminum chloride, aqueous	x		110/43
Aluminum chloride, dry			
Aluminum fluoride			
Aluminum hydroxide	70/21		
Aluminum nitrate			
Aluminum oxychloride			
Aluminum sulfate	x		
Ammonia gas	212/100		
Ammonium bifluoride			
Ammonium carbonate	70/21		
Ammonium chloride 10%			200/93
Ammonium chloride 50%			
Ammonium chloride, sat.			
Ammonium fluoride 10%			
Ammonium fluoride 25%			
Ammonium hydroxide 25%	70/21		
Ammonium hydroxide, sat.	70/21		
Ammonium nitrate	212/100		
Ammonium persulfate 5%	70/21		
Ammonium phosphate	70/21		
Ammonium sulfate 10–40%	x		
Ammonium sulfide			
Ammonium sulfite			
Amyl acetate	70/21		
Amyl alcohol			
Amyl chloride	x		
Aniline	70/21		

Table 6-13 Continued

Chemical	430 (°F/°C)	444 (°F/°C)	XM-27 (°F/°C)
Antimony trichloride	x		
Aqua regia 3:1			x
Barium carbonate	70/21		
Barium chloride	70/21[a]		
Barium hydroxide			
Barium sulfate	70/21		
Barium sulfide	70/21		
Benzaldehyde			210/99
Benzene	70/21		
Benzene sulfonic acid 10%			
Benzoic acid	70/21		
Benzyl alcohol			
Benzyl chloride			
Borax 5%	200/93		
Boric acid	200/93[a]		
Bromine gas, dry	x		
Bromine gas, moist	x		
Bromine liquid	x		
Butadiene			
Butyl acetate			
Butyl alcohol			
n-Butylamine			
Butyl phthalate			
Butyric acid	200/93		
Calcium bisulfide			
Calcium bisulfite			
Calcium carbonate	200/93		
Calcium chlorate			
Calcium chloride	x		
Calcium hydroxide 10%			
Calcium hydroxide, sat.			
Calcium hypochlorite	x		
Calcium nitrate			
Calcium oxide			
Calcium sulfate	70/21		
Caprylic acid			
Carbon bisulfide	70/21		
Carbon dioxide, dry	70/21		
Carbon dioxide, wet			
Carbon disulfide			
Carbon monoxide	1600/871		
Carbon tetrachloride, dry	212/100		
Carbonic acid	x		
Cellosolve			
Chloracetic acid, 50% water	x		
Chloracetic acid	x		
Chlorine gas, dry	x		

Table 6-13 Continued

Chemical	430 (°F/°C)	444 (°F/°C)	XM-27 (°F/°C)
Chlorine gas, wet	x		
Chlorine, liquid			
Chlorobenzene			
Chloroform, dry	70/21		
Chlorosulfonic acid			
Chromic acid 10%	70/21		120/49
Chromic acid 50%	x		x
Chromyl chloride			
Citric acid 15%	70/21	200/93	200/93
Citric acid, concentrated	x		
Copper acetate	70/21		
Copper carbonate	70/21		
Copper chloride	x		x
Copper cyanide	212/100		
Copper sulfate	212/100		
Cresol			
Cupric chloride 5%	x		
Cupric chloride 50%	x		
Cyclohexane			
Cyclohexanol			
Dichloroacetic acid			
Dichloroethane			
Ethylene glycol	70/21		
Ferric chloride	x		80/27
Ferric chloride 10% in water			75/25
Ferric nitrate 10–50%	70/21		
Ferrous chloride	x		
Ferrous nitrate			
Fluorine gas, dry	x		
Fluorine gas, moist	x		
Hydrobromic acid, dilute	x		
Hydrobromic acid 20%	x		
Hydrobromic acid 50%	x		
Hydrochloric acid 20%	x		
Hydrochloric acid 38%	x		
Hydrocyanic acid 10%	x		
Hydrofluoric acid 30%	x		x
Hydrofluoric acid 70%	x		x
Hydrofluoric acid 100%	x		x
Hypochlorous acid			
Iodine solution 10%	x		
Ketones, general			
Lactic acid 20%	x	200/93	200/93
Lactic acid, concentrated	x		
Magnesium chloride			200/93
Malic acid	200/93		
Manganese chloride			
Methyl chloride			

Table 6-13 Continued

Chemical	430 (°F/°C)	444 (°F/°C)	XM-27 (°F/°C)
Methyl ethyl ketone			
Methyl isobutyl ketone			
Muriatic acid	x		
Nitric acid 5%	70/21	200/93	320/160
Nitric acid 20%	200/93	200/93	320/160
Nitric acid 70%	70/21	x	210/99
Nitric acid, anhydrous	x	x	
Nitrous acid 5%	70/21		
Oleum			
Perchloric acid 10%			
Perchloric acid 70%			
Phenol	200/93		
Phosphoric acid 50–80%	x	200/93	200/93
Picric acid	x		
Potassium bromide 30%			
Salicylic acid			
Silver bromide 10%	x		
Sodium carbonate			
Sodium chloride	70/21[a]		
Sodium hydroxide 10%	70/21	212/100	200/93
Sodium hydroxide 50%		x	180/82
Sodium hydroxide, concentrated		x	
Sodium hypochlorite 30%			90/32
Sodium hypochlorite, conc			
Sodium sulfide to 50%	x		
Stannic chloride	x		
Stannous chloride 10%			90/32
Sulfuric acid 10%	x	x	x
Sulfuric acid 50%	x	x	x
Sulfuric acid 70%		x	x
Sulfuric acid 90%		x	x
Sulfuric acid 98%		x	280/138
Sulfuric acid 100%	70/21	x	
Sulfuric acid, fuming		x	
Sulfurous acid 5%	x		360/182
Thionyl chloride			
Toluene			210/99
Trichloroacetic acid	x		
White liquor			
Zinc chloride 20%	70/21[a]		200/93

[a]Pitting may occur.
Source: Schweitzer, Philip A. (1991). *Corrosion Resistance Tables,* Marcel Dekker, Inc., New York, Vols. 1 and 2.

Table 6-14 Typical Mechanical Properties of Stainless Steels at Cryogenic Temperatures

Stainless steel type	Temp. (°F/°C)	Yield strength 0.2% offset (ksi)	Tensile strength (ksi)	Elongation in 2 in. (%)	Izod impact (ft/lb)
304	−40/−40	34	150	47	110
	−80/−62	34	170	39	110
	−320/−196	39	225	40	110
	−423/−252	50	243	40	110
310	−40/−40	39	95	57	110
	−80/−62	40	100	55	110
	−320/−196	74	152	54	85
	−423/−252	108	176	56	110
316	−40/−40	41	104	59	110
	−80/−62	44	116	57	
	−320/−196	75	155	59	
	−423/−252	84	210	52	
347	−40/−40	44	117	63	110
	−80/−62	45	130	57	110
	−320/−196	47	200	46	95
	−423/−252	55	228	34	
410	−40/−40	90	122	23	25
	−80/−62	94	128	22	25
	−320/−196	148	156	10	5
430	−40/−40	41	76	36	10
	−80/−62	44	81	36	8
	−320/−196	87	92	2	2

Table 6-15 Mechanical and Physical Properties of Types 304 and 304L Stainless Steel

Property	Type of alloy	
	304	304L
Modulus of elasticity $\times 10^6$, psi	28.0	28.0
Tensile strength $\times 10^3$, psi	85	80
Yield strength 0.2% offset $\times 10^3$, psi	35	30
Elongation in 2 in., %	55	55
Hardness, Rockwell	B-80	B-80
Density, lb/in.^{3}	0.29	0.29
Specific gravity	8.02	8.02
Specific heat (32–212°F), Btu/lb°F	0.12	0.12
Thermal conductivity, Btu/hr ft^2 °F at 212°F	9.4	9.4
Thermal expansion coefficient (32–212°F) $\times 10^{-6}$ in/in.°F	9.6	9.6
Izod impact, ft-lb	110	110

Table 6-16 Allowable Design Stress for Types 304 and 304L Stainless Steel

Temperature (°F/°C)	Allowable design stress (psi)	
	304	304L
−325 to 100/− 198 to 38	18750	15600
200/93	16650	15300
300/149	15000	13100
400/204	13650	11000
500/260	12500	9700
600/316	11600	9000
650/343	11200	8750
700/371	10800	8500
750/399	10400	8300
800/427	10000	8100
850/454	9700	
900/482	9400	
950/510	9100	
1000/538	8800	
1050/560	8500	
1100/593	7500	
1150/621	5750	
1200/649	4500	
1250/677	3250	
1300/704	2450	
1350/732	1800	
1400/760	1400	
1450/788	1000	
1500/816	750	

Table 6-17 Maximum Allowable Operating Pressure of Seamless Schedule 5S Type 304 Stainless Steel Pipe

Nominal pipe size (in.)	Maximum operating pressure (psi) at °F/°C								
	−325 to 100/−198 to 38	200/ 93	300/ 149	400/ 204	500/ 260	600/ 316	700/ 371	800/ 427	1000/ 538
½	2150	1910	1720	1570	1440	1340	1240	1150	1010
¾	1750	1510	1360	1240	1140	1060	980	910	800
1	1350	1200	1080	980	900	840	780	720	640
1½	920	820	740	670	620	580	540	490	432
2	740	650	590	540	488	454	422	390	344
3	640	570	510	460	422	392	365	338	298
4	491	435	392	356	327	304	283	262	231
6	437	387	349	317	291	271	252	233	205
8	335	297	268	243	223	207	193	178	157
10	330	293	264	240	220	205	190	176	155
12	324	287	259	235	216	201	187	173	152

These operating pressures are based on 85% of the allowable stress values in Table 6-16.

Table 6-18   Maximum Allowable Operating Pressure of Seamless Schedule 10S Type 304 Stainless Steel Pipe

Nominal pipe size (in.)	Maximum operating pressure (psi) at °F/°C								
	−325 to 100/−198 to 38	200/ 93	300/ 149	400/ 204	500/ 260	600/ 316	800/ 427	1000/ 538	
½	2790	2480	2230	2030	1860	1730	1610	1490	1310
¾	2200	1950	1760	1600	1470	1370	1270	1180	1040
1	2310	2050	1850	1650	1540	1440	1340	1240	1090
1½	1570	1400	1260	1150	1050	980	910	840	740
2	1250	1110	1000	910	830	780	720	670	590
3	930	820	740	680	620	580	540	492	434
4	720	640	580	520	476	443	412	381	336
6	540	477	431	391	359	334	311	287	253
8	456	404	365	331	304	283	263	243	214
10	408	362	326	296	272	253	235	218	192
12	375	333	300	273	250	233	216	200	176

These operating pressures are based on 85% of the allowable stress values in Table 6-16.

Table 6-19   Maximum Allowable Operating Pressure of Seamless Schedule 40S Type 304 Stainless Steel Pipe, Plain End

Nominal pipe size (in.)	Maximum operating pressure (psi) at °F/°C							
	−325 to 100/−198 to 38	200/ 93	300/ 149	400/ 204	500/ 260	600/ 316	800/ 427	1000/ 538
½	4690	4160	3750	3410	3150	2900	2500	2200
¾	3820	3400	3060	2780	2550	2370	2040	1800
1	3580	3180	2860	2600	2390	2210	1910	1680
1½	2650	2350	2120	1930	1770	1640	1420	1250
2	2230	1980	1790	1630	1490	1380	1200	1050
3	2120	1880	1700	1550	1420	1310	1130	1000
4	1800	1600	1440	1310	1200	1110	960	850
6	1430	1270	1150	1040	960	890	770	680
8	1260	1120	1010	920	840	780	680	590
10	1150	1020	920	840	770	710	610	540
12	990	880	790	720	660	610	530	462

These operating pressures are based on 85% of the allowable stress values in Table 6-16.

Table 6-20 Maximum Allowable Operating Pressure of Seamless Schedule 80S Type 304 Stainless Steel Pipe, Plain End

Nominal pipe size (in.)	Maximum operating pressure (psi) at °F/°C							
	−325 to 100/−198 to 38	200/ 93	300/ 149	400/ 204	500/ 260	600/ 316	800/ 427	1000/ 538
½	6550	5820	5240	4770	4370	4050	3490	3080
¾	5370	4770	4300	3910	3580	3320	2860	2520
1	4940	4390	3950	3600	3300	3060	2646	2320
1½	3730	3320	2990	2720	2490	2310	1990	1750
2	3220	2860	2580	2350	2150	2000	1720	1520
3	3000	2660	2400	2180	2000	1860	1600	1410
4	2600	2310	2080	1890	1730	1610	1390	1220
6	2250	2000	1800	1640	1500	1390	1200	1060
8	1990	1770	1590	1450	1330	1230	1060	940
10	1580	1410	1270	1150	1060	980	850	740
12	1330	1180	1060	970	890	820	710	630

These operating pressures are based on 85% of the allowable stress values in table 6-16.

Table 6-21 Maximum Allowable Operating Pressure of Seamless Schedule 40S and Schedule 80S Type 304 Stainless Steel Pipe, Threaded Ends

Nominal pipe size (in.)	Maximum operating pressure (psi) at °F/°C									
	Schedule 40S					Schedule 80S				
	−325 to 100/ −198 to 38	200/93	300/149	400/204	500/260	−325 to 100/ −198 to 38	200/93	300/149	400/204	500/260
⅛	898	800	720	655	600	3270	2900	2615	2380	2180
¼	2120	1880	1695	1540	1412	4070	3610	3255	2960	2710
⅜	1758	1560	1405	1280	1170	3590	3190	2875	2615	2395
½	1549	1370	1238	1125	1030	3200	2840	2560	2330	2130
¾	1362	1210	1090	994	910	2820	2590	2340	2130	1950
1	1180	1050	945	860	787	2420	2150	1940	1765	1615
1½	1040	925	835	760	695	2050	1825	1645	1495	1370
2	960	855	770	700	640	1885	1675	1510	1370	1259
3	885	785	710	645	590	1720	1525	1372	1250	1142
4	840	745	672	612	560	1605	1421	1281	1165	1070
6	780	693	626	570	470	1590	1410	1270	1157	1060

These operating pressures are based on 85% of the allowable stress values in table 6-16.

Table 6-22 Factors for Conversion of Working Pressures of Type 304 Stainless Steels to Other Stainless Alloys

Operating temperature (°F/°C)	Alloy								
	304L	316	316L	317	321	347	20Cb3	20Mo6	A16XN
−325 to 100/ −198 to 38	0.83	1.00	0.83	1.00	1.00	1.00	0.77	0.91	
200/93	0.92	1.13	0.94	1.13	1.13	1.13	0.87	1.02	1.33
300/149	0.87	1.19	0.97	1.19	1.13	1.13	0.95	1.07	1.37
400/204	0.81	1.28	0.88	1.28	1.16	1.16	0.99	1.09	1.41
500/260	0.78	1.38	0.88	1.38	1.22	1.22	1.05	1.10	1.42
600/316	0.78	1.47	0.88	1.47	1.28	1.28	1.11	1.12	1.46
700/371	0.79	1.57	0.88	1.57	1.37	1.37	1.16	1.15	1.52
800/427	0.81	1.68	0.88	1.68	1.46	1.46	1.22	1.18	1.56
1000/538		1.59		1.69	1.59	1.53			

Table 6-23 Compatibility of Types 304, 304L, and 347 Stainless Steel with Selected Corrodents
The chemicals listed are in the pure state or in a saturated solution unless otherwise indicated. Compatibility is shown to the maximum allowable temperature for which data are available. Incompatibility is shown by an x. A blank space indicates that data are unavailable. When compatible, the corrosion rate is <20 mpy.

Chemical	Maximum temp. °F	Maximum temp. °C	Chemical	Maximum temp. °F	Maximum temp. °C
Acetaldehyde	200	93	Barium carbonate	80	27
Acetamide	100	38	Barium chloride	x	x
Acetic acid 10%	200	93	Barium hydroxide	230	110
Acetic acid 50%	170	77	Barium sulfate	210	99
Acetic acid 80%	170	77	Barium sulfide	210	99
Acetic acid, glacial	210	99	Benzaldehyde	210	99
Acetic anhydride	220	104	Benzene	230	110
Acetone	190	88	Benzene sulfonic acid 10%	210	99
Acetyl chloride	100	38	Benzoic acid	400	204
Acrylic acid	130	54	Benzyl alcohol	90	32
Acrylonitrile	210	99	Benzyl chloride	210	99
Adipic acid	210	99	Borax	150	66
Allyl alcohol	220	104	Boric acid[a]	400	204
Allyl chloride	120	49	Bromine gas, dry	x	x
Alum	x	x	Bromine gas, moist	x	x
Aluminum acetate	210	99	Bromine liquid	x	x
Aluminum chloride, aqueous	x	x	Butadiene	180	82
Aluminum chloride, dry	150	66	Butyl acetate	80	27
Aluminum fluoride	x	x	Butyl alcohol	200	93
Aluminum hydroxide	80	27	n-Butylamine		
Aluminum nitrate	80	27	Butyl phthalate	210	99
Aluminum oxychloride			Butyric acid	180	82
Aluminum sulfate[a]	210	99	Calcium bisulfide		
Ammonia gas	90	32	Calcium bisulfite[c]	300	149
Ammonium bifluoride			Calcium carbonate	210	99
Ammonium carbonate	200	93	Calcium chlorate 10%	210	99
Ammonium chloride 10%	230	110	Calcium chloride[a,b]	80	27
Ammonium chloride 50%	x	x	Calcium hydroxide 10%	210	99
Ammonium chloride, sat.	x	x	Calcium hydroxide, sat.	200	93
Ammonium fluoride 10%	x	x	Calcium hypochlorite	x	x
Ammonium fluoride 25%	x	x	Calcium nitrate	90	32
Ammonium hydroxide 25%	230	110	Calcium oxide	90	32
Ammonium hydroxide, sat.	210	99	Calcium sulfate	210	99
Ammonium nitrate[b]	210	99	Caprylic acid[a]	210	99
Ammonium persulfate	x	x	Carbon bisulfide	210	99
Ammonium phosphate 40%	130	54	Carbon dioxide, dry	210	99
Ammonium sulfate 10–40%	x	x	Carbon dioxide, wet	200	93
Ammonium sulfide	210	99	Carbon disulfide	210	99
Ammonium sulfite	210	99	Carbon monoxide	570	299
Amyl acetate	300	149	Carbon tetrachloride	210	99
Amyl alcohol	80	27	Carbonic acid	210	99
Amyl chloride	150	66	Cellosolve	210	99
Aniline	500	260	Chloracetic acid, 50% water	x	x
Antimony trichloride	x	x	Chloracetic acid	x	x
Aqua regia 3:1	x	x	Chlorine gas, dry	x	x

280

Table 6-23 *Continued*

Chemical	°F	°C	Chemical	°F	°C
Chlorine gas, wet	x	x	Magnesium chloride	x	x
Chlorine, liquid[a]	110	43	Malic acid 50%	120	49
Chlorobenzene	210	99	Manganese chloride	x	x
Chloroform[b]	210	99	Methyl chloride[a]	210	99
Chlorosulfonic acid	x	x	Methyl ethyl ketone	200	93
Chromic acid 10%[a]	200	93	Methyl isobutyl ketone	200	93
Chromic acid 50%	90	32	Muriatic acid	x	x
Chromyl chloride	210	99	Nitric acid 5%	210	99
Citric acid 15%	210	99	Nitric acid 20%	190	88
Citric acid, concentrated[a]	80	27	Nitric acid 70%	170	77
Copper acetate	210	99	Nitric acid, anhydrous	80	27
Copper carbonate 10%	80	27	Nitrous acid, concentrated	80	27
Copper chloride	x	x	Oleum	100	38
Copper cyanide	210	99	Perchloric acid 10%	x	x
Copper sulfate[c]	210	99	Perchloric acid 70%	x	x
Cresol	160	71	Phenol[a]	560	293
Cupric chloride 5%	x	x	Phosphoric acid 50–80%[c]	120	49
Cupric chloride 50%	x	x	Picric acid[a]	300	149
Cyclohexane	100	38	Potassium bromide 30%	210	99
Cyclohexanol	80	27	Salicylic acid	210	99
Dichloroacetic acid			Silver bromide 10%	x	x
Dichloroethane (ethylene di-chloride)	210	99	Sodium carbonate 30%	210	99
			Sodium chloride to 30%[a]	210	99
Ethylene glycol	210	99	Sodium hydroxide 10%	210	99
Ferric chloride	x	x	Sodium hydroxide 50%	210	99
Ferric chloride 50% in water	x	x	Sodium hydroxide, con-centrated	90	32
Ferric nitrate 10–50%	210	99	Sodium hypochlorite 20%	x	x
Ferrous chloride	x	x	Sodium hypochlorite, con-centrated	x	x
Ferrous nitrate					
Fluorine gas, dry	470	243	Sodium sulfide to 50%[a]	210	99
Fluorine gas, moist	x	x	Stannic chloride	x	x
Hydrobromic acid, dilute	x	x	Stannous chloride	x	x
Hydrobromic acid 20%	x	x	Sulfuric acid 10%	x	x
Hydrobromic acid 50%	x	x	Sulfuric acid 50%	x	x
Hydrochloric acid 20%	x	x	Sulfuric acid 70%	x	x
Hydrochloric acid 38%	x	x	Sulfuric acid 90%[c]	80	27
Hydrocyanic acid 10%	210	99	Sulfuric acid 98%[c]	80	27
Hydrofluoric acid 30%	x	x	Sulfuric acid 100%[c]	80	27
Hydrofluoric acid 70%	x	x	Sulfuric acid, fuming	90	32
Hydrofluoric acid 100%	x	x	Sulfurous acid	x	x
Hypochlorous acid	x	x	Thionyl chloride	x	x
Iodine solution 10%	x	x	Toluene	210	99
Ketones, general	200	93	Trichloroacetic acid	x	x
Lactic acid 25%[a,c]	120	49	White liquor	100	38
Lactic acid, concentrated[a,c]	80	27	Zinc chloride	x	x

[a]Subject to pitting.
[b]Subject to stress cracking.
[c]Subject to intergranular attack (type 304).
[d]Subject to stress cracking when wet.
Source: Schweitzer, Philip A. (1991). *Corrosion Resistance Tables,* Marcel Dekker, Inc., New York, Vols. 1 and 2.

Table 6-24   Mechanical and Physical Properties of Type 316 and 316L Stainless Steels

Property	Alloy type	
	316	316L
Modulus of elasticity $\times 10^6$, psi	28	28
Tensile strength $\times 10^3$, psi	75	70
Yield strength 0.2% offset $\times 10^3$, psi	30	25
Elongation in 2 in., %	50	50
Hardness, Rockwell	B-80	B-80
Density, lb/in.^{3}	0.286	0.286
Specific gravity	7.95	7.95
Specific heat (32–212°F), Btu/lb°F	0.12	0.12
Thermal conductivity, Btu/hr ft^2 °F		
at 70°F	9.3	9.3
at 1500°F	12.4	12.4
Thermal expansion coefficient (32–212°F) $\times 10^{-6}$, in./in.°F	8.9	8.9
Izod impact, ft-lb	110	110

Table 6-25   Maximum Allowable Design Stresses for Type 316 and 316L Stainless Steel Pipe

Temp (°F/°C)	Allowable design stress (psi)	
	316	316L
−325 to 100/ −198 to 38	18750	15600
200/93	18750	15600
300/149	17900	14500
400/204	17500	12000
500/260	17200	11000
600/316	17100	10150
650/343	17050	9800
700/371	17000	9450
750/399	16900	9100
800/427	16750	8800
850/450	16500	
900/482	16000	
950/510	15100	
1000/538	14000	
1050/560	12200	
1100/593	10400	
1150/621	8500	
1200/649	6800	
1250/677	5300	
1300/704	4000	
1350/732	3000	
1400/760	2350	
1450/788	1850	
1500/816	1500	

Table 6-26 Compatibility of Types 316, 316L Stainless Steels with Selected Corrodents
The chemicals listed are in the pure state or in a saturated solution unless otherwise indicated. Compatibility is shown to the maximum allowable temperature for which data are available. Incompatibility is shown by an x. A blank space indicates that data are unavailable. When compatible, the corrosion rate is <20 mpy.

Chemical	Maximum temp.		Chemical	Maximum temp.	
	°F	°C		°F	°C
Acetaldehyde	210	99	Barium carbonate	80	27
Acetamide	340	171	Barium chloride[b]	210	99
Acetic acid 10%	420	216	Barium hydroxide	400	204
Acetic acid 50%	400	204	Barium sulfate	210	99
Acetic acid 80%	230	110	Barium sulfide	210	99
Acetic acid, glacial	400	204	Benzaldehyde	400	204
Acetic anhydride	380	193	Benzene	400	204
Acetone	400	204	Benzene sulfonic acid 10%	210	99
Acetyl chloride	400	204	Benzoic acid	400	204
Acrylic acid	120	49	Benzyl alcohol	400	204
Acrylonitrile	210	99	Benzyl chloride	210	99
Adipic acid	210	99	Borax	400	204
Allyl alcohol	400	204	Boric acid	400	204
Allyl chloride	100	38	Bromine gas, dry	x	x
Alum	200	93	Bromine gas, moist	x	x
Aluminum acetate	200	93	Bromine liquid	x	x
Aluminum chloride, aqueous	x	x	Butadiene	400	204
Aluminum chloride, dry	150	66	Butyl acetate	380	193
Aluminum fluoride	90	32	Butyl alcohol	400	204
Aluminum hydroxide	400	204	n-Butylamine	400	204
Aluminum nitrate	200	93	Butyl phthalate	210	99
Aluminum oxychloride			Butyric acid	400	204
Aluminum sulfate[a]	210	99	Calcium bisulfide	60	16
Ammonia gas	90	32	Calcium bisulfite	350	177
Ammonium bifluoride 10%	90	32	Calcium carbonate	205	96
Ammonium carbonate	400	204	Calcium chlorate		
Ammonium chloride 10%	230	110	Calcium chloride[a]	210	99
Ammonium chloride 50%	x	x	Calcium hydroxide 10%	210	99
Ammonium chloride, sat.	x	x	Calcium hydroxide, sat.		
Ammonium fluoride 10%	90	32	Calcium hypochlorite	80	27
Ammonium fluoride 25%	x	x	Calcium nitrate	350	177
Ammonium hydroxide 25%	230	110	Calcium oxide	80	27
Ammonium hydroxide, sat.	210	99	Calcium sulfate	210	99
Ammonium nitrate[a]	300	149	Caprylic acid	400	204
Ammonium persulfate 10%	360	182	Carbon bisulfide	400	204
Ammonium phosphate 40%	130	54	Carbon dioxide, dry	570	299
Ammonium sulfate 10–40%	400	204	Carbon dioxide, wet	200	93
Ammonium sulfide	390	171	Carbon disulfide	400	204
Ammonium sulfite	210	99	Carbon monoxide	570	299
Amyl acetate	300	149	Carbon tetrachloride[a,b]	400	204
Amyl alcohol	400	204	Carbonic acid	350	177
Amyl chloride	150	66	Cellosolve	400	204
Aniline	500	260	Chloracetic acid, 50% water	x	x
Antimony trichloride	x	x	Chloracetic acid	x	x
Aqua regia 3:1	x	x	Chlorine gas, dry	400	204

283

Table 6-26 *Continued*

Chemical	°F	°C	Chemical	°F	°C
Chlorine gas, wet	x	x	Magnesium chloride 50%[a,b]	210	99
Chlorine, liquid dry	120	49	Malic acid	250	121
Chlorobenzene, ELC only	260	127	Manganese chloride 30%	210	99
Chloroform[a]	210	99	Methyl chloride, dry	350	177
Chlorosulfonic acid	x	x	Methyl ethyl ketone	330	166
Chromic acid 10%[c]	400	204	Methyl isobutyl ketone	350	177
Chromic acid 50%[c]	150	49	Muriatic acid	x	x
Chromyl chloride	210	99	Nitric acid 5%[d]	210	99
Citric acid 15%[b]	200	93	Nitric acid 20%[d]	270	132
Citric acid, concentrated[b]	380	193	Nitric acid 70%[d]	400	204
Copper acetate	210	99	Nitric acid, anhydrous[d]	110	43
Copper carbonate 10%	80	27	Nitrous acid, concentrated	80	27
Copper chloride	x	x	Oleum	80	27
Copper cyanide	210	99	Perchloric acid 10%	x	x
Copper sulfate	400	204	Perchloric acid 70%	x	x
Cresol	100	38	Phenol	570	299
Cupric chloride 5%	x	x	Phosphoric acid 50–80%[d]	400	204
Cupric chloride 50%	x	x	Picric acid	400	204
Cyclohexane	400	204	Potassium bromide 30%[b]	350	177
Cyclohexanol	80	27	Salicylic acid	350	177
Dichloroacetic acid			Silver bromide 10%	x	x
Dichloroethane (ethylene di-chloride)	400	204	Sodium carbonate	350	177
			Sodium chloride to 30%[a]	350	177
Ethylene glycol	340	171	Sodium hydroxide 10%	350	177
Ferric chloride	x	x	Sodium hydroxide 50%[a]	350	177
Ferric chloride 50% in water	x	x	Sodium hydroxide, con-centrated	350	177
Ferric nitrate 10–50%	350	177	Sodium hypochlorite 20%	x	x
Ferrous chloride	x	x	Sodium hypochlorite, con-centrated	x	x
Ferrous nitrate					
Fluorine gas, dry	420	216	Sodium sulfide to 50%	190	88
Fluorine gas, moist	x	x	Stannic chloride	x	x
Hydrobromic acid, dilute	x	x	Stannous chloride 10%	210	99
Hydrobromic acid 20%	x	x	Sulfuric acid 10%	x	x
Hydrobromic acid 50%	x	x	Sulfuric acid 50%	x	x
Hydrochloric acid 20%	x	x	Sulfuric acid 70%	x	x
Hydrochloric acid 38%	x	x	Sulfuric acid 90%[d]	80	27
Hydrocyanic acid 10%	210	99	Sulfuric acid 98%[d]	210	99
Hydrofluoric acid 30%	x	x	Sulfuric acid 100%[d]	210	99
Hydrofluoric acid 70%	x	x	Sulfuric acid, fuming	210	99
Hydrofluoric acid 100%	80	27	Sulfurous acid[d]	150	66
Hypochlorous acid	x	x	Thionyl chloride	x	x
Iodine solution 10%	x	x	Toluene	350	177
Ketones, general	250	121	Trichloroacetic acid	x	x
Lactic acid 25%	210	99	White liquor	100	38
Lactic acid, concentrated[b,d]	300	149	Zinc chloride	200	93

[a]Subject to stress cracking.
[b]Subject to pitting.
[c]Subject to crevice attack.
[d]Subject to intergranular corrosion.
Source: Schweitzer, Philip A. (1991). *Corrosion Resistance Tables,* Marcel Dekker, Inc., New York, Vols. 1 and 2.

Table 6-27 Mechanical and Physical Properties of Types 317 and 317-L Stainless Steel

Property	Alloy type	
	317	317-L
Modulus of elasticity $\times 10^6$, psi	28.0	28.0
Tensile strength $\times 10^3$, psi	75	75
Yield strength 0.2% offset $\times 10^3$, psi	30	30
Elongation in 2 in., %	35	35
Hardness, Rockwell	B-85	B-85
Density, lb/in.^{3}	0.286	0.286
Specific gravity		
Specific heat (32–212°F), Btu/lb°F	0.12	0.12
Thermal conductivity, Btu/hr ft^2 °F		
at 70°F	9.3	9.3
at 1500°F	12.4	12.4
Thermal expansion coefficient (32–212°F) $\times 10^{-6}$, in./in. °F	9.2	9.2
Izod impact, ft-lb	110	110

Table 6-28 Maximum Allowable Design Stress for Type 317 Stainless Steel Pipe

Temp (°F/°C)	Allowable stress (psi)
−325 to 100/ −198 to 38	18750
200/93	18750
300/149	17900
400/204	17500
500/260	17200
600/316	17100
650/343	17050
700/371	17000
750/399	16900
800/427	16750
850/450	16500
900/482	16000
950/516	15100
1000/538	14000
1050/560	12200
1100/593	10400
1150/621	8500
1200/649	6800
1250/677	5600
1300/704	4000
1350/732	3000
1400/760	2350
1450/788	1850
1500/816	1500

Table 6-29 Compatibility of Types 317 and 317-L Stainless Steel with Selected Corrodents The chemicals listed are in the pure state or in a saturated solution unless otherwise indicated. Compatibility is shown to the maximum allowable temperature for which data is available. Incompatibility is shown by an X. A blank space indicates that data is unavailable. When compatible the corrosion rate is <20 MPY.

Chemical	Maximum temp. (°F/°C)	Chemical	Maximum temp. (°F/°C)
Acetaldehyde	150/66	Ferric chloride	70/21
Acetic acid 10%	232/111	Hydrochloric acid 20%	x
Acetic acid 50%	232/111	Hydrochloric acid 38%	x
Acetic acid 80%	240/116	Hydrofluoric acid 30%	x
Acetic acid, glacial	240/116	Hydrofluoric acid 70%	x
Acetic anhydride	70/21	Hydrofluoric acid 100%	x
Acetone	70/21	Iodine solution 10%	70/21
Aluminum chloride, aqueous	x	Lactic acid 25%	70/21
Aluminum chloride, dry	x	Lactic acid, concentrated	330/166
Aluminum sulfate 50–55%	225/107	Magnesium chloride	70/21
Ammonium nitrate 66%	70/21	Nitric acid 5%	70/21
Ammonium phosphate	80/27	Nitric acid 20%	210/99
Ammonium sulfate 10–40%	100/38	Nitric acid 70%	210/99
Benzene	100/38	Phenol	70/21
Boric acid	210/99	Phosphoric acid 50–80%	140/60
Bromine gas, dry	x	Sodium carbonate	210/99
Bromine gas, moist	x	Sodium chloride	x
Bromine liquid	x	Sodium hydroxide 10%	210/99
Butyl alcohol 5%	195/91	Sodium hydroxide 50%	70/21
Calcium chloride	210/99	Sodium hydochlorite 20%	70/21
Calcium Hypochlorite	70/21	Sodium hypochlorite, concentrated	70/21
Carbon tetrachloride	70/21		
Carbonic acid	70/21	Sodium sulfide to 50%	210/99
Chloracetic acid 78%	122/50	Sulfuric acid 10%	120/49
Chlorine, liquid	x	Sulfuric acid 50%	x
Chlorobenzene	265/129	Sulfuric acid 70%	x
Chromic acid 10%	x	Sulfuric acid 90%	x
Chromic acid 50%	x	Sulfuric acid 98%	x
Citric acid 15%	210/99	Sulfuric acid 100%	x
Citric acid, concentrated	210/99	Sulfurous acid	x
Copper sulfate	70/21		

Table 6-30 Mechanical and Physical Properties of Type 321 Stainless Steel

Modulus of elasticity × 10^6, psi	29
Tensile strength × 10^3, psi	75
Yield strength 0.2% offset × 10^3, psi	30
Elongation in 2 in., %	35
Hardness, Rockwell	B-85
Density, lb/in.^{3}	0.286
Specific gravity	7.92
Specific heat (32–212°F), Btu/lb°F	0.12
Thermal conductivity, Btu/hr ft^2 °F	
at 70°F	9.3
at 1500°F	12.8
Thermal expansion coefficient (312–212°F) × 10^{-6}, in./in. °F	9.3
Izod impact, ft-lb	110

Table 6-31 Maximum Allowable Design Stress for
Type 321 Stainless Steel Pipe

Temperature (°F/°C)	Allowable design stress (psi)
−325 to 100/−198 to 38	18750
200/93	18750
300/149	17000
400/204	15800
500/260	15200
600/316	14900
700/371	14800
750/399	14700
800/427	14550
850/450	14300
900/482	14100
950/510	13850
1000/538	13500

Table 6-32 Compatibility of Type 321 Stainless Steel with Selected Corrodents
The chemicals listed are in the pure state or in a saturated solution unless otherwise indicated.
Compatibility is shown to the maximum allowable temperature for which data are available.
Incompatibility is shown by an x. A blank space indicates that data are unavailable. When
compatible, the corrosion rate is <20 mpy.

Chemical	Maximum temp. (°F/°C)	Chemical	Maximum temp. (°F/°C)
Acetic acid 10%	x	Ferric chloride	x
Acetic acid 50%	x	Hydrochloric acid 20%	x
Acetic acid 80%	x	Hydrochloric acid 38%	x
Acetic acid, glacial	x	Hydrofluoric acid 30%	x
Acetic anhydride	70/21	Hydrofluoric acid 70%	x
Alum	x	Hydrofluoric acid 100%	x
Aluminum chloride, aqueous	x	Iodine solution 10%	x
Aluminum chloride, dry	x	Lactic acid 25%	70/21
Aluminum sulfate	70/21	Lactic acid, concentrated	70/21
Ammonium phosphate	70/21	Magnesium chloride	x
Ammonium sulfate 10–40%	70/21	Nitric acid 5%	70/21
Benzene	100/38	Nitric acid 20%	210/99
Boric acid	210/99	Nitric acid 70%	210/99
Bromine gas, dry	x	Phenol	x
Bromine gas, moist	x	Phosphoric acid 50–80%	70/21
Bromine liquid	x	Sodium carbonate	70/21
Calcium chloride	x	Sodium chloride	x
Calcium hypochlorite	x	Sodium hydroxide 10%	70/21
Carbon tetrachloride	x	Sodium hydroxide 50%	70/21
Carbonic acid	70/21	Sodium hypochlorite 20%	x
Chloracetic acid	x	Sodium hypochlorite, concentrated	x
Chlorine, liquid	x		
Chromic acid 10%	x	Sodium sulfide to 50%	70/21
Chromic acid 50%	x	Sulfuric acid 98%	x
Citric acid 15%	70/21	Sulfuric acid 100%	x
Citric acid, concentrated	70/21	Sulfurous acid	x
Copper sulfate	70/21		

Table 6-33 Mechanical and Physical Properties of Type 347 Stainless Steel

Modulus of elasticity $\times$ 10^6, psi	29.0
Tensile strength $\times$ 10^3, psi	75
Yield strength 0.2% offset $\times$ 10^3, psi	30
Elongation in 2 in., %	35
Hardness, Rockwell	B-85
Density, lb/in.^{3}	0.285
Specific gravity	7.92
Specific heat (32–212°F), Btu/lb°F	
Thermal conductivity, Btu/hr ft^2 °F	
at 70–212°F	9.3
at 1500°F	12.8
Thermal expansion coefficient (32–212°F) $\times$ 10^{-6}, in./in. °F	9.3
Izod impact, ft-lb	110

Table 6-34 Maximum Allowable Design Stress for Type 347 Stainless Steel Pipe

Temperature (°F/°C)	Allowable design stress (psi)
–325 to 100/–198 to 38	18750
200/93	18750
300/149	17000
400/204	15800
500/260	15200
600/316	14900
650/343	14850
700/371	14800
750/399	14700
800/427	14550
850/450	14300
900/482	14100
950/510	13850
1000/538	13500

Table 6-35 Mechanical and Physical Properties of 20Cb-3 Stainless Steel

Modulus of elasticity $\times 10^6$, psi	28
Tensile strength $\times 10^3$, psi	91
Yield strength 0.2% offset $\times 10^3$, psi	45
Elongation in 2 in., %	45
Hardness, Rockwell	B-86
Density, lb/in.^{3}	0.292
Specific gravity	8.08
Specific heat, Btu/lb °F	0.12
Thermal conductivity, Btu/hr ft^2 °F	
at 122°F	7.05
at 212°F	7.57
at 392°F	8.50
at 572°F	9.53
752°F	10.50
Thermal expansion coefficient, in./in. °F	
at 77–212°F	8.16×10^{-6}
at 77–842°F	8.84×10^{-6}
at 77–1652°F	9.53×10^{-6}
Charpy V-notch impact strength, ft-lb	200

Table 6-36 Elevated Temperature Tensile Properties of 20Cb-3 Stainless Steel

Temperature (°F/°C)	0.2% yield strength (ksi)	Ultimate tensile strength (ksi)
Room	45	91
200/93	40	86
400/204	35	83
600/316	33	80
800/427	30	79
1000/538	28	77
1400/760	26	45
1600/871	19	29

Table 6-37 Maximum Allowable Design Stress Values for 20Cb3 Pipe

Temperature (°F/°C)	Stress value (psi)
100/38	14450
200/93	14450
300/199	14280
400/204	13515
500/260	13175
600/316	12835
650/343	12665
700/371	12495
750/399	12325
800/427	12155

Table 6-38 Compatibility of Types 20Cb3 Stainless Steel with Selected Corrodents
The chemicals listed are in the pure state or in a saturated solution unless otherwise indicated.
Compatibility is shown to the maximum allowable temperature for which data are available.
Incompatibility is shown by an x. A blank space indicates that data are unavailable. When
compatible, the corrosion rate is <20 mpy.

Chemical	Maximum temp. °F	Maximum temp. °C	Chemical	Maximum temp. °F	Maximum temp. °C
Acetaldehyde	200	93	Barium carbonate	90	32
Acetamide	60	16	Barium chloride 40%	210	99
Acetic acid 10%	220	104	Barium hydroxide 50%	230	110
Acetic acid 50%	300	149	Barium sulfate	210	99
Acetic acid 80%	300	149	Barium sulfide	210	99
Acetic acid, glacial	300	149	Benzaldehyde	210	99
Acetic anhydride	180	82	Benzene	230	110
Acetone	220	104	Benzene sulfonic acid 10%	210	99
Acetyl chloride	210	99	Benzoic acid	400	204
Acrylic acid			Benzyl alcohol	210	99
Acrylonitrile	210	99	Benzyl chloride	230	110
Adipic acid	210	99	Borax	100	38
Allyl alcohol	300	149	Boric acid	130	54
Allyl chloride	200	93	Bromine gas, dry	80	27
Alum	200	93	Bromine gas, moist	x	x
Aluminum acetate	60	16	Bromine liquid		
Aluminum chloride, aqueous	120	43	Butadiene	180	82
Aluminum chloride, dry	120	43	Butyl acetate	300	149
Aluminum fluoride	x	x	Butyl alcohol	90	32
Aluminum hydroxide	80	27	n-Butylamine		
Aluminum nitrate	80	27	Butyl phthalate	210	99
Aluminum oxychloride			Butyric acid	300	149
Aluminum sulfate	210	99	Calcium bisulfide		
Ammonia gas	90	32	Calcium bisulfite	300	149
Ammonium bifluoride	90	32	Calcium carbonate	210	99
Ammonium carbonate	310	154	Calcium chlorate	90	32
Ammonium chloride 10%	230	110	Calcium chloride	210	99
Ammonium chloride 50%	170	77	Calcium hydroxide 10%	210	99
Ammonium chloride, sat.[a]	210	99	Calcium hydroxide, sat.	210	99
Ammonium fluoride 10%	90	32	Calcium hypochlorite	90	32
Ammonium fluoride 25%	90	32	Calcium nitrate		
Ammonium hydroxide 25%	90	32	Calcium oxide	80	27
Ammonium hydroxide, sat.	210	99	Calcium sulfate	210	99
Ammonium nitrate[b]	210	99	Caprylic acid	400	204
Ammonium persulfate	210	99	Carbon bisulfide	210	99
Ammonium phosphate	210	99	Carbon dioxide, dry	570	299
Ammonium sulfate 10–40%	210	99	Carbon dioxide, wet	400	204
Ammonium sulfide	210	99	Carbon disulfide	210	99
Ammonium sulfite	210	99	Carbon monoxide	570	299
Amyl acetate	310	154	Carbon tetrachloride	210	99
Amyl alcohol	160	71	Carbonic acid	570	299
Amyl chloride	130	54	Cellosolve	210	99
Aniline	500	260	Chloracetic acid, 50% water		
Antimony trichloride	200	93	Chloracetic acid	80	27
Aqua regia 3 : 1	x	x	Chlorine gas, dry	400	204

Table 6-38 *Continued*

Chemical	°F	°C	Chemical	°F	°C
Chlorine gas, wet	x	x			
Chlorine, liquid			Magnesium chloride	200	93
Chlorobenzene, dry	100	38	Malic acid 50%	160	71
Chloroform	210	99	Manganese chloride 40%	210	99
Chlorosulfonic acid	130	54	Methyl chloride	210	99
Chromic acid 10%	130	54	Methyl ethyl ketone	200	93
Chromic acid 50%	140	60	Methyl isobutyl ketone	210	99
Chromyl chloride	210	99	Muriatic acid	x	x
Citric acid 15%	210	99	Nitric acid 5%	210	99
Citric acid, concentrated	210	99	Nitric acid 20%	210	99
Copper acetate	100	38	Nitric acid 70%	210	99
Copper carbonate	90	32	Nitric acid, anhydrous	80	27
Copper chloride	x	x	Nitrous acid, concentrated	90	32
Copper cyanide	210	99	Oleum	110	43
Copper sulfate	210	99	Perchloric acid 10%	100	38
Cresol			Perchloric acid 70%	110	43
Cupric chloride 5%	60	16	Phenol	570	299
Cupric chloride 50%	x	x	Phosphoric acid 50–80%	210	99
Cyclohexane	200	93	Picric acid	300	149
Cyclohexanol	80	27	Potassium bromide 30%	210	99
Dichloroacetic acid			Salicylic acid	210	99
Dichloroethane (ethylene di-chloride)	210	99	Silver bromide 10%	90	32
			Sodium carbonate	570	299
Ethylene glycol	210	99	Sodium chloride to 30%[a]	210	99
Ferric chloride	x	x	Sodium hydroxide 10%	300	149
Ferric chloride 50% in water	x	x	Sodium hydroxide 50%[b]	300	149
Ferric nitrate 10–50%	210	99	Sodium hydroxide, con-centrated	200	93
Ferrous chloride	x	x	Sodium hypochlorite 30%	90	32
Ferrous nitrate			Sodium hypochlorite, con-centrated		
Fluorine gas, dry	570	299			
Fluorine gas, moist	x	x	Sodium sulfide to 50%	200	93
Hydrobromic acid, dilute	x	x	Stannic chloride	x	x
Hydrobromic acid 20%	x	x	Stannous chloride 10%	90	32
Hydrobromic acid 50%	x	x	Sulfuric acid 10%	200	93
Hydrochloric acid 20%	x	x	Sulfuric acid 50%	110	43
Hydrochloric acid 38%	x	x	Sulfuric acid 70%	120	49
Hydrocyanic acid 10%	210	99	Sulfuric acid 90%	100	38
Hydrofluoric acid 30%	190	88	Sulfuric acid 98%	300	149
Hydrofluoric acid 70%	x	x	Sulfuric acid 100%	300	149
Hydrofluoric acid 100%	80	27	Sulfuric acid, fuming	210	99
Hypochlorous acid			Sulfurous acid	360	182
Iodine solution 10%	x	x	Thionyl chloride		
Ketones, general	100	38	Toluene	210	99
Lactic acid 25%[a]	210	49	Trichloroacetic acid		
Lactic acid, concentrated, air free	300	149	White liquor	100	38
			Zinc chloride	210	99

[a]Material subject to intergranular corrosion.
[b]Material subject to stress cracking.
Source: Schweitzer, Philip A. (1991). *Corrosion Resistance Tables,* Marcel Dekker, Inc., New York, Vols. 1 and 2.

Table 6-39 Mechanical and Physical Properties of 904-L Stainless Steel

Modulus of elasticity $\times 10^6$, psi	28.5
Tensile strength $\times 10^3$, psi	75
Yield strength 0.2% offset $\times 10^3$, psi	32
Elongation in 2 in., %	35
Hardness, Rockwell	B-90
Density, lb/in.^{3}	0.29
Specific gravity	7.9
Specific heat, Btu/lb°F	0.105
Thermal conductivity, Btu-in./ft^2 hr °F	
at 68°F	90
at 200°F	94
at 400°F	97
at 600°F	105
at 800°F	113
at 1000°F	121
Thermal expansion coefficient, in./in.°F $\times 10^{-6}$	
at 200°F	8.4
at 600°F	8.8
at 800°F	9.0
at 1000°F	9.2
at 1500°F	10.0
at 1800°F	10.3
Charpy V-notch impact, ft-lb	125

Table 6-40 Mechanical and Physical Properties of 20Mo-6 Stainless Steel

Modulus of elasticity $\times 10^6$, psi	27.0
Tensile strength $\times 10^3$, psi	88
Yield strength 0.2% offset $\times 10^3$ psi	45
Elongation in 2 in., %	50
Hardness, Rockwell	B-85
Density, lb/in.^{3}	0.294
Specific gravity	8.133
Specific heat, Btu/hr °F (32–12°F)	0.11
Thermal conductivity, Btu/ft hr °F	
at 122°F	6.99
at 212°F	7.51
at 392°F	8.55
at 572°F	9.53
at 752°F	10.52
Thermal expansion coefficient, in./in. °F $\times 10^{-6}$	
at 77–212°F	8.22
at 77–392°F	8.29
at 77–752°F	8.73
at 77–1112°F	9.29
at 77–1652°F	9.86

Table 6-41 Maximum Allowable
Design Stress Values for 20Mo-6
Stainless Steel

Temperature (°F/°C)	Allowable stress (psi)
100/38	17000
200/93	17000
300/149	16100
400/204	14900
500/260	13800
600/316	13000
700/371	12400
800/427	11800

Table 6-42 Mechanical and Physical Properties of AL-6XN[a] Stainless
Steel Pipe

Modulus of elasticity $\times 10^6$, psi	27
Tensile strength $\times 10^3$, psi	107.5
Yield strength 0.2% offset $\times 10^3$, psi	54.7
Elongation in 2 in., %	50
Density, lb/in.^{3}	0.291
Specific gravity	8.06
Hardness, Rockwell	B-90
Thermal conductivity at 68–212°F, Btu/hr °F	7.9
Thermal expansion coefficient, in./in. °F $\times 10^{-6}$	
at 68–212°F	8.5
at 68–932°F	8.9
at 68–1472°F	10.0
Izod impact ft-lb	140

[a]Registered trademark of Allegheny Ludlum Corporation.

Table 6-43 Maximum Allowable
Design Stress Values for AL-6XN
Stainless Steel Pipe

Temperature (°F/°C)	Allowable stress values (psi)
200/93	22100
300/149	20600
400/204	19200
500/260	17700
600/316	16900
700/371	16400
750/399	15800
800/427	15600

Table 6-44 Compatibility of AL-6XN Stainless Steel with Selected Corrodents
Compatibility is shown to the maximum allowable temperature for which data are available. Incompatibility is shown by an x. When compatible, the corrosion rate is <20 mpy

Chemical	Maximum temperature (°F/°C)
Acetic acid 20%	210/99
Acetic acid 80%	217/103
Formic acid 45%	220/104
Formic acid 50%	220/104
Nitric acid 10%	194/90
Nitric acid 65%	241/116
Oxalic acid 10%	210/99
Phosphoric acid 20%	210/99
Phosphoric acid 85%	158/76
Sulfamic acid 10%	210/99
Sulfuric acid 10%	x/x
Sulfuric acid 60%	122/50
Sulfuric acid 95%	86/30
Sodium bisulfate 10%	210/99
Sodium hydroxide 50%	210/99

Table 6-45 Mechanical and Physical Properties of Alloy 350 Stainless Steel

Modulus of elasticity $\times$ 10^6, psi	(aged) 29.4
Tensile strength $\times$ 10^3, psi	(aged) 200
Yield strength 0.2% offset $\times$ 10^3, psi	(aged) 585
Elongation in 2 in., %	(aged) 12
Hardness, Rockwell	(aged) C-30
Density, lb/in.^{3}	0.286
Thermal conductivity, Btu/ft hr °F	
at 70°F (20°C)	8.4
at 1500°F (815°C)	12.2

Table 6-46   Mechanical and Physical Properties of Alloy 410 Stainless Steel

Modulus of elasticity $\times\ 10^6$, psi	29
Tensile strength $\times\ 10^3$, psi	
annealed	75
heat-treated	150
Yield strength 0.2% offset $\times\ 10^3$, psi	
annealed	40
heat-treated	115
Elongation in 2 in., %	
annealed	30
heat-treated	15
Hardness, Brinell	
annealed	150
heat-treated	410
Density, lb/in.^{3}	0.28
Specific gravity	7.75
Specific heat, (32–212°F) Btu/lb °F	0.11
Thermal expansion coefficient (32–212°F), in./in. °F $\times\ 10^{-6}$	173

Table 6-47 Compatibility of Type 410 Stainless Steel with Selected Corrodents
The chemicals listed are in the pure state or in a saturated solution unless otherwise indicated. Compatibility is shown to the maximum allowable temperature for which data are available. Incompatibility is shown by an x. A blank space indicates that data are unavailable. When compatible the corrosion rate is <20 mpy.

Chemical	Maximum temp. °F	Maximum temp. °C	Chemical	Maximum temp. °F	Maximum temp. °C
Acetaldehyde	60	16	Barium carbonate 10%	210	99
Acetamide	60	16	Barium chloride[a]	60	16
Acetic acid 10%	70	21	Barium hydroxide	230	110
Acetic acid 50%	70	21	Barium sulfate	210	99
Acetic acid 80%	70	21	Barium sulfide	70	21
Acetic acid, glacial	x	x	Benzaldehyde		
Acetic anhydride	x	x	Benzene	230	110
Acetone	210	99	Benzene sulfonic acid 10%		
Acetyl chloride			Benzoic acid	210	99
Acrylic acid			Benzyl alcohol	130	54
Acrylonitrile	110	43	Benzyl chloride		
Adipic acid			Borax	150	66
Allyl alcohol	90	27	Boric acid	130	54
Allyl chloride			Bromine gas, dry	x	x
Alum	x	x	Bromine gas, moist	x	x
Aluminum acetate			Bromine liquid	x	x
Aluminum chloride, aqueous	x	x	Butadiene	60	16
Aluminum chloride, dry	150	66	Butyl acetate	90	32
Aluminum fluoride	x	x	Butyl alcohol	60	16
Aluminum hydroxide 10%	60	16	*n*-Butylamine		
Aluminum nitrate	210	99	Butyl phthalate		
Aluminum oxychloride	x	x	Butyric acid	150	66
Aluminum sulfate	x	x	Calcium bisulfide		
Ammonia gas			Calcium bisulfite	x	x
Ammonium bifluoride	x	x	Calcium carbonate	210	99
Ammonium carbonate	210	99	Calcium chlorate		
Ammonium chloride 10%[a]	230	110	Calcium chloride[a]	150	66
Ammonium chloride 50%	x	x	Calcium hydroxide 10%	210	99
Ammonium chloride, sat.	x	x	Calcium hydroxide, sat.		
Ammonium fluoride 10%			Calcium hypochlorite	x	x
Ammonium fluoride 25%			Calcium nitrate		
Ammonium hydroxide 25%			Calcium oxide		
Ammonium hydroxide, sat.	70	21	Calcium sulfate	210	99
Ammonium nitrate	210	99	Caprylic acid		
Ammonium persulfate 5%	60	16	Carbon bisulfide	60	16
Ammonium phosphate 5%	90	32	Carbon dioxide, dry	570	299
Ammonium sulfate 10–40%	60	16	Carbon dioxide, wet	570	299
Ammonium sulfide			Carbon disulfide	60	16
Ammonium sulfite	x	x	Carbon monoxide	570	299
Amyl acetate[a]	60	16	Carbon tetrachloride[a]	210	99
Amyl alcohol	110	43	Carbonic acid	60	16
Amyl chloride	x	x	Cellosolve		
Aniline	210	99	Chloracetic acid, 50% water		
Antimony trichloride	x	x	Chloracetic acid	x	x
Aqua regia 3:1			Chlorine gas, dry	x	x

296

Table 6-47 *Continued*

Chemical	Maximum temp. °F	Maximum temp. °C	Chemical	Maximum temp. °F	Maximum temp. °C
Chlorine gas, wet	x	x	Magnesium chloride 50%	210	99
Chlorine, liquid	x	x	Malic acid	210	99
Chlorobenzene, dry	60	16	Manganese chloride		
Chloroform	150	66	Methyl chloride, dry	210	99
Chlorosulfonic acid	x	x	Methyl ethyl ketone	60	16
Chromic acid 10%	x	x	Methyl isobutyl ketone		
Chromic acid 50%	x	x	Muriatic acid	x	x
Chromyl chloride			Nitric acid 5%	90	32
Citric acid 15%	210	99	Nitric acid 20%	160	71
Citric acid 50%	140	60	Nitric acid 70%	60	16
Copper acetate	90	32	Nitric acid, anhydrous	x	x
Copper carbonate	80	27	Nitrous acid, concentrated	60	16
Copper chloride	x	x	Oleum		
Copper cyanide	210	99	Perchloric acid 10%	x	x
Copper sulfate	210	99	Perchloric acid 70%	x	x
Cresol			Phenol[a]	210	99
Cupric chloride 5%	x	x	Phosphoric acid 50–80%	x	x
Cupric chloride 50%	x	x	Picric acid	60	16
Cyclohexane	80	27	Potassium bromide 30%	210	99
Cyclohexanol	90	32	Salicylic acid	210	99
Dichloroacetic acid			Silver bromide 10%	x	x
Dichloroethane (ethylene di-chloride)			Sodium carbonate 10–30%	210	99
			Sodium chloride[a]	210	99
Ethylene glycol	210	99	Sodium hydroxide 10%	210	99
Ferric chloride	x	x	Sodium hydroxide 50%	60	16
Ferric chloride 50% in water	x	x	Sodium hydroxide, con-centrated		
Ferric nitrate 10–50%	60	16	Sodium hypochlorite 20%	x	x
Ferrous chloride	x	x	Sodium hypochlorite, con-centrated	x	x
Ferrous nitrate					
Fluorine gas, dry	570	299	Sodium sulfide to 50%	x	x
Fluorine gas, moist	x	x	Stannic chloride	x	x
Hydrobromic acid, dilute	x	x	Stannous chloride	x	x
Hydrobromic acid 20%	x	x	Sulfuric acid 10%	x	x
Hydrobromic acid 50%	x	x	Sulfuric acid 50%	x	x
Hydrochloric acid 20%	x	x	Sulfuric acid 70%	x	x
Hydrochloric acid 38%	x	x	Sulfuric acid 90%	x	x
Hydrocyanic acid 10%	210	99	Sulfuric acid 98%	x	x
Hydrofluoric acid 30%	x	x	Sulfuric acid 100%	x	x
Hydrofluoric acid 70%	x	x	Sulfuric acid, fuming		
Hydrofluoric acid 100%	x	x	Sulfurous acid	x	x
Hypochlorous acid			Thionyl chloride		
Iodine solution 10%			Toluene	210	99
Ketones, general	60	16	Trichloroacetic acid	x	x
Lactic acid 25%	60	16	White liquor		
Lactic acid, concentrated	60	16	Zinc chloride	x	x

[a]Material is subject to pitting.

Source: Schweitzer, Philip A. (1991). *Corrosion Resistance Tables*, Marcel Dekker, Inc., New York, Vols. 1 and 2.

Table 6-48 Mechanical and Physical Properties of Custom 450[a]
Stainless Steel

Modulus of elasticity $\times 10^3$, psi	
annealed	28
aged	29
Tensile strength $\times 10^3$, psi	
annealed	142
aged	196
Yield strength 0.2% offset $\times 10^3$, psi	
annealed	118
aged	188
Elongation in 2 in., %	
annealed	13
aged	14
Hardness, Rockwell	
annealed	C-28
aged	C-42.5
Density, lb/in.^{3}	0.28

[a]Custom 450 is the registered trademark of Carpenter Technology Corp.

Table 6-49 Mechanical and Physical Properties of Custom 455[a]
Stainless Steel

Modulus of elasticity $\times 10^6$, psi	
aged	29
Tensile strength $\times 10^3$, psi	
annealed	140
aged	230
Yield strength 0.2% offset $\times 10^3$, psi	
annealed	115
aged	220
Elongation in 2 in., %	
annealed	12
aged	10
Hardness, Rockwell	
annealed	C-31
aged	C-48
Density, lb/in.^{3}	0.28
Thermal conductivity, Btu/ft hr °F	
at 70°F (20°C)	10.4
at 1500°F (815°C)	14.3

[a]Custom 455 is the registered trademark of Carpenter Technology Corp.

Table 6-50 Mechanical and Physical Properties of Alloy 718 Stainless Steel

Modulus of elasticity $\times$ 10^6, psi	
aged	29
Tensile strength $\times$ 10^3, psi	
annealed	140
aged	180
Yield strength 0.2% offset $\times$ 10^3, psi	
annealed	80
aged	150
Elongation in 2 in., %	
annealed	30
aged	15
Hardness, Rockwell	
annealed	C-20
aged	C-36
Density, lb/in.^{3}	0.296
Thermal conductivity, Btu/ft hr °F	
at 70°F (20°C)	6.6
at 1500°F (815°C)	13.9

Table 6-51 Mechanical and Physical Properties of 17-7PH Stainless Steel

Modulus of elasticity $\times$ 10^6, psi	
annealed	30.5
aged	32.5
Tensile strength $\times$ 10^3, psi	
annealed	133
aged	210
Yield strength 0.2% offset $\times$ 10^3, psi	
annealed	42
aged	190
Elongation in 2 in., %	
annealed	19
Hardness, Rockwell	
annealed	B-85
aged	C-48
Density, lb/in.^{3}	0.282
Thermal conductivity, Btu/ft hr °F	
at 70°F (20°C)	9.75
at 1500°F (815°C)	12.2

Table 6-52 Mechanical and Physical Properties of Type 7 Mo Plus[a] Stainless Steel

Modulus of elasticity $\times$ 10^6, psi	29.0
Tensile strength $\times$ 10^3, psi	90
Yield strength 0.2% offset $\times$ 10^3, psi	70
Elongation in 2 in., %	20
Hardness, Rockwell	C30.5
Density, lb/in.^{3}	0.280
Specific gravity	7.74
Specific heat, (75–212°F) Btu/lb °F	0.114
Thermal conductivity, Btu/hr °F	
at 70°F (20°C)	8.8
at 1500°F (815°C)	12.5
Thermal expansion coefficient, in./in. °F $\times$ 10^{-6}	
at 75–400°F	6.39
at 75–600°F	6.94
at 75–800°F	7.49
at 75–1000°F	7.38
Charpy V-notch impact at 75°F (20°C), ft-lb	101

[a]Registered trademark of Carpenter Technology Corp.

Table 6-53 Mechanical and Physical Properties of Alloy 2205 Duplex Stainless Steel

Modulus of elasticity $\times$ 10^6, psi	29.0
Tensile strength $\times$ 10^3, psi	90
Yield strength 0.2% offset $\times$ 10^3, psi	65
Elongation in 2 in., %	25
Hardness, Rockwell	C30.5
Density, lb/in.^{3}	0.283
Specific gravity	7.83
Thermal conductivity at 70°F (20°C), Btu/hr °F	10
Thermal expansion coefficient at 68–212°F, in./in. °F $\times$ 10^{-6}	7.5

Table 6-54 Maximum Nonshock Ratings of Forged Stainless Steel Flanges

Operating temperature (°F/°C)	Primary service rating					
	150 psi	300 psi				
	304, 304L, 316, 316L, 347	304	304L[a]	316	316L[b]	347
−20 to 100/−29 to 38	275	615	515	720	515	720
150/66	255	585	510	710	515	710
200/93	240	550	505	700	515	700
250/121	225	520	465	690	495	690
300/149	210	495	430	680	475	680
350/177	195	470	395	675	435	675
400/204	180	450	360	665	395	665
450/232	165	430	340	650	380	650
500/260	150	410	320	625	360	625
550/288	140	395	310	590	350	590
600/316	130	380	300	555	335	555
650/343	120	370	290	515	325	515
700/371	110	355	280	495	310	495
750/399	100	340	275	470	300	470
800/427	92	330	265	450	290	450
850/454	82	320		425	280	425
875/468	75	315		415		415
900/482	70	310		400		400
925/496	60	305		390		390
950/510	55	305		380		380
975/524	50	300		370		370
1000/538	40	300		355		355
1025/552		295		345		345
1050/566		290		335		335
1075/579		275		325		325
1100/593		255		310		310
1125/607		225		300		300
1150/621		195		290		260

[a]Maximum temperature rating of 800°F (427°C).
[b]Maximum temperature rating of 850°F (454°C).

Table 6-55   Maximum Nonshock Ratings of Forged Stainless Steel Flanges with Primary Service Rating of 600 psi

Operating temperature (°F/°C)	Type of stainless steel				
	304	304L[a]	316	316L[b]	347
–20 to 100/–29 to 38	1235	1030	1440	1030	1440
150/66	1165	1020	1420	1030	1420
200/93	1095	1005	1400	1030	1400
250/121	1040	935	1380	990	1380
300/149	985	860	1365	955	1365
350/177	945	795	1350	870	1350
400/204	900	725	1330	790	1330
450/232	860	680	1305	755	1305
500/260	825	640	1250	725	1250
550/288	795	615	1180	695	1180
600/316	765	600	1110	670	1110
650/343	735	575	1030	645	1030
700/371	710	560	985	620	985
750/399	685	545	940	600	940
800/427	660	535	895	580	895
850/454	640		850	560	850
875/468	630		825		825
900/482	620		805		805
925/496	615		780		780
950/510	610		760		760
975/524	605		735		735
1000/538	600		715		715
1025/552	595		690		690
1050/566	585		670		670
1075/579	550		645		645
1100/593	515		625		625
1125/607	455		600		600
1150/621	395		585		520

[a]Type 304L has a maximum temperature rating of 800°F (427°C).
[b]Type 316L has a maximum temperature rating of 850°F (454°C).

Table 6-56 Maximum Nonshock Ratings of Forged Stainless Steel Flanges with Primary Service Rating of 900 psi

Operating temperature (°F/°C)	Allowable pressure (psi)				
	304[a]	304L	316[a]	316L	347[a]
−20 to 100/−29 to 38	1850	1545	1850	1545	1850
200/93	1545	1315	1590	1300	1635
300/149	1390	1180	1440	1165	1510
400/204	1275	1080	1370	1060	1425
500/260	1195	1005	1230	980	1335
600/316	1125	955	1160	925	1265
650/343	1105	940	1140	900	1235
700/371	1090	920	1115	885	1205
750/399	1070	905	1100	865	1185
800/427	1035	890	1086	845	1175
850/454	1020	875	1075	825	1140
900/482	1000		1070		1110
950/516	980		1055		1105
1000/538	965		1050		1050
1050/566	965		1050		1050
1100/593	905		1050		935
1150/621	790		955		625
1200/649	615		760		455
1250/677	485		555		340
1300/709	380		430		225
1350/733	300		310		155
1400/760	225		225		125
1450/788	175		175		95
1500/816	125		125		70

[a]At temperatures above 1000°F (538°C), only use when carbon is 0.04% or higher.

Table 6-57 Maximum Nonshock Ratings of Forged Stainless Steel Flanges with Primary Service Rating of 1500 psi

Operating temperature (°F/°C)	Allowable pressure (psi)				
	304[a]	304L	316[a]	316L	347[a]
–20 to 100/–29 to 38	3085	2570	3085	2570	3085
200/93	2570	2190	2655	2170	2725
300/149	2315	1965	2395	1945	2520
400/204	2130	1800	2200	1770	2375
500/260	1995	1675	2045	1635	2220
600/316	1870	1595	1935	1545	2110
650/343	1840	1565	1905	1500	2055
700/371	1820	1535	1860	1470	2005
750/399	1780	1510	1830	1440	1975
800/427	1730	1480	1810	1410	1955
850/454	1695	1460	1790	1380	1905
900/482	1665		1780		1850
950/516	1635		1760		1840
1000/538	1605		1750		1750
1050/566	1605		1750		1750
1100/593	1510		1750		1560
1150/621	1320		1595		1045
1200/649	1030		1270		755
1250/677	805		925		565
1300/709	635		705		375
1350/733	495		515		255
1400/760	375		375		205
1450/788	290		290		155
1500/816	205		205		120

[a]At temperatures above 1000°F (538°C), only use when carbon is 0.04% or higher.

Table 6-58 Maximum Nonshock Ratings of Forged Stainless Steel Flanges with Primary Service Rating of 2500 psi

Operating temperature (°F/°C)	Allowable pressure (psi)				
	304[a]	304L	316[a]	316L	347[a]
−20 to 100/−29 to 38	5145	4285	5145	4285	5145
200/93	4285	3650	4425	3615	4545
300/149	3855	3275	3995	3240	4200
400/204	3550	3000	3670	2950	3960
500/260	3325	2795	3410	2725	3705
600/316	3120	2655	3225	2570	3515
650/343	3070	2605	3170	2505	3430
700/371	3035	2555	3105	2450	3345
750/399	2965	2520	3050	2400	3290
800/427	2880	2470	3015	2350	3255
850/454	2830	2435	2985	2293	3170
900/482	2775		2965		3085
950/516	2725		2930		3070
1000/538	2675		2915		2915
1050/566	2675		2915		2915
1100/593	2515		2915		2600
1150/621	2200		2655		1745
1200/649	1715		2115		1255
1250/677	1345		1545		945
1300/709	1055		1170		630
1350/733	830		853		430
1400/760	630		630		345
1450/788	485		485		255
1500/816	345		345		200

[a]At temperatures over 1000°F (538°C), only use when carbon is 0.04% or higher.

Table 6-59 Representative Pressure Ratings of Van Stone Flanges Types 304 and 316L

Temperature (°F/°C)	Schedule 10S (psi)	Schedule 40S (psi)
100/38	275	300
200/93	240	300
300/149	210	300
400/204	180	300
500/260	150	300
600/316	130	300
650/343	120	300

Table 6-60 Filler Metals Suggested for Welding
Stainless Steel Pipe

Type of stainless pipe to be welded	Electrode or filler rod material to be used
430	308
444	In-82, 316L
XM-27	XM-27
304	308
304L	347 or 308L
316	310
316L	316Cb or 316L
317	317
317L	317Cb
321	347
347	347
20Cb-3	ER 320LR
904L	Alloy 625
20Mo-6	Alloy 625
A1-6XN	Alloy 625

Table 6-61 Support Spacing for Stainless Steel Pipe

Nominal pipe size (in.)	Support Spacing (ft)		
	5S	10S	40S
½	4.0	4.5	5.0
¾	5.0	5.5	6.0
1	6.0	6.5	7.0
1½	7.0	8.0	9.0
2	8.0	9.0	10.0
3	9.5	10.75	12.0
4	10.5	12.25	14.0
6	14.0	15.5	17.0
8	15.5	17.25	19.0
10	18.0	20.0	22.0
12	19.0	21.0	23.0

Based on fluids having a specific gravity of 1.35 and the lines being
uninsulated and a maximum operating temperature of 750°F (399°C).

Table 6-62 Mechanical and Physical Properties of Nickel 200 and Nickel 201

Property	Nickel 200	Nickel 201
Modulus of elasticity $\times 10^6$, psi	28	30
Tensile strength $\times 10^3$, psi	27000	58500
Yield strength 0.2% offset $\times 10^3$, psi	21500	15000
Elongation in 2 in., %	47	50
Hardness, Brinell	105	87
Density, lb/in.^{3}	0.321	0.321
Specific gravity	8.89	8.89
Specific heat, Btu/lb °F	0.109	0.109
Thermal conductivity, Btu/ hr ft^2/°F/in.		
at 0–70°F	500	569
at 70–200°F	465	512
at 70–400°F	425	460
at 70–600°F	390	408
at 70–800°F	390	392
at 70–1000°F	405	410
at 70–1200°F	420	428
Thermal expansion coefficient, in./in./°F $\times 10^{-6}$		
at 0–70°F	6.3	
at 70–200°F	7.4	7.3
at 70–400°F	7.7	
at 70–600°F	8.0	
at 70–800°F	8.3	
at 70–1000°F	8.5	
at 70–1200°F	8.7	

Table 6-63 Allowable Design Stress for Nickel 200 and Nickel 201 Pipe

	Allowable stress (psi)			
	4-in.-diameter pipe		5- to 12-inch-diameter pipe	
Operating temperature (°F/°C)	Nickel 200	Nickel 201	Nickel 200	Nickel 201
100/38	10000	8000	8000	6700
200/93	10000	7700	8000	6400
300/149	10000	7500	8000	6300
400/204	10000	7500	8000	6200
500/260	10000	7500	8000	6200
600/316	10000	7500	8000	6200
700/371		7400		6200
800/427		7200		5900
900/482		4500		4500
1000/538		3000		3000
1100/593		2000		2000
1200/649		1200		1200

Table 6-64 Maximum Allowable Operating Pressure of Seamless Nickel 200 pipe, Plain End[a]

Nominal pipe size (in.)	Maximum allowable operating pressure (psi)			
	5S	10S	40S	80S
½	1854	1858	2498	3490
¾	1132	1464	2037	2860
1	896	1500	1905	2630
1½	613	1046	1411	1989
2	488	830	1189	1717
3	422	615	1129	1596
4	327	476	957	1383
6	233	287	609	965
8	178	243	536	845
10	176	217	486	673
12	172	199	420	364

[a]Temperature range 100–600°F (38–316°C).

Table 6-65 Maximum Allowable Operating Pressure of Seamless Nickel 200 Pipe, Threaded Ends

Nominal pipe size (in.)	Maximum allowable operating pressure (psi)	
	40S	80S
½	825	1705
¾	727	1507
1	630	1290
1½	556	1095
2	512	1005
3	472	915
4	418	855
6	333	676

[a]Temperature range 100–600°F (38–316°C).

Table 6-66 Maximum Allowable Operating Pressure of Seamless Nickel 201 Pipe Schedule 5S, Plain End

Nominal pipe size (in.)	Maximum allowable operating pressure (psi) at °F/°C									
	100/38	200/93	300/149	400/204	700/371	800/427	900/482	1000/538	1100/593	1200/649
½	1485	1430	1390	1390	1370	1335	835	556	371	222
¾	905	870	850	850	836	815	509	340	226	135
1	717	690	672	672	664	645	403	279	179	107
1½	490	472	460	460	454	441	276	184	122	73
2	390	375	365	365	360	351	219	146	97	58
3	338	325	317	317	312	304	190	127	84	50
4	262	252	245	245	242	235	147	98	65	39
6	195	186	183	180	180	171	131	87	58	35
8	149	142	140	138	138	131	100	67	44	26
10	147	141	138	136	136	130	99	66	44	26
12	153	146	144	142	142	135	103	68	45	27

Table 6-67 Maximum Allowable Operating Pressure of Seamless Nickel 201 Pipe Schedule 10S, Plain End

Nominal pipe size (in.)	Maximum allowable operating pressure (psi) at °F/°C									
	100/38	200/93	300/149	400/204	700/371	800/427	900/482	1000/538	1100/593	1200/649
½	1483	1429	1390	1390	1370	1335	835	557	371	222
¾	1173	1130	1100	1100	1085	1055	660	430	294	176
1	1200	1160	1130	1130	1113	1081	677	450	301	180
1½	836	806	785	785	775	759	471	314	209	125
2	665	640	623	623	615	598	374	249	166	99
3	492	473	460	460	455	443	276	184	123	74
4	370	366	357	357	352	343	214	143	95	57
6	240	230	226	222	222	212	161	107	71	43
8	203	194	191	183	183	179	136	91	60	36
10	182	174	171	168	168	160	122	81	54	32
12	167	160	157	155	155	147	112	75	50	30

Table 6-68　Maximum Allowable Operating Pressure of Seamless Nickel 201 Pipe Schedule 40S, Plain End

Nominal pipe size (in.)	Maximum allowable operating pressure (psi) at °F/°C									
	100/38	200/93	300/149	400/204	700/371	800/427	900/482	1000/538	1100/593	1200/649
1/2	1995	1920	1870	1870	1845	1795	1120	750	498	299
3/4	1629	1567	1525	1525	1505	1465	915	610	407	244
1	1525	1468	1430	1430	1410	1371	859	572	381	229
1 1/2	1130	1085	1059	1059	1045	1015	635	356	282	169
2	950	915	891	891	891	856	535	356	238	142
3	904	870	845	845	845	813	507	338	225	135
4	765	735	716	716	716	690	430	286	191	115
6	510	487	480	472	472	442	343	228	152	91
8	450	430	423	416	416	396	302	201	134	80
10	407	390	384	377	377	359	274	182	121	75
12	353	337	332	326	326	315	237	158	105	63

Table 6-69　Maximum Allowable Operating Pressure of Seamless Nickel 201 Pipe Schedule 80S, Plain End

Nominal pipe size (in.)	Maximum allowable operating pressure (psi) at °F/°C									
	100/38	200/93	300/149	400/204	700/371	800/427	900/482	1000/538	1100/593	1200/649
1/2	2790	2690	2620	2620	2580	2515	1570	1045	698	420
3/4	2265	2200	2140	2140	2115	2060	1285	859	571	343
1	2100	2025	1970	1970	1950	1895	1182	790	525	338
1 1/2	1590	1580	1490	1490	1470	1430	895	595	398	238
2	1370	1320	1285	1285	1270	1235	773	515	344	206
3	1252	1210	1175	1175	1160	1130	705	470	314	188
4	1080	1065	1040	1040	1025	996	624	415	276	169
6	800	765	754	741	741	705	540	359	239	143
8	710	676	666	655	655	625	476	318	211	127
10	565	539	530	522	522	497	359	253	168	111
12	473	452	445	437	437	416	317	212	141	85

Table 6-70 Maximum Allowable Operating Pressure of Seamless Nickel 201 Pipe, Threaded End

| | Maximum allowable operating pressure (psi) at °F/°C | | | | | | | |
| | Schedule 40S | | | | Schedule 80S | | | |
Nominal pipe size (in.)	100/38	200/93	300/149	400/204 to 600/316	100/38	200/93	300/149	400/204 to 600/316
½	660	635	620	620	1365	1315	1280	1280
¾	583	570	545	545	1205	1160	1130	1130
1	505	485	473	473	1032	995	970	970
1½	445	429	417	417	876	845	823	823
2	410	395	384	384	805	775	755	755
3	378	364	354	354	732	705	685	685
4	358	345	336	336	684	659	640	640
6	279	267	262	258	567	542	533	525

Table 6-71 Support Spacing for Nickel Pipe

| | Support spacing (ft) | | | | | |
| | 5S | | 10S | | 40S and 80S | |
Nominal pipe size (in.)	Nickel 200	Nickel 201	Nickel 200	Nickel 201	Nickel 200	Nickel 201
½	4.0	3.00	4.50	3.40	5.0	3.75
¾	5.0	3.75	5.50	4.00	6.0	4.50
1	6.0	4.50	6.50	4.80	7.0	5.25
1½	7.0	5.25	8.00	6.00	9.0	6.75
2	8.0	6.0	9.00	6.75	10.0	7.50
3	9.5	7.0	10.75	8.00	12.0	9.00
4	10.5	7.50	12.25	9.20	14.0	10.50
6	11.0	9.00	12.00	9.75	13.5	11.00
8	12.0	9.75	14.00	11.50	15.0	12.50
10	14.0	11.50	16.00	13.00	17.5	14.5
12	15.0	12.50	16.50	13.50	18.0	15.00

Table 6-72 Maximum Nonshock Ratings of Forged Nickel Flanges

Operating temperature (°F/°C)	Nickel 200						Nickel 201					
	150	300	600	900	1500	2500	150	300	600	900	1500	2500
−20 to 100/−29 to 38	120	310	615	925	1545	2570	80	205	410	615	1030	1715
150/66	120	310	615	925	1545	2570	75	200	405	610		
200/93	120	310	615	925	1545	2570	75	200	400	600	1000	1665
250/121	120	310	615	925	1545	2570	75	195	390	590	965	1610
300/149	120	310	615	925	1545	2570	75	195	385	580	960	1590
350/177	120	310	615	925	1545	2570	75	190	380	575	955	1590
400/204	120	310	615	925	1545	2570	75	190	380	575	955	1590
450/232	120	310	615	925	1545	2570	75	190	380	575	955	1590
500/260	120	310	615	925	1545	2570	75	190	380	575	955	1590
550/288	120	310	615	925	1545	2570	75	190	380	575	955	1590
600/316	120	310	615	925	1545	2570	75	190	380	575	955	1590
650/343	120	310	615	925	1545	2570	75	190	380	575	955	1590
700/371							70	190	380	575	945	1575
750/390							70	185	370	570	925	1545
800/427							70	180	360	555	905	1510
850/454							70	175	355	545	885	1475
900/481							60	155	310	530	770	1285
950/510							50	125	255	465	635	1060
1000/538							40	105	205	380	515	855
1050/566								85	165	310	410	685
1100/593								70	135	245	345	570
1150/621								50	105	155	255	430
1200/649										125	205	345

Maximum nonshock rating (psi)

Table 6-73 Compatibility of Nickel 200 and Nickel 201 with Selected Corrodents
The chemicals listed are in the pure state or in a saturated solution unless otherwise indicated. Compatibility is shown to the maximum allowable temperature for which data are available. Incompatibility is shown by an x. A blank space indicates that data are unavailable. When compatible, corrosion rate is <20 mpy.

Chemical	Maximum temp. °F	Maximum temp. °C	Chemical	Maximum temp. °F	Maximum temp. °C
Acetaldehyde	200	93	Barium carbonate	210	99
Acetamide			Barium chloride	80	27
Acetic acid 10%	90	32	Barium hydroxide	90	32
Acetic acid 50%	90	32	Barium sulfate	210	99
Acetic acid 80%	120	49	Barium sulfide	110	43
Acetic acid, glacial	x	x	Benzaldehyde	210	99
Acetic anhydride	170	77	Benzene	210	99
Acetone	190	88	Benzene sulfonic acid 10%	190	88
Acetyl chloride	100	38	Benzoic acid	400	204
Acrylic acid			Benzyl alcohol	210	99
Acrylonitrile	210	99	Benzyl chloride	210	99
Adipic acid	210	99	Borax	200	93
Allyl alcohol	220	104	Boric acid	210	99
Allyl chloride	190	88	Bromine gas, dry	60	16
Alum	170	77	Bromine gas, moist	x	x
Aluminum acetate			Bromine liquid		
Aluminum chloride, aqueous	300	149	Butadiene	80	27
Aluminum chloride, dry	60	16	Butyl acetate	80	27
Aluminum fluoride	90	32	Butyl alcohol	200	93
Aluminum hydroxide	80	27	n-Butylamine		
Aluminum nitrate			Butyl phthalate	210	99
Aluminum oxychloride			Butyric acid	x	x
Aluminum sulfate	210	99	Calcium bisulfide		
Ammonia gas	90	32	Calcium bisulfite	x	x
Ammonium bifluoride			Calcium carbonate		
Ammonium carbonate	190	88	Calcium chlorate	140	60
Ammonium chloride 10%	230	110	Calcium chloride	80	27
Ammonium chloride 50%	170	77	Calcium hydroxide 10%	210	99
Ammonium chloride, sat.	570	299	Calcium hydroxide, sat.	200	93
Ammonium fluoride 10%	210	99	Calcium hypochlorite	x	x
Ammonium fluoride 25%	200	93	Calcium nitrate		
Ammonium hydroxide 25%	x	x	Calcium oxide	90	32
Ammonium hydroxide, sat.	320	160	Calcium sulfate	210	99
Ammonium nitrate	90	32	Caprylic acid[a]	210	99
Ammonium persulfate	x	x	Carbon bisulfide	x	x
Ammonium phosphate 30%	210	99	Carbon dioxide, dry	210	99
Ammonium sulfate 10–40%	210	99	Carbon dioxide, wet	200	93
Ammonium sulfide			Carbon disulfide	x	x
Ammonium sulfite	x	x	Carbon monoxide	570	290
Amyl acetate	300	149	Carbon tetrachloride	210	99
Amyl alcohol			Carbonic acid	80	27
Amyl chloride	90	32	Cellosolve	210	99
Aniline	210	99	Chloracetic acid, 50% water		
Antimony trichloride	210	99	Chloracetic acid	210	99
Aqua regia 3:1	x	x	Chlorine gas, dry	200	93

313

Table 6-73 *Continued*

Chemical	°F	°C	Chemical	°F	°C
Chlorine gas, wet	x	x	Magnesium chloride	300	149
Chlorine, liquid			Malic acid	210	99
Chlorobenzene, dry	120	49	Manganese chloride 37%	90	32
Chloroform	210	99	Methyl chloride	210	99
Chlorosulfonic acid	80	27	Methyl ethyl ketone		
Chromic acid 10%	100	38	Methyl isobutyl ketone	200	93
Chromic acid 50%	x	x	Muriatic acid	x	x
Chromyl chloride	210	99	Nitric acid 5%	x	x
Citric acid 15%	210	99	Nitric acid 20%	x	x
Citric acid, concentrated	80	27	Nitric acid 70%	x	x
Copper acetate	100	38	Nitric acid, anhydrous	x	x
Copper carbonate	x	x	Nitrous acid, concentrated	x	x
Copper chloride	x	x	Oleum		
Copper cyanide	x	x	Perchloric acid 10%	x	x
Copper sulfate	x	x	Perchloric acid 70%		
Cresol	100	38	Phenol, sulfur free	570	299
Cupric chloride 5%	x	x	Phosphoric acid 50–80%	x	x
Cupric chloride 50%	x	x	Picric acid	80	27
Cyclohexane	80	27	Potassium bromide 30%		
Cyclohexanol	80	27	Salicylic acid	80	27
Dichloroacetic acid			Silver bromide 10%		
Dichloroethane (ethylene di-chloride)	x	x	Sodium carbonate to 30%	210	99
			Sodium chloride to 30%	210	99
Ethylene glycol	210	99	Sodium hydroxide 10%[b]	210	99
Ferric chloride	x	x	Sodium hydroxide 50%[b]	300	149
Ferric chloride 50% in water	x	x	Sodium hydroxide, con-centrated	200	93
Ferric nitrate 10–50%	x	x	Sodium hypochlorite 20%	x	x
Ferrous chloride	x	x	Sodium hypochlorite, con-centrated	x	x
Ferrous nitrate					
Fluorine gas, dry	570	299	Sodium sulfide to 50%	x	x
Fluorine gas, moist	60	16	Stannic chloride	x	x
Hydrobromic acid, dilute	x	x	Stannous chloride, dry	570	299
Hydrobromic acid 20%	x	x	Sulfuric acid 10%	x	x
Hydrobromic acid 50%	x	x	Sulfuric acid 50%	x	x
Hydrochloric acid 20%	80	27	Sulfuric acid 70%	x	x
Hydrochloric acid 38%	x	x	Sulfuric acid 90%	x	x
Hydrocyanic acid 10%			Sulfuric acid 98%	x	x
Hydrofluoric acid 30%[b]	170	77	Sulfuric acid 100%	x	x
Hydrofluoric acid 70%[b]	100	38	Sulfuric acid, fuming	x	x
Hydrofluoric acid 100%[b]	120	49	Sulfurous acid	x	x
Hypochlorous acid	x	x	Thionyl chloride	210	99
Iodine solution 10%			Toluene	210	99
Ketones, general	100	38	Trichloroacetic acid	80	27
Lactic acid 25%	x	x	White liquor		
Lactic acid, concentrated	x	x	Zinc chloride to 80%	200	93

[a]Material subject to pitting.
[b]Material subject to stress cracking.

Source: Schweitzer, Philip A. (1991). *Corrosion Resistance Tables*, Marcel Dekker, Inc., New York, Vols. 1 and 2.

Table 6-74 Mechanical and Physical Properties of Monel 400

Modulus of elasticity $\times$ 10^6, psi	26
Tensile strength $\times$ 10^3, psi	70
Yield strength 0.2% offset $\times$ 10^3, psi	25–28
Elongation in 2 in., %	48
Hardness, Brinell	130
Density, lb/in.^{3}	0.318
Specific gravity	8.84
Specific heat, at 32–212°F Btu/lb °F	0.102
Thermal conductivity, Btu/hr/ft^2/in./°F	
at 70°F	151
at 200°F	167
at 500°F	204
Thermal expansion coefficient, in/in °F $\times$ 10^{-6}	
at 70–200°F	7.7
at 70–400°F	8.6
at 70–500°F	8.7
at 70–1000°F	9.1

Table 6-75 Allowable Design Stress for Monel 400 Pipe

	Allowable design stress $\times$ 10^3 (psi)	
	≤4 inch diameter	5–12 inch diameter
100/38	17.5	16.6
200/93	16.5	14.6
300/149	15.5	13.6
400/204	14.8	13.2
500/260	14.7	13.1
600/316	14.7	13.1
700/371	14.7	13.1
800/427	14.5	13.1
900/482	8.0	8.0

Table 6-76 Maximum Allowable Operating Pressure of Seamless, Schedule 5S Monel 400 Pipe, Plain End

Nominal pipe size (in.)	Maximum allowable operating pressure (psi) at °F/°C				500/260 to		
	100/38	200/93	300/149	400/204	700/371	800/427	900/482
½	2462	2362	2219	2118	2104	2075	1145
¾	1947	1868	1755	1675	1664	1641	905
1	1541	1478	1388	1326	1317	1299	716
1½	1055	1012	950	907	901	889	490
2	839	805	756	722	717	708	390
3	725	696	654	624	620	611	337
4	562	539	506	483	480	474	261
5	578	508	473	459	456	456	261
6	483	425	396	384	381	381	233
8	370	325	303	294	292	292	178
10	365	321	299	290	288	288	176
12	358	315	293	285	282	282	172

Table 6-77 Maximum Allowable Operating Pressure of Seamless, Schedule 10S Monel 400 Pipe, Plain End

Nominal pipe size (in.)	Maximum allowable operating pressure (psi) at °F/°C						
	100/38	200/93	300/149	400/204	500/260 to 700/371	800/427	900/482
½	3195	3065	2879	2749	2730	2491	1486
¾	2518	2416	2269	2167	2152	2123	1171
1	2648	2540	2386	2279	2263	2232	1231
1½	1799	1725	1621	1548	1537	1516	836
2	1427	1369	1286	1228	1219	1203	663
3	1057	1014	952	909	903	891	491
4	817	784	737	703	699	689	380
5	712	626	583	566	562	562	343
6	596	524	488	473	470	470	287
8	504	443	413	401	398	398	243
10	450	396	369	358	355	355	217
12	414	364	339	329	326	326	199

Table 6-78 Maximum Allowable Operating Pressure of Seamless, Schedule 40S Monel 400 Pipe, Plain End

Nominal pipe size (in.)	Maximum allowable operating pressure (psi) at °F/°C					500/260 to		
	100/38	200/93	300/149	400/204	700/371	800/427	900/482
½	4296	4121	3871	3696	3671	3621	1998
¾	3503	3360	3156	3014	2994	2953	1629
1	3276	3142	2952	2819	2800	2761	1523
1½	2426	2327	2186	2088	2074	2045	1128
2	2044	1961	1842	1759	1747	1723	950
3	1941	1862	1749	1670	1659	1636	903
4	1645	1578	1483	1416	1406	1387	765
5	1393	1225	1141	1107	1099	1099	671
6	1265	1112	1036	1006	998	998	609
8	1113	979	912	885	878	878	536
10	1010	888	827	803	797	797	486
12	872	767	714	693	688	688	420

Table 6-79 Maximum Allowable Operating Pressure of Seamless, Schedule 80S Monel 400 Pipe, Plain End

Nominal pipe size (in.)	Maximum allowable operating pressure (psi) at °F/°C							
	100/38	200/93	300/149	400/204	500/260 to 700/371	800/427	900/482
½	6002	5758	5409	5165	5130	5060	2792
¾	4919	4719	4433	4233	4204	4147	2288
1	4528	4344	4081	3896	3870	3817	2106
1½	3420	3281	3082	2943	2923	2883	1590
2	2952	2832	2660	2540	2523	2489	1373
3	2744	2632	2473	2361	2345	2313	1276
4	2378	2282	2143	2046	2033	2005	1106
5	2055	1807	1684	1634	1622	1622	990
6	1984	1745	1626	1578	1566	1566	956
8	1755	1543	1438	1395	1385	1385	845
10	1396	1228	1144	1110	1102	1102	673
12	1171	1030	959	931	924	924	564

Table 6-80 Maximum Allowable Operating Pressure of Seamless, Schedule 40S Monel 400 Pipe, Threaded

Nominal pipe size (in.)	Maximum allowable operating pressure (psi) at °F/°C						
	100/38	200/93	300/149	400/204	500/260 to 700/371	800/427	900/482
½	1420	1360	1280	1220	1210	1195	660
¾	1250	1200	1125	1075	1070	1055	582
1	1082	1040	975	933	925	915	505
1½	956	917	863	825	819	806	495
2	880	845	795	760	555	745	410
3	810	780	731	699	699	685	377
4	770	740	695	665	660	650	358
6	692	610	567	550	545	545	333

Table 6-81 Maximum Allowable Operating Pressure of Seamless, Schedule 80S Monel 400 Pipe, Threaded

Nominal pipe size (in.)	Maximum allowable operating pressure (psi) at °F/°C						
	100/38	200/93	300/149	400/204	500/260 to 700/371	800/427	900/482
½	2940	2835	2645	2525	2510	2480	1362
¾	2580	2480	2335	2230	2210	2180	1207
1	2220	2130	2000	1910	1900	1870	1030
1½	1885	1810	1700	1620	1610	1590	875
2	1730	1660	1560	1490	1480	1460	805
3	1572	1510	1420	1353	1345	1328	732
4	1470	1410	1325	1265	1260	1240	685
6	1408	1239	1150	1120	1110	1110	678

Table 6-82 Maximum Nonshock Ratings of Forged Monel 400 Flanges

Operating temperature (°F/°C)	Maximum nonshock ratings (psi)					
	150	300	600	900	1500	2500
–20 to 100/–29 to 38	195	515	1030	1545	2570	4285
150/66	180	475	945			
200/93	175	455	905	1360	2265	3770
250/121	165	435	870			
300/147	160	420	845	1265	2110	3515
350/177	155	410	825			
400/204	155	410	815	1225	2040	3400
450/232	155	405	810			
500/260	155	405	805	1210	2015	3360
550/288	155	405	805	1210	2015	3360
600/316	155	405	805	1210	2015	3360
650/343	155	405	805	1210	2015	3360
700/371	155	405	805	1210	2015	3360
750/399	155	405	805	1210	2015	3360
800/427	155	405	805	1210	2015	3360
850/454	145	375	755	1130	1590	3145
875/468	120	310	615			
900/482	105	275	555	825	1520	2285

Table 6-83 Support Spacing for Monel 400 Pipe

Nominal pipe size (in.)	Support spacing (ft)		
	5S	10S	40S
½	5.50	6.3	7.00
¾	7.00	7.5	8.40
1	8.40	9.0	9.75
1½	9.75	11.0	12.50
2	11.00	12.5	14.00
3	13.00	15.0	16.50
4	14.50	21.0	23.50
6	17.50	19.5	21.50
8	19.00	21.5	24.00
10	22.50	25.0	27.50
12	24.00	26.5	29.00

Based on lines being uninsulated, handling a liquid with a maximum specific gravity of 1.35 and operating at a maximum temperature of 800°F (427°C)

Table 6-84 Compatibility of Monel 400 with Selected Corrodents

The chemicals listed are in the pure state or in a saturated solution unless otherwise indicated. Compatibility is shown to the maximum allowable temperature for which data are available. Incompatibility is shown by an x. A blank space indicates that data are unavailable. When compatible, corrosion rate is <20 mpy.

Chemical	Maximum temp. °F	Maximum temp. °C	Chemical	Maximum temp. °F	Maximum temp. °C
Acetaldehyde	170	77	Barium carbonate	210	99
Acetamide	340	171	Barium chloride	210	99
Acetic acid 10%	80	27	Barium hydroxide	80	27
Acetic acid 50%	200	93	Barium sulfate	210	99
Acetic acid 80%	200	93	Barium sulfide	x	x
Acetic acid, glacial	290	143	Benzaldehyde	210	99
Acetic anhydride	190	88	Benzene	210	99
Acetone	190	88	Benzene sulfonic acid 10%	210	99
Acetyl chloride	400	204	Benzoic acid	210	99
Acrylic acid			Benzyl alcohol	400	204
Acrylonitrile	210	99	Benzyl chloride	210	99
Adipic acid	210	99	Borax	90	32
Allyl alcohol	400	204	Boric acid	210	99
Allyl chloride	200	93	Bromine gas, dry	120	49
Alum	100	38	Bromine gas, moist	x	x
Aluminum acetate	80	27	Bromine liquid		
Aluminum chloride, aqueous	x	x	Butadiene	180	82
Aluminum chloride, dry	150	66	Butyl acetate	380	193
Aluminum fluoride	90	32	Butyl alcohol	200	93
Aluminum hydroxide	80	27	n-Butylamine		
Aluminum nitrate			Butyl phthalate	210	99
Aluminum oxychloride			Butyric acid	210	99
Aluminum sulfate	210	99	Calcium bisulfide	60	16
Ammonia gas	x	x	Calcium bisulfite	x	x
Ammonium bifluoride	400	204	Calcium carbonate	200	93
Ammonium carbonate	190	88	Calcium chlorate	140	60
Ammonium chloride 10%	230	110	Calcium chloride	350	177
Ammonium chloride 50%	170	77	Calcium hydroxide 10%	210	99
Ammonium chloride, sat.	570	299	Calcium hydroxide, sat.	200	93
Ammonium fluoride 10%	400	204	Calcium hypochlorite	x	x
Ammonium fluoride 25%	400	204	Calcium nitrate		
Ammonium hydroxide 25%	x	x	Calcium oxide	90	32
Ammonium hydroxide, sat.	x	x	Calcium sulfate	80	27
Ammonium nitrate	x	x	Caprylic acid[a]	210	99
Ammonium persulfate	x	x	Carbon bisulfide	x	x
Ammonium phosphate 30%	210	99	Carbon dioxide, dry	570	299
Ammonium sulfate 10–40%	400	204	Carbon dioxide, wet[a]	400	204
Ammonium sulfide			Carbon disulfide	x	x
Ammonium sulfite	90	32	Carbon monoxide	570	299
Amyl acetate	300	149	Carbon tetrachloride	400	204
Amyl alcohol	180	82	Carbonic acid	x	x
Amyl chloride	400	204	Cellosolve	210	99
Aniline	210	99	Chloracetic acid, 50% water	180	82
Antimony trichloride	350	177	Chloracetic acid	x	x
Aqua regia 3:1	x	x	Chlorine gas, dry	570	299

322

Table 6-84 *Continued*

Chemical	°F	°C	Chemical	°F	°C
Chlorine gas, wet	x	x	Malic acid	210	99
Chlorine, liquid	150	66	Manganese chloride 40%	100	38
Chlorobenzene, dry	400	204	Methyl chloride	210	99
Chloroform	210	99	Methyl ethyl ketone	200	93
Chlorosulfonic acid	80	27	Methyl isobutyl ketone	200	93
Chromic acid 10%	130	54	Muriatic acid	x	x
Chromic acid 50%	x	x	Nitric acid 5%	x	x
Chromyl chloride	210	99	Nitric acid 20%	x	x
Citric acid 15%	210	99	Nitric acid 70%	x	x
Citric acid, concentrated	80	27	Nitric acid, anhydrous	x	x
Copper acetate	x	x	Nitrous acid, concentrated	x	x
Copper carbonate	x	x	Oleum	x	x
Copper chloride	x	x	Perchloric acid 10%	x	x
Copper cyanide	x	x	Perchloric acid 70%	x	x
Copper sulfate	x	x	Phenol	570	299
Cresol	100	38	Phosphoric acid 50–80%	x	x
Cupric chloride 5%	x	x	Picric acid	x	x
Cupric chloride 50%	x	x	Potassium bromide 30%, air free	210	99
Cyclohexane	180	82			
Cyclohexanol	80	27	Salicylic acid	210	99
Dichloroacetic acid			Silver bromide 10%	80	27
Dichloroethane (ethylene dichloride)	200	93	Sodium carbonate	210	99
			Sodium chloride to 30%	210	99
Ethylene glycol	210	99	Sodium hydroxide 10%[b]	350	177
Ferric chloride	x	x	Sodium hydroxide 50%[b]	300	149
Ferric chloride 50% in water	x	x	Sodium hydroxide, concentrated	350	177
Ferric nitrate 10–50%	x	x	Sodium hypochlorite 20%	x	x
Ferrous chloride	x	x	Sodium hypochlorite, concentrated	x	x
Ferrous nitrate					
Fluorine gas, dry	570	299	Sodium sulfide to 50%	210	99
Fluorine gas, moist	x	x	Stannic chloride	x	x
Hydrobromic acid, dilute	x	x	Stannous chloride, dry	570	299
Hydrobromic acid 20%	x	x	Sulfuric acid 10%	x	x
Hydrobromic acid 50%	x	x	Sulfuric acid 50%	80	27
Hydrochloric acid 20%	80	27	Sulfuric acid 70%	80	27
Hydrochloric acid 38%	x	x	Sulfuric acid 90%	x	x
Hydrocyanic acid 10%	80	27	Sulfuric acid 98%	x	x
Hydrofluoric acid 30%[b]	400	204	Sulfuric acid 100%	x	x
Hydrofluoric acid 70%[b]	400	204	Sulfuric acid, fuming	x	x
Hydrofluoric acid 100%[b]	210	99	Sulfurous acid	x	x
Hypochlorous acid	x	x	Thionyl chloride	300	149
Iodine solution 10%	x	x	Toluene	210	99
Ketones, general	100	38	Trichloroacetic acid	170	77
Lactic acid 25%	x	x	White liquor	x	x
Lactic acid, concentrated	x	x	Zinc chloride to 80%	200	93
Magnesium chloride 50%	350	177			

[a]Not for use with carbonated beverages.
[b]Material is subject to stress cracking.
Source: Schweitzer, Philip A. (1991). *Corrosion Resistance Tables*, Marcel Dekker, Inc., New York, Vols. 1 and 2.

Table 6-85 Mechanical and Physical Properties of Alloy 600 Pipe

Modulus of elasticity $\times 10^6$, psi	30–31
Tensile strength $\times 10^3$, psi	80
Yield strength 0.2% offset $\times 10^3$, psi	30–35
Elongation in 2 in., %	40
Hardness, Rockwell	B- 120–170
Density, lb/in.^{3}	0.306
Specific gravity	8.42
Specific heat, Btu/lb°F	0.106
Thermal conductivity, Btu/hr ft^2/°F/in.	
at 70°F	103
at 200°F	109
at 400°F	121
at 600°F	133
at 800°F	145
at 1000°F	158
at 1200°F	172
Thermal expansion coefficient, in./in./°F $\times 10^{-6}$	
at 70–200°F	7.4
at 70–400°F	7.7
at 70–600°F	7.9
at 70–800°F	8.1
at 70–1000°F	8.4
at 70–1200°F	8.6

Table 6-86 Allowable Design Stress for Alloy 600 Pipe

Operating temperature (°F/°C)	Allowable design stress (psi)	
	≤4-inch pipe	≥5-inch pipe
100/38	20000	20000
200/93	19300	19100
300/149	18800	18200
400/204	18500	17450
500/260	18500	16850
600/316	18500	16100
700/371	18500	15600
800/427	18500	15300
900/482	16000	14900
1000/538	7000	7000
1100/593	3000	3000
1200/649	2000	2000

Table 6-87 Maximum Allowable Operating Pressure of Seamless Alloy 600 Pipe Schedule 5S

Nominal pipe size (in.)	Maximum allowable operating pressure (psi) at °F/°C							
	100/ 38	200/ 93	300/ 149	400/204 to 800/427	900/ 482	1000/ 538	1100/ 593	1200/ 649
½	2820	2730	2640	2610	2260	990	425	282
¾	2265	2190	2130	2100	1812	795	340	226
1	1793	1730	1685	1660	1433	626	268	179
1½	1223	1181	1150	1131	980	430	184	122
2	966	944	919	905	781	342	146	96
3	845	815	792	780	675	296	126	84
4	655	630	615	605	522	228	98	65

Table 6-88 Maximum Allowable Operating Pressure of Seamless Alloy 600 Pipe Schedule 10S

Nominal pipe size (in.)	Maximum allowable operating pressure (psi) at °F/°C							
	100/ 38	200/ 93	300/ 149	400/204 to 800/427	900/ 482	1000/ 538	1100/ 593	1200/ 649
¼	4600	4450	4330	4250	3680	1610	690	460
⅜	3620	3490	3400	3340	2890	1265	543	362
½	3690	3560	3470	3400	2950	1290	555	369
¾	2930	2830	2755	2710	2340	1025	440	293
1	3010	2900	2830	2780	2410	1050	452	301
1½	2095	2020	1970	1035	1675	733	314	209
2	1660	1600	1560	1535	1330	580	248	166
3	1230	1185	1155	1138	983	430	184	123
4	955	920	895	850	762	334	143	95

Table 6-89 Maximum Allowable Operating Pressure of Seamless Alloy 600 Pipe, Schedule 40S, Plain End

Nominal pipe size (in.)	Maximum allowable operating pressure (psi) at °F/°C											
	100/38	200/93	300/149	400/204	500/260	600/316	700/371	800/427	900/482	1000/538	1100/593	1200/649
¼	6430	6200	6100	5950	5950	5950	5950	5950	5150	2250	965	463
⅜	5200	5030	4900	4820	4820	4820	4820	4820	4160	1820	780	520
½	5000	4820	4700	4620	4620	4620	4620	4610	3990	1745	750	500
¾	4070	3930	3830	3760	3760	3760	3760	3760	3260	1425	610	407
1	3810	3780	3550	3520	3520	3520	3520	3520	3045	1330	572	381
1½	2815	2725	2655	2615	2615	2615	2615	2615	2260	990	424	281
2	2380	2300	2230	2200	2200	2200	2200	2200	1905	832	356	238
3	2260	2180	2120	2090	2090	2090	2090	2090	1805	790	339	226
4	1910	1850	1800	1770	1770	1770	1770	1770	1530	670	287	191
6	1525	1455	1390	1330	1285	1225	1190	1165	1135	534	229	152
8	1341	1281	1220	1171	1130	1080	1048	1029	1000	470	202	134

Table 6-90 Maximum Allowable Operating Pressure of Seamless Alloy 600 Pipe, Schedule 80S, Plain End

Nominal pipe size (in.)	Maximum allowable operating pressure (psi) at °F/°C											
	100/ 38	200/ 93	300/ 149	400/ 204	500/ 260	600/ 316	700/ 371	800/ 427	900/ 482	1000/ 538	1100/ 593	1200/ 649
³⁄₈	7500	7240	7050	6950	6950	6950	6950	6950	6000	2620	1120	750
¹⁄₂	6980	6750	6560	6450	6450	6450	6450	6450	5600	2440	1048	698
³⁄₄	5720	5520	5380	5300	5300	5300	5300	5300	4570	2000	860	572
1	5260	5080	4950	4870	4870	4870	4870	4870	4210	1840	790	526
1¹⁄₂	3970	3840	3740	3680	3680	3680	3680	3680	3180	1390	595	397
2	3430	3315	3225	3175	3175	3175	3175	3175	2745	1220	515	343
3	3140	3025	2945	2900	2900	2900	2900	2900	2510	1095	470	314
4	2770	2670	2600	2540	2540	2540	2540	2540	2220	970	415	277

Table 6-91 Maximum Allowable Operating Pressure of Seamless
Alloy 600 Pipe, Threaded

Nominal pipe size (in.)	Maximum allowable operating pressure (psi) at °F/°C					
	100/ 38	200/ 93	300/ 149	400/ 204	500/ 260	600/ 316
			Schedule 40S			
¼	2260	2180	2125	2090	2090	2090
⅜	1872	1805	1760	1730	1730	1730
½	1650	1590	1550	1525	1525	1525
¾	1452	1400	1365	1345	1345	1345
1	1260	1215	1182	1165	1165	1165
1½	1130	1075	1045	1030	1030	1030
2	1023	990	965	950	950	950
3	945	913	890	875	875	875
4	896	866	845	830	830	830
6	835	797	760	728	703	672
			Schedule 80S			
⅜	3840	3700	3600	3540	3540	3540
½	3410	3290	3200	3150	3150	3150
¾	3010	2910	2830	2785	2785	2785
1	2580	2490	2425	2390	2390	2390
1½	2190	2115	2060	2015	2015	2015
2	2010	1940	1890	1860	1860	1860
3	1828	1765	1715	1690	1690	1690
4	1710	1650	1610	1580	1580	1580

Table 6-92 Maximum Nonshock Ratings of Forged Alloy 600 Flanges at Different Primary Service Ratings

Operating temperature (°F/°C)	Maximum nonshock rating (psi)		
	150	300	600
−20 to 100/−29 to 38	235	615	1235
150/66	225	595	1185
200/93	220	580	1160
250/121	220	575	1145
300/149	215	560	1115
350/177	210	555	1110
400/204	210	555	1110
500/260	210	555	1110
550/288	210	555	1110
600/316	210	555	1110
650/343	210	555	1110
700/371	210	550	1095
750/399	205	540	1075
800/427	205	530	1065
850/454	200	520	1035
875/468	195	510	1020
900/482	190	505	1010
925/496	170	445	890
950/510	140	360	720
975/524	110	290	585
1000/538	90	240	480
1025/552		205	410
1050/566		170	345
1075/579		150	295
1100/593		105	205
1125/607		100	200
1150/621		85	170
1175/635		75	150
1200/649		70	135

Table 6-93 Maximum Nonshock Ratings of Forged Alloy 600 Flanges at Different Primary Service Ratings

Operating temperature (°F/°C)	Maximum nonshock rating (psi)		
	900	1500	2500
−20 to 100/−29 to 38	1850	3085	5140
200/93	1740	2900	4830
300/149	1675	2795	4655
350/177	1670		
400/204	1665	2775	4625
450/232	1665	2775	4625
500/260	1665	2775	4625
550/288	1665	2775	4625
600/316	1665	2775	4625
650/343	1665	2775	4625
700/371	1645	2745	4570
750/399	1615	2690	4490
800/427	1595	2655	4430
850/454	1555	2590	4315
900/482	1510	2520	4200
950/510	1080	1800	3000
1000/538	720	1200	2000
1050/566	515	855	1430
1100/593	310	515	860
1150/621	255	430	715
1200/649	205	340	570

Table 6-94 Support Spacing for Alloy 600 Pipe

Nominal pipe size (in.)	Support spacing (ft)		
	5S	10S	40S
¼		5.0	6.0
½	6.0	6.5	7.5
¾	7.5	8.5	9.0
1	9.0	10.0	10.5
1½	10.5	12.0	13.5
2	12.0	13.5	15.0
3	14.5	16.5	18.0
4	16.0	18.5	21.0
6			24.0
8			27.0

Table is based on fluid having a specific gravity of 1.35, the lines being uninsulated, and a maximum operating temperature of 900°F (482°C). If lines are insulated, the spacing should be reduced by 30%.

Table 6-95 Compatibility of Alloy 600 and Alloy 625 with Selected Corrodents
The chemicals listed are in the pure state or in a saturated solution unless otherwise indicated. Compatibility is shown to the maximum allowable temperature for which data are available. Incompatibility is shown by an x. A blank space indicates that data are unavailable. When compatible, corrosion rate is <20 mpy.

Chemical	Maximum temp. °F	°C	Chemical	Maximum temp. °F	°C
Acetaldehyde	140	60	Barium carbonate	80	27
Acetamide			Barium chloride	570	299
Acetic acid 10%	80	27	Barium hydroxide	90	32
Acetic acid 50%	x	x	Barium sulfate	210	99
Acetic acid 80%	x	x	Barium sulfide		
Acetic acid, glacial	220	104	Benzaldehyde	210	99
Acetic anhydride	200	93	Benzene	210	99
Acetone	190	88	Benzene sulfonic acid 10%		
Acetyl chloride	80	27	Benzoic acid 10%	90	32
Acrylic acid			Benzyl alcohol	210	99
Acrylonitrile	210	99	Benzyl chloride	210	99
Adipic acid	210	99	Borax	90	32
Allyl alcohol	200	93	Boric acid	80	27
Allyl chloride	150	66	Bromine gas, dry	60	16
Alum	200	93	Bromine gas, moist	x	x
Aluminum acetate	80	27	Bromine liquid		
Aluminum chloride, aqueous	x	x	Butadiene	80	27
Aluminum chloride, dry	x	x	Butyl acetate	80	27
Aluminum fluoride	80	27	Butyl alcohol	80	27
Aluminum hydroxide	80	27	*n*-Butylamine		
Aluminum nitrate			Butyl phthalate	210	99
Aluminum oxychloride			Butyric acid	x	x
Aluminum sulfate	x	x	Calcium bisulfide		
Ammonia gas			Calcium bisulfite	x	x
Ammonium bifluoride			Calcium carbonate	90	32
Ammonium carbonate	190	88	Calcium chlorate	80	27
Ammonium chloride 10%[a]	230	110	Calcium chloride	80	27
Ammonium chloride 50%	170	77	Calcium hydroxide 10%	210	99
Ammonium chloride, sat.	200	93	Calcium hydroxide, sat.	90	32
Ammonium fluoride 10%	90	32	Calcium hypochlorite	x	x
Ammonium fluoride 25%	90	32	Calcium nitrate		
Ammonium hydroxide 25%	80	27	Calcium oxide		
Ammonium hydroxide, sat.	90	32	Calcium sulfate[b]	210	99
Ammonium nitrate	x	x	Caprylic acid	230	110
Ammonium persulfate	80	27	Carbon bisulfide	80	27
Ammonium phosphate 10%	210	99	Carbon dioxide, dry	210	99
Ammonium sulfate 10–40%[b]	210	99	Carbon dioxide, wet	200	93
Ammonium sulfide			Carbon disulfide	80	27
Ammonium sulfite	90	32	Carbon monoxide	570	299
Amyl acetate	300	149	Carbon tetrachloride	210	99
Amyl alcohol			Carbonic acid	210	99
Amyl chloride	x	x	Cellosolve	210	99
Aniline	210	99	Chloracetic acid, 50% water		
Antimony trichloride	90	32	Chloracetic acid	x	x
Aqua regia 3:1	x	x	Chlorine gas, dry	90	32

331

Table 6-95 *Continued*

Chemical	Maximum temp. °F	Maximum temp. °C	Chemical	Maximum temp. °F	Maximum temp. °C
Chlorine gas, wet	x	x	Magnesium chloride 50%	130	54
Chlorine, liquid			Malic acid	210	99
Chlorobenzene	210	99	Manganese chloride 37%	x	x
Chloroform	210	99	Methyl chloride	210	99
Chlorosulfonic acid			Methyl ethyl ketone	210	99
Chromic acid 10%	130	54	Methyl isobutyl ketone	200	93
Chromic acid 50%	90	32	Muriatic acid	x	x
Chromyl chloride	210	99	Nitric acid 5%	90	32
Citric acid 15%	210	99	Nitric acid 20%	80	27
Citric acid, concentrated	210	99	Nitric acid 70%	x	x
Copper acetate	100	38	Nitric acid, anhydrous	x	x
Copper carbonate	80	27	Nitrous acid, concentrated	x	x
Copper chloride	x	x	Oleum	x	x
Copper cyanide	80	27	Perchloric acid 10%		
Copper sulfate	80	27	Perchloric acid 70%		
Cresol	100	38	Phenol	570	299
Cupric chloride 5%	x	x	Phosphoric acid 50–80%	190	88
Cupric chloride 50%	x	x	Picric acid	x	x
Cyclohexane			Potassium bromide 30%	210	99
Cyclohexanol	80	27	Salicylic acid	80	27
Dichloroacetic acid			Silver bromide 10%		
Dichloroethane (ethylene dichloride)	200	93	Sodium carbonate to 30%	210	99
			Sodium chloride to 30%	210	99
Ethylene glycol	210	99	Sodium hydroxide 10%	300	149
Ferric chloride	x	x	Sodium hydroxide 50%[a]	300	149
Ferric chloride 50% in water	x	x	Sodium hydroxide, concentrated	80	27
Ferric nitrate 10–50%	x	x	Sodium hypochlorite 20%	x	x
Ferrous chloride	x	x	Sodium hypochlorite, concentrated	x	x
Ferrous nitrate					
Fluorine gas, dry	570	299	Sodium sulfide to 50%	210	99
Fluorine gas, moist	60	16	Stannic chloride	x	x
Hydrobromic acid, dilute	90	32	Stannous chloride, dry	570	299
Hydrobromic acid 20%	80	27	Sulfuric acid 10%	x	x
Hydrobromic acid 50%	x	x	Sulfuric acid 50%	x	x
Hydrochloric acid 20%	80	27	Sulfuric acid 70%	x	x
Hydrochloric acid 38%	x	x	Sulfuric acid 90%	x	x
Hydrocyanic acid 10%			Sulfuric acid 98%	x	x
Hydrofluoric acid 30%	x	x	Sulfuric acid 100%	x	x
Hydrofluoric acid 70%	x	x	Sulfuric acid, fuming	x	x
Hydrofluoric acid 100%	120	49	Sulfurous acid	90	32
Hypochlorous acid			Thionyl chloride		
Iodine solution 10%			Toluene	210	99
Ketones, general			Trichloroacetic acid	80	27
Lactic acid 25%	210	99	White liquor		
Lactic acid, concentrated	90	32	Zinc chloride, dry	80	27

[a]Material is subject to stress cracking.
[b]Material subject to pitting.
Source: Schweitzer, Philip A. (1991). *Corrosion Resistance Tables*, Marcel Dekker, Inc., New York, Vols. 1 and 2.

Table 6-96 Mechanical and Physical Properties of Alloy 625 Pipe

Modulus of elasticity $\times 10^6$, psi	30.1
Tensile strength $\times 10^3$, psi	100–120
Yield strength 0.2% offset $\times 10^3$, psi	60
Elongation in 2 in., %	30
Hardness, Brinell	192
Density, lb/in.^{3}	0.305
Specific gravity	8.44
Specific heat, Btu/lb°F	0.098
Thermal conductivity, Btu-in/ft^2 hr °F	
at −250°F	50
at −100°F	58
at 0°F	64
at 70°F	68
at 100°F	70
at 200°F	75
at 400°F	87
at 600°F	98
at 1000°F	121
at 1400°F	144
Thermal expansion coefficient in./in. °F $\times 10^{-6}$	
at 70–200°F	7.1
at 70–400°F	7.3
at 70–600°F	7.4
at 70–800°F	7.6
at 70–1000°F	7.8
at 70–1200°F	8.2
at 70–1400°F	8.5
at 70–1600°F	8.8

Table 6-97 Mechanical and Physical Properties of Alloy 800 Pipe

Modulus of elasticity $\times$ 10^6, psi	28.5
Tensile strength $\times$ 10^3, psi	75
Yield strength 0.2% offset $\times$ 10^3, psi	26.6
Elongation in 2 in., %	30
Hardness, Brinell	152
Density, lb/in.^{3}	0.287
Specific gravity	7.94
Specific heat, Btu/lb°F	0.11
Thermal conductivity, Btu-in/hr/ft^2/°F/in.	
at 0°F	75
at 70°F	80
at 100°F	83
at 200°F	89
at 400°F	103
at 600°F	115
at 1200°F	152
at 1600°F	181
Coefficient of thermal expansion, in./in. °F $\times$ 10^{-6}	
at 70–200°F	7.9
at 70–400°F	8.8
at 70–600°F	9.0
at 70–800°F	9.2
at 70–1000°F	9.4
at 70–1200°F	9.6
at 70–1400°F	9.9
at 70–1600°F	10.2

Table 6-98 Allowable Design Stress for Alloy 800 Pipe

Operating temperature (°F/°C)	Allowable design stress (psi)
100/38	15600
200/93	13400
300/149	12100
400/204	11100
500/260	10400
600/316	10000
650/343	9800
700/371	9600
750/399	9500
800/427	9300
850/454	9200
900/482	9100
950/510	9000
1000/538	8900
1050/566	8800
1100/593	8800
1150/621	8700
1200/649	7100
1250/677	5400
1300/704	4150
1350/732	3250
1400/760	2500
1450/788	1900
1500/816	1500

Table 6-99 Maximum Allowable Operating Pressure of Schedule 5S Alloy 800 Seamless Pipe for Different Pipe Sizes (inches)

Operating temperature (°F/°C)	Maximum allowable operating pressure (psi)						
	½	¾	1	1½	2	3	4
100/38	2900	1765	1400	950	760	656	507
200/93	2490	1520	1205	820	655	567	437
300/149	2240	1370	1080	740	593	510	396
400/204	2060	1260	998	679	541	469	363
500/260	1935	1180	935	639	507	440	340
600/316	1850	1132	895	613	489	422	326
650/343	1815	1110	880	600	478	413	320
700/371	1773	1085	860	587	468	405	313
750/399	1765	1080	850	583	464	402	311
800/427	1720	1055	835	570	454	393	304
900/482	1690	1030	815	557	445	384	298
950/510	1670	1021	805	551	439	379	294
1000/538	1650	1008	795	545	434	375	290
1050/566	1630	1000	787	538	430	371	288
1100/593	1630	995	790	558	430	371	287

Table 6-100 Maximum Allowable Operating Pressure of Schedule 10S Alloy 800 Seamless Pipe at Different Nominal Pipe Sizes (inches)

Operating temperature (°F/°C)	Maximum allowable operating pressure (psi)							
	¼	½	¾	1	1½	2	3	4
100/38	3600	2910	2290	2340	1630	1290	950	745
200/93	3080	2500	1965	2020	1400	1110	835	890
300/149	2790	2250	1775	1820	1260	1050	745	577
400/204	2560	2070	1630	1670	1160	920	684	529
500/260	2390	1935	1530	1570	1090	865	640	495
600/316	2300	1855	1465	1500	1045	830	615	471
650/343	2260	1815	1439	1470	1025	810	604	462
700/371	2200	1775	1410	1440	1005	795	589	452
750/399	2180	1765	1405	1430	995	790	584	448
800/427	2140	1725	1365	1400	975	771	571	439
850/454	2120	1710	1350	1390	965	764	565	435
900/482	2090	1690	1331	1370	952	756	560	430
950/510	2070	1680	1319	1355	942	745	552	425
1000/538	2050	1650	1300	1340	930	736	547	420
1050/566	2020	1635	1290	1320	918	730	540	415
1100/593	2020	1635	1290	1320	918	730	540	415

Table 6-101 Maximum Allowable Operating Pressure of Schedule 40S Alloy 800 Seamless Pipe, Plain End, at Different Nominal Pipe Sizes (inches)

Operating temperature (°F/°C)	Maximum allowable operating pressure (psi)									
	¼	½	¾	1	1½	2	3	4	6	8
100/38	5030	3900	3160	2960	2200	1850	1765	1500	1190	1050
200/93	4390	3350	2720	2560	1890	1590	1510	1280	1023	900
300/149	3900	3310	2460	2310	1710	1435	1365	1160	920	813
400/204	3480	2880	2265	2110	1560	1320	1255	1062	845	745
500/260	3330	2590	2115	1980	1468	1235	1170	996	795	698
600/316	3220	2500	2030	1905	1410	1185	1125	956	760	673
650/343	3150	2440	1990	1870	1380	1165	1105	937	745	655
700/371	3080	2390	1950	1825	1350	1140	1080	916	733	644
750/399	3060	2370	1930	1810	1340	1130	1075	910	725	637
800/427	2980	2320	1890	1770	1310	1105	1049	890	710	625
850/454	2970	2300	1875	1750	1300	1095	1042	881	703	619
900/482	2920	2280	1850	1735	1281	1082	1030	872	695	611
950/510	2890	2250	1830	1715	1270	1070	1015	861	688	604
1000/538	2860	2220	1810	1690	1255	1055	1005	850	680	596
1050/566	2820	2200	1790	1675	1240	1048	992	840	670	590
1100/593	2820	2200	1790	1675	1240	1048	992	840	670	590

Table 6-102 Maximum Allowable Operating Pressure of Schedule 80S Alloy 800 Seamless Pipe, Plain End, at Different Nominal Pipe Sizes (inches)

Operating temperature (°F/°C)	Maximum allowable operating pressure (psi)							
	¼	½	¾	1	1½	2	3	4
100/38	7110	5450	4470	4080	3100	2770	2450	2150
200/93	6140	4680	3830	3520	2660	2300	2100	1855
300/149	5530	4220	3460	3190	2400	2080	1890	1675
400/204	5070	3890	3160	2810	2220	1905	1740	1540
500/260	4750	3630	2980	2740	2065	1780	1625	1440
600/316	4560	3480	2850	2630	1985	1715	1565	1380
650/343	4480	3420	2800	2580	1945	1685	1540	1355
700/371	4370	3340	2740	2520	1915	1645	1505	1325
750/399	4350	3320	2720	2500	1889	1620	1489	1317
800/427	4250	3240	2660	2440	1845	1590	1460	1285
850/454	4200	3220	2640	2430	1830	1585	1445	1275
900/482	4160	3180	2600	2395	1810	1561	1430	1260
950/510	4110	3150	2570	2370	1785	1545	1410	1245
1000/538	4050	3100	2540	2340	1765	1525	1390	1230
1050/566	4020	3070	2520	2310	1745	1510	1380	1215
1100/593	4020	3070	2520	2310	1745	1510	1380	1215

Table 6-103 Support Spacing for Alloy 800 Pipe

Nominal pipe size (in.)	Support spacing (ft)		
	5S	10S	40S
¼		3.00	3.50
½	3.5	4.00	4.50
¾	4.5	5.00	5.50
1	5.5	5.50	6.40
1½	6.4	7.30	8.25
2	7.3	8.25	9.00
3	8.5	10.00	11.00
4	9.5	11.00	12.50
6			15.50
8			17.00

Table is based on pipe being uninsulated and carrying a fluid having a specific gravity of 1.35 and operating at a maximum temperature of 750°F (399°C). If the lines are to be insulated, reduce the spans by 30%.

Table 6-104 Compatibility of Alloy 800 and Alloy 825 with Selected Corrodents
The chemicals listed are in the pure state or in a saturated solution unless otherwise indicated. Compatibility is shown to the maximum allowable temperature for which data are available. Incompatibility is shown by an x. A blank space indicates that data are unavailable. When compatible, corrosion rate is <20 mpy.

Chemical	Maximum temp.		Chemical	Maximum temp.	
	°F	°C		°F	°C
Acetaldehyde			Barium carbonate	90	32
Acetamide			Barium chloride		
Acetic acid 10%[a]	200	93	Barium hydroxide		
Acetic acid 50%[a]	220	104	Barium sulfate	90	32
Acetic acid 80%[a]	210	99	Barium sulfide		
Acetic acid, glacial[a]	220	104	Benzaldehyde		
Acetic anhydride	230	110	Benzene	190	88
Acetone	210	99	Benzene sulfonic acid 10%		
Acetyl chloride	210	99	Benzoic acid 5%	90	32
Acrylic acid			Benzyl alcohol		
Acrylonitrile			Benzyl chloride		
Adipic acid			Borax	190	88
Allyl alcohol			Boric acid 5%	210	99
Allyl chloride			Bromine gas, dry[a]	90	32
Alum			Bromine gas, moist		
Aluminum acetate	60	16	Bromine liquid		
Aluminum chloride, aqueous	60	16	Butadiene		
Aluminum chloride, dry			Butyl acetate[a]	90	32
Aluminum fluoride 5%	80	27	Butyl alcohol		
Aluminum hydroxide	80	27	n-Butylamine		
Aluminum nitrate			Butyl phthalate		
Aluminum oxychloride			Butyric acid 5%	90	32
Aluminum sulfate	210	99	Calcium bisulfide		
Ammonia gas			Calcium bisulfite		
Ammonium bifluoride			Calcium carbonate	90	32
Ammonium carbonate	190	88	Calcium chlorate	80	27
Ammonium chloride 10%[a]	230	110	Calcium chloride[a,b]	60	16
Ammonium chloride 50%			Calcium hydroxide 10%	200	93
Ammonium chloride, sat.	200	93	Calcium hydroxide, sat.		
Ammonium fluoride 10%			Calcium hypochlorite	x	x
Ammonium fluoride 25%			Calcium nitrate		
Ammonium hydroxide 25%			Calcium oxide		
Ammonium hydroxide, sat.	110	43	Calcium sulfate	90	32
Ammonium nitrate	90	32	Caprylic acid		
Ammonium persulfate	90	32	Carbon bisulfide		
Ammonium phosphate			Carbon dioxide, dry		
Ammonium sulfate 10–40%	210	99	Carbon dioxide, wet		
Ammonium sulfide			Carbon disulfide		
Ammonium sulfite	210	99	Carbon monoxide	570	299
Amyl acetate[a]	200	93	Carbon tetrachloride	90	32
Amyl alcohol			Carbonic acid	90	32
Amyl chloride	90	32	Cellosolve		
Aniline	90	32	Chloracetic acid, 50% water		
Antimony trichloride	90	32	Chloracetic acid	x	x
Aqua regia 3 : 1			Chlorine gas, dry[a]	90	x

338

Table 6-104 *Continued*

Chemical	Maximum temp. °F	Maximum temp. °C	Chemical	Maximum temp. °F	Maximum temp. °C
Chlorine gas, wet	x	x	Magnesium chloride 1–5%	170	77
Chlorine, liquid			Malic acid	170	77
Chlorobenzene	90	32	Manganese chloride 10–50%	210	99
Chloroform	90	32	Methyl chloride		
Chlorosulfonic acid	x	x	Methyl ethyl ketone		
Chromic acid 10%[a]	210	99	Methyl isobutyl ketone		
Chromic acid 50%	x	x	Muriatic acid[a]	90	32
Chromyl chloride			Nitric acid 5%	90	32
Citric acid 15%	210	99	Nitric acid 20%	60	16
Citric acid, concentrated[a]	210	99	Nitric acid 70%		
Copper acetate	90	32	Nitric acid, anhydrous	210	99
Copper carbonate	90	32	Nitrous acid, concentrated		
Copper chloride 5%[a]	80	27	Oleum		
Copper cyanide	210	99	Perchloric acid 10%		
Copper sulfate	210	99	Perchloric acid 70%		
Cresol			Phenol	90	32
Cupric chloride 5%	x	x	Phosphoric acid 50–80%		
Cupric chloride 50%			Picric acid	90	32
Cyclohexane			Potassium bromide 5%	90	32
Cyclohexanol			Salicylic acid	90	32
Dichloroacetic acid			Silver bromide 10%[a]	90	32
Dichloroethane (ethylene di-chloride)			Sodium carbonate	90	32
			Sodium chloride[b]	200	93
Ethylene glycol			Sodium hydroxide 10%	90	32
Ferric chloride	x	x	Sodium hydroxide 50%		
Ferric chloride 50% in water	x	x	Sodium hydroxide, con-centrated	90	32
Ferric nitrate 10–50%	90	32	Sodium hypochlorite 20%		
Ferrous chloride[a,b]	90	32	Sodium hypochlorite, con-centrated		
Ferrous nitrate					
Fluorine gas, dry	x	x	Sodium sulfide to 50%	90	32
Fluorine gas, moist	x	x	Stannic chloride	x	x
Hydrobromic acid, dilute			Stannous chloride 5%	90	32
Hydrobromic acid 20%	x	x	Sulfuric acid 10% [a]	230	110
Hydrobromic acid 50%	x	x	Sulfuric acid 50%[a]	210	99
Hydrochloric acid 20%[a]	90	32	Sulfuric acid 70%[a]	150	66
Hydrochloric acid 38%	x	x	Sulfuric acid 90%[a]	180	82
Hydrocyanic acid 10%	60	16	Sulfuric acid 98%[a]	220	104
Hydrofluoric acid 30%	x	x	Sulfuric acid 100%[a]	230	110
Hydrofluoric acid 70%	x	x	Sulfuric acid, fuming	x	x
Hydrofluoric acid 100%	x	x	Sulfurous acid[a]	370	188
Hypochlorous acid			Thionyl chloride		
Iodine solution 10%			Toluene		
Ketones, general			Trichloroacetic acid		
Lactic acid 25%			White liquor		
Lactic acid, concentrated			Zinc chloride 5%	140	60

[a]Applicable to alloy 825 only.
[b]Material subject to pitting.
Source: Schweitzer, Philip A. (1991). *Corrosion Resistance Tables*, Marcel Dekker, Inc., New York, Vols. 1 and 2.

Table 6-105 Mechanical and Physical Properties of Alloy 825 Pipe

Modulus of elasticity $\times 10^6$, psi	28.3
Tensile strength $\times 10^3$, psi	85
Yield strength 0.2% offset $\times 10^3$, psi	35
Elongation in 2 in., %	30
Hardness, Brinell	150
Density, lb/in.^{3}	0.294
Specific gravity	8.14
Coefficient of thermal expansion, in./in. °F $\times 10^{-6}$	
at 80– 200°F	7.8
at 80– 400°F	8.3
at 80– 600°F	8.5
at 80– 800°F	8.7
at 80–1000°F	8.8
at 80–1200°F	9.1
at 80–1400°F	9.5
at 80–1600°F	9.7

Table 6-106 Mechanical and Physical Properties of Alloy B-2 Pipe

Modulus of elasticity $\times 10^6$, psi	31.4
Tensile strength $\times 10^3$, psi	110
Yield strength 0.2% offset $\times 10^3$, psi	60
Elongation in 2 in., %	60
Hardness, Brinell	210
Density, lb/in.^{3}	0.333
Specific gravity	9.22
Specific heat, at 212°F, Btu/lb °F	0.093
Thermal conductivity, Btu/ft^2/in. hr °F	
at 32°F	77
at 212°F	85
at 392°F	93
at 572°F	102
at 752°F	111
at 932°F	120
at 1112°F	130
Coefficient of thermal expansion, in./in. °F $\times 10^{-6}$	
at 68–200°F	5.7
at 68–600°F	6.2
at 68–1000°F	6.5

Table 6-107 Allowable Design Stress for Seamless Alloy
B-2 Pipe

Operating temperature (°F/°C)	Allowable design stress (psi)
100/38	25000
200/93	25000
300/149	24750
400/204	22750
500/260	21450
600/316	20750
650/343	20100

Table 6-108 Maximum Allowable Operating Pressure of Seamless Alloy B-2 Pipe, Schedule 5S

Nominal pipe size (in.)	Maximum allowable operating pressure (psi) at °F/°C						
	100/38	200/93	300/149	400/204	500/260	600/316	650/343
½	3520	3520	3480	3210	3020	2920	2835
¾	2840	2840	2810	2580	2440	2350	2290
1	2240	2240	2220	2040	1930	1860	1800
1½	1530	1530	1515	1390	1313	1270	1230
2	1220	1220	1209	1110	1095	1011	980
3	1055	1055	1045	910	905	875	850
4	815	815	809	745	700	679	655
6	727	727	720	660	627	604	585
8	558	558	552	507	480	463	447
10	550	550	545	500	475	457	443
12	574	574	567	522	493	476	460

Table 6-109 Maximum Allowable Operating Pressure of Seamless Alloy B-2 Pipe, Schedule 10S

Nominal pipe size (in.)	Maximum allowable operating pressure (psi) at °F/°C						
	100/38	200/93	300/149	400/204	500/260	600/316	650/343
⅜	4560	4560	4470	4110	3880	3750	3618
½	4620	4620	4560	4200	3960	3840	3720
¾	3660	3660	3620	3339	3140	3060	2940
1	3770	3770	3730	3430	3230	3130	3030
1½	2620	2620	2590	2381	2250	2178	2103
2	2075	2075	2050	1890	1770	1720	1665
3	1535	1535	1520	1400	1320	1275	1235
4	1190	1190	1179	1083	1020	988	956
6	900	900	890	820	768	746	722
8	760	760	753	693	653	631	610
10	680	680	674	620	584	565	546
12	625	625	620	569	537	519	502

Table 6-110 Maximum Allowable Operating Pressure of Seamless Alloy B-2 Pipe, Schedule 40S, Plain End

Nominal pipe size (in.)	Maximum allowable operating pressure (psi) at °F/°C						
	100/38	200/93	300/149	400/204	500/260	600/316	650/343
⅜	6510	6510	6450	5930	5580	5400	5240
½	6250	6250	6190	5680	5360	5180	5020
¾	5100	5100	5040	4640	4370	4230	4100
1	4760	4760	4710	4340	4090	3955	3830
1½	3530	3530	3500	3210	3030	2930	2840
2	2980	2980	2940	2710	2555	2470	2390
3	2820	2820	2790	2565	2420	2340	2270
4	2390	2390	2365	2165	2050	1990	1920
6	1905	1905	1885	1735	1635	1580	1530
8	1679	1679	1660	1525	1440	1390	1350
10	1520	1520	1510	1385	1308	1265	1220
12	1315	1315	1300	1188	1130	1090	1058

Table 6-111 Maximum Allowable Operating Pressure of Seamless Alloy B-2 Pipe, Schedule 80S, Plain End

Nominal pipe size (in.)	Maximum allowable operating pressure (psi) at °F/°C						
	100/38	200/93	300/149	400/204	500/260	600/316	650/343
1½	4960	4960	4910	4510	4260	4130	4000
2	4300	4300	4250	3920	3690	3570	3550
3	3920	3920	3885	3580	3370	3260	3150
4	3460	3460	3430	3155	2970	2875	2780
6	3000	3000	2970	2730	2570	2490	2410
8	2641	2641	2620	2410	2270	2200	2125
10	2105	2105	2080	1915	1805	1750	1690
12	1765	1765	1750	1610	1515	1470	1420

Table 6-112 Maximum Allowable Operating Pressure of Seamless Alloy B-2 Pipe, Schedule 40S, Threaded

Nominal pipe size (in.)	Maximum allowable operating pressure (psi) at °F/°C				
	100/38	200/93	300/149	400/204	500/260
⅜	2340	2340	2320	2130	2010
½	2060	2060	2040	1880	1770
¾	1820	1820	1800	1660	1560
1	1575	1575	1560	1433	1350
1½	1390	1390	1378	1267	1192
2	1280	1280	1268	1165	1100
3	1180	1180	1170	1075	1013
4	1120	1120	1110	1020	960
6	1044	1044	1034	950	895

Table 6-113 Maximum Allowable Operating Pressure of Seamless Alloy B-2 Pipe, Schedule 80S, Threaded

Nominal pipe size (in.)	Maximum allowable operating pressure (psi) at °F/°C				
	100/38	200/93	300/149	400/204	500/260
1½	2740	2740	2710	2490	2310
2	2515	2515	2490	2290	2160
3	2285	2285	2265	2080	1960
4	2135	2135	2115	1950	1835
6	2115	2115	2095	1925	1845

Table 6-114 Maximum Nonshock Ratings of Forged Alloy B-2 and Alloy C-276 Flanges

Operating temperature (°F/°C)	Maximum nonshock ratings (psi)					
	150 psi		300 psi		600 psi	
	B-2	C-276	B-2	C-276	B-2	C-276
−20 to 100/−29 to 38	350	345	910	900	1825	1805
150/66	345	335	900	875	1805	1750
200/93	335	230	880	845	1765	1685
250/121	330	310	865	815	1730	1625
300/149	320	300	845	790	1685	1575
350/177	315	290	825	765	1645	1530
400/204	305	285	800	740	1605	1480
450/232	295	280	770	725	1545	1455
500/260	285	275	745	715	1490	1435
550/288	280	265	725	695	1455	1390
600/316	275	260	715	680	1435	1360
650/343	265	260	700	680	1400	1360
700–1000/371–538		260		680		1360

Table 6-114 Maximum Nonshock Ratings of Forged Alloy B-2 and Alloy C-276 Flanges

Operating temperature (°F/°C)	Maximum nonshock ratings (psi)					
	150 psi		300 psi		600 psi	
	B-2	C-276	B-2	C-276	B-2	C-276
−20 to 100/−29 to 38	350	345	910	900	1825	1805
150/66	345	335	900	875	1805	1750
200/93	335	230	880	845	1765	1685
250/121	330	310	865	815	1730	1625
300/149	320	300	845	790	1685	1575
350/177	315	290	825	765	1645	1530
400/204	305	285	800	740	1605	1480
450/232	295	280	770	725	1545	1455
500/260	285	275	745	715	1490	1435
550/288	280	265	725	695	1455	1390
600/316	275	260	715	680	1435	1360
650/343	265	260	700	680	1400	1360
700–1000/371–538		260		680		1360

Table 6-115 Support Spacing for Alloy B-2 and Alloy C-276 Pipe

Nominal pipe size (in.)	Support spacing (ft)					
	Alloy B-2			Alloy C-276		
	55	105	405	55	105	405
⅜	5.0	5.5	6.5	4.0	4.5	5.5
½	5.5	6.0	7.0	4.5	5.0	6.0
¾	7.0	7.5	8.5	6.0	6.5	7.0
1	8.5	9.0	9.5	7.0	7.5	8.0
1½	9.5	11.0	12.0	8.0	9.5	10.5
2	11.0	12.5	14.0	9.5	11.0	12.0
3	13.0	15.0	16.5	11.0	13.0	14.5
4	14.5	17.0	19.5	12.5	14.5	17.0
6	14.5	21.5	24.0	17.0	18.5	21.0
8	21.5	24.0	26.5	18.5	21.0	23.0
10	25.5	28.0	31.0	22.0	24.5	27.0
12	26.5	29.5	32.0	23.0	25.5	28.0

Table is based on a maximum operating temperature of 650°F (343°C) for alloy B-2 and 1000°F (538°C) for alloy C-276 when lines are conveying a fluid having a specific gravity of 1.35 and are uninsulated. No allowance has been made for concentrated loads.

Table 6-116 Compatibility of Alloy B-2 and Alloy C-276 with Selected Corrodents

The chemicals listed are in the pure state or in a saturated solution unless otherwise indicated. Compatibility is shown to the maximum allowable temperature for which data are available. Incompatibility is shown by an x. A blank space indicates that data are unavailable. When compatible, corrosion rate is <20 mpy.

Chemical	Maximum temperature (°F/°C)	
	Alloy B-2	Alloy C-276
Acetaldehyde	80/27	140/60
Acetamide		60/16
Acetic acid 10%	300/149	300/149
Acetic acid 50%	300/149	300/149
Acetic acid 80%	300/149	300/149
Acetic acid, glacial	560/293	560/293
Acetic anhydride	280/138	280/138
Acetone	200/93	200/93
Acetyl chloride	80/27	
Acrylic acid	210/99	
Acrylonitrile	210/99	210/99
Adipic acid		210/99
Allyl alcohol		570/299
Allyl chloride	200/93	
Alum	150/66	150/66
Aluminum acetate	60/16	60/16
Aluminum chloride, aqueous	300/149	210/99
Aluminum chloride, dry	210/99	210/99
Aluminum fluoride	80/27	80/27
Aluminum hydroxide		
Aluminum nitrate		
Aluminum oxychloride		
Aluminum sulfate	210/99	210/99
Ammonia gas	200/93	200/93
Ammonium bifluoride		380/193
Ammonium carbonate	300/149	300/149
Ammonium chloride 10%	210/99	210/99
Ammonium chloride 50%	210/99	210/99
Ammonium chloride, sat.	570/299	570/299
Ammonium fluoride 10%	210/99	210/99
Ammonium fluoride 25%		210/99
Ammonium hydroxide 25%	210/99	570/299
Ammonium hydroxide, sat.	210/99	570/299
Ammonium nitrate		
Ammonium persulfate	x	
Ammonium phosphate		
Ammonium sulfate 10–40%	80/27	200/93
Ammonium sulfide		
Ammonium sulfite		100/38
Amyl acetate	340/171	340/171
Amyl alcohol		180/82

Table 6-116 Continued

Chemical	Maximum temperature (°F/°C)	
	Alloy B-2	Alloy C-276
Amyl chloride	210/99	90/32
Aniline	570/299	570/299
Antimony trichloride	210/99	210/99
Aqua regia 3:1	x	x
Barium carbonate	570/299	570/299
Barium chloride	570/299	210/99
Barium hydroxide	270/132	270/132
Barium sulfate	80/27	
Barium sulfide		
Benzaldehyde	210/99	210/99
Benzene	210/99	210/99
Benzene sulfonic acid 10%	210/99	210/99
Benzoic acid	210/99	
Benzyl alcohol	210/99	210/99
Benzyl chloride	210/99	
Borax	120/49	120/49
Boric acid	570/299	570/299
Bromine gas, dry	60/16	60/16
Bromine gas, moist		60/16
Bromine liquid		180/82
Butadiene	300/149	300/149
Butyl acetate	200/93	200/93
Butyl alcohol	210/99	200/93
n-Butylamine	210/99	210/99
Butyl phthalate		
Butyric acid	280/138	280/138
Calcium bisulfide		
Calcium bisulfite		80/27
Calcium carbonate	210/99	210/99
Calcium chlorate		210/99
Calcium chloride	350/177	350/177
Calcium hydroxide 10%	210/99	170/77
Calcium hydroxide, sat.	210/99	
Calcium hypochlorite	x	
Calcium nitrate	210/99	210/99
Calcium oxide		90/32
Calcium sulfate 10%	320/160	320/160
Caprylic acid	300/149	300/149
Carbon bisulfide	180/82	210/99
Carbon dioxide, dry	570/299	570/299
Carbon dioxide, wet	570/299	200/93
Carbon disulfide	180/82	300/149
Carbon monoxide	570/299	570/299
Carbon tetrachloride	300/149	300/149
Carbonic acid	80/27	80/27
Cellosolve	210/99	210/99
Chloracetic acid, 50% water		210/99

Table 6-116 Continued

	Maximum temperature (°F/°C)	
Chemical	Alloy B-2	Alloy C-276
Chloracetic acid	370/188	300/149
Chlorine gas, dry	200/93	570/299
Chlorine gas, wet	x	220/104
Chlorine, liquid		110/43
Chlorobenzene	350/177	350/177
Chloroform	210/99	210/99
Chlorosulfonic acid	230/110	230/110
Chromic acid 10%	130/54	210/99
Chromic acid 50%	x	210/99
Chromyl chloride	210/99	210/99
Citric acid 15%	210/99	210/99
Citric acid, concentrated	210/99	210/99
Copper acetate	100/38	100/38
Copper carbonate	90/32	90/32
Copper chloride	200/93	200/93
Copper cyanide	150/66	150/66
Copper sulfate	210/99	210/99
Cresol	210/99	210/99
Cupric chloride 5%	60/16	210/99
Cupric chloride 50%	210/99	210/99
Cyclohexane	210/99	210/99
Cyclohexanol	80/27	80/27
Dichloroacetic acid		
Dichloroethane	230/110	230/110
Ethylene glycol	570/299	570/299
Ferric chloride	90/32	90/32
Ferric chloride 50% in water	x	
Ferric nitrate 10–50%	x	
Ferrous chloride	280/138	280/138
Ferrous nitrate		
Fluorine gas, dry	80/27	150/66
Fluorine gas, moist		570/299
Hydrobromic acid, dilute	210/99	
Hydrobromic acid 20%	210/99	90/32
Hydrobromic acid 50%	260/127	90/32
Hydrochloric acid 20%	140/60	150/66
Hydrochloric acid 38%	140/60	90/32
Hydrocyanic acid 10%		
Hydrofluoric acid 30%	140/60	210/99
Hydrofluoric acid 70%	110/43	200/93
Hydrofluoric acid 100%	80/27	210/99
Hypochlorous acid	90/32	80/27
Iodine solution 10%		180/82
Ketones, general	180/82	100/38
Lactic acid 25%	250/121	210/99
Lactic acid, concentrated	250/121	210/99
Magnesium chloride	300/149	300/149

Table 6-116   Continued

Chemical	Maximum temperature (°F/°C)	
	Alloy B-2	Alloy C-276
Malic acid	210/99	210/99
Manganese chloride 40%	210/99	210/99
Methyl chloride	210/99	90/32
Methyl ethyl ketone	210/99	210/99
Methyl isobutyl ketone	200/93	200/93
Muriatic acid	90/32	90/32
Nitric acid 5%	x	210/99
Nitric acid 20%	x	160/71
Nitric acid 70%	x	200/93
Nitric acid, anhydrous	x	80/27
Nitrous acid, concentrated	x	x
Oleum to 25%	110/43	140/60
Perchloric acid 10%		
Perchloric acid 70%		220/104
Phenol	570/299	570/299
Phosphoric acid 50–80%	210/99	210/99
Picric acid	220/104	300/149
Potassium bromide 30%	90/32	90/32
Salicylic acid	80/27	250/121
Silver bromide 10%	90/32	90/32
Sodium carbonate	570/299	210/99
Sodium chloride to 30%	210/99	210/99
Sodium hydroxide 10%[a]	240/116	230/110
Sodium hydroxide 50%	250/121	210/99
Sodium hydroxide, concentrated	200/93	120/49
Sodium hypochlorite 20%	x	x
Sodium hypochlorite, concentrated	x	x
Sodium sulfide to 50%	210/99	210/99
Stannic chloride to 50%	210/99	210/99
Stannous chloride[b]	570/299	210/99
Sulfuric acid 10%	210/99	200/93
Sulfuric acid 50%	230/110	230/110
Sulfuric acid 70%	290/143	290/143
Sulfuric acid 90%	190/88	190/88
Sulfuric acid 98%	280/138	210/99
Sulfuric acid 100%	290/143	190/88
Sulfuric acid, fuming	210/99	90/32
Sulfurous acid	210/99	370/188
Thionyl chloride		
Toluene	210/99	210/99
Trichloroacetic acid	210/99	210/99
White liquor	100/38	100/38
Zinc chloride	60/16	250/121

[a]Alloy B-2 is subject to stress cracking.
[b]Alloy B-2 is subject to pitting.
Source: Schweitzer, Philip A. (1991). *Corrosion Resistance Tables*, Marcel Dekker, Inc., New York, Vols. 1 and 2.

Table 6-117 Mechanical and Physical Properties of Alloy C-276

Modulus of elasticity $\times$ 10^6, psi	29.8
Tensile strength $\times$ 10^3, psi	100
Yield strength 0.2% offset $\times$ 10^3, psi	41
Elongation in 2 in., %	40
Hardness, Brinell	190
Density, lb/in.^{3}	0.321
Specific gravity	8.89
Specific heat, Btu/lb°F	0.102
Thermal conductivity, Btu/ft^2/hr°F/in.	
at $-$ 270°F	50
at 0°F	65
at 100°F	71
at 200°F	77
at 400°F	90
at 600°F	104
at 800°F	117
at 1000°F	132
at 1200°F	145
Coefficient of thermal expansion, in./in. °F $\times$ 10^{-6}	
at 75– 200°F	6.2
at 75– 400°F	6.7
at 75– 600°F	7.1
at 75– 800°F	7.3
at 75–1000°F	7.4
at 75–1200°F	7.8
at 75–1400°F	8.3
at 75–1600°F	8.8

Table 6-118 Allowable Design Stress for Alloy C-276
Seamless Pipe

Operating temperature (°F/°C)	Allowable design stress (psi)
100/38	25,000
200/93	24,500
300/149	24,000
400/204	22,500
500/260	20,750
600/316	19,400
700/371	18,750
800/427	16,750
900/482	16,750
1000/538	15,400

Table 6-119 Maximum Allowable Operating Pressure of Seamless Alloy C-276 Pipe, Schedule 5S

Nominal pipe size (in.)	Maximum allowable operating pressure (psi) at °F/°C									
	100/ 38	200/ 93	300/ 149	400/ 204	500/ 260	600/ 316	700/ 371	800/ 427	900/ 482	1000/ 538
½	3520	3450	3380	3170	2920	2740	2640	2360	2360	2170
¾	2830	2760	2720	2550	2350	2200	2130	1900	1900	1745
1	2240	2195	2150	2020	1860	1740	1680	1500	1500	1380
1½	1530	1500	1470	1375	1270	1185	1150	1025	1025	943
2	1218	1195	1170	1095	1010	945	915	818	818	750
3	1053	1032	1011	950	875	820	790	706	706	650
4	819	802	785	736	678	635	614	548	548	504
6	728	714	700	655	604	565	545	488	488	449
8	558	547	535	503	463	433	419	374	374	344
10	550	540	529	495	456	427	414	368	368	319
12	573	560	550	575	475	445	430	382	386	351

Table 6-120 Maximum Allowable Operating Pressure of Seamless Alloy C-276 Pipe, Schedule 10S

Nominal pipe size (in.)	Maximum allowable operating pressure (psi) at °F/°C									
	100/ 38	200/ 93	300/ 149	400/ 204	500/ 260	600/ 316	700/ 371	800/ 427	900/ 482	1000/ 538
⅜	4510	4430	4340	4060	3750	3500	3390	3030	3030	2780
½	4620	4530	4440	4150	3830	3580	3460	3080	3080	2840
¾	3660	3590	3520	3300	3040	2840	2750	2460	2460	2260
1	3760	3690	3610	3390	3120	2920	2820	2520	2520	2320
1½	2620	2570	2510	2360	2170	2030	1965	1755	1755	1610
2	2080	2040	1995	1870	1723	1610	1560	1390	1390	1280
3	1535	1505	1475	1382	1275	1190	1150	1030	1030	945
4	1190	1165	1140	1070	985	925	892	798	798	734
6	898	830	863	810	746	697	674	603	603	554
8	760	745	730	684	630	590	570	509	509	469
10	680	667	655	612	565	528	510	466	466	420
12	635	613	600	562	519	485	479	419	419	385

Table 6-121 Maximum Allowable Operating Pressure of Seamless Alloy C-276 Pipe, Schedule 40S, Plain End

Nominal pipe size (in.)	Maximum allowable operating pressure (psi) at °F/°C									
	100/ 38	200/ 93	300/ 149	400/ 204	500/ 260	600/ 316	700/ 371	800/ 427	900/ 482	1000/ 538
⅜	6500	6390	6250	5860	5400	5050	4890	4360	4360	4010
½	6250	6100	6000	5610	5190	4850	4670	4180	4180	3840
¾	5090	4970	4870	4570	4220	3940	3810	3410	3410	3130
1	4770	4670	4570	4300	3960	3700	3570	3190	3190	1940
1½	3530	3460	3395	3180	2940	2740	2650	2370	2370	2180
2	2980	2910	2850	2675	2470	2310	2230	1990	1990	1830
3	2820	2760	2700	2540	2340	2185	2110	1890	1890	1735
4	2390	2345	2300	2155	1990	1860	1795	1605	1605	1475
6	1905	1865	1830	1713	1580	1480	1430	1278	1278	1173
8	1680	1643	1611	1510	1393	1300	1260	1125	1125	1033
10	1520	1490	1460	1370	1261	1180	1140	1020	1020	936
12	1315	1290	1261	1180	1090	1020	986	880	880	820

Table 6-122 Maximum Allowable Operating Pressure of Seamless Alloy C-276 Pipe, Schedule 80S, Plain End

Nominal pipe size (in.)	Maximum allowable operating pressure (psi) at °F/°C									
	100/ 38	200/ 93	300/ 149	400/ 204	500/ 260	600/ 316	700/ 371	800/ 427	900/ 482	1000/ 538
1½	4970	4860	4770	4480	4130	3860	3730	3330	3330	3060
2	4300	4200	4120	3860	3560	3330	3220	2875	2875	2640
3	3920	3840	3760	3520	3250	3040	2940	2620	2620	2420
4	3460	3400	3320	3120	2875	2690	2600	2320	2320	2135
6	2990	2930	2870	2690	2480	2320	2240	2000	2000	1840
8	2640	2585	2540	2375	2190	2050	1980	1770	1770	1630
10	2100	2060	2020	1890	1748	1632	1578	1410	1410	1295
12	1765	1730	1695	1590	1465	1370	1324	1185	1185	1090

Table 6-123 Maximum Allowable Operating Pressure of Seamless Alloy C-276 Pipe, Schedule 40S, Threaded

Nominal pipe size (in.)	Maximum allowable operating pressure (psi) at °F/°C				
	100/38	200/93	300/149	400/204	500/260
⅜	2390	2240	2250	2110	1940
½	2060	2025	1980	1860	1715
¾	1820	1780	1745	1635	1510
1	1575	1543	1510	1415	1309
1½	1390	1360	1335	1250	1155
2	1280	1255	1230	1150	1060
3	1180	1160	1135	1061	980
4	1120	1095	1075	1008	932
6	1040	1020	1000	936	865

Table 6-124 Maximum Allowable Operating Pressure of Seamless Alloy C-276 Schedule 80S Pipe, Threaded

Nominal pipe size (in.)	Maximum allowable operating pressure (psi) at °F/°C				
	100/38	200/93	300/149	400/204	500/260
1½	2740	2680	2630	2465	2275
2	2520	2470	2420	2270	2090
3	2290	2240	2195	2060	1900
4	2135	2090	2050	1920	1772
6	2115	2075	2030	1905	1760

Table 6-125 Mechanical and Physical Properties of Alloy G and Alloy G-3

Property	G	G-3
Modulus of elasticity $\times 10^6$, psi	27.8	27.8
Tensile strength $\times 10^3$, psi	90	90
Yield strength 0.2% offset $\times 10^3$, psi	35	35
Elongation in 2 in., %	35	45
Hardness, Brinell	169	885(Rb)
Density, lb/in.^{3}	0.30	0.30
Specific gravity	8.31	8.31
Specific heat, J/kg K	456	464
Thermal conductivity, W/mK	10.1	10.0
Coefficient of thermal expansion, in./in. °F $\times 10^{-6}$		
at 70– 200°F	7.5	7.5
at 70– 400°F	7.7	7.7
at 70– 600°F	7.9	7.9
at 70– 800°F	8.3	8.3
at 70–1000°F	8.7	8.7
at 70–1200°F	9.1	9.1

Table 6-126 Mechanical and Physical Properties of Aluminum Alloys

Property	Aluminum alloy			
	3003-3	5052-0	6061-T6	6063-T6
Modulus of elasticity $\times$ 10^6, psi	10	10.2	10	10
Tensile strength $\times$ 10^3, psi	17	41	45	35
Yield strength 0.2% offset $\times$ 10^3, psi	8	36	40	31
Elongation in 2 in., %	40	25	12	12
Density, lb/in.^{3}	.099	.097		.098
Specific gravity	2.73	2.68		2.70
Specific heat, Btu/hr °F	0.23	0.23		
Thermal conductivity, Btu/hr/ft^2/°F/in.	1070	960	900	1090
Coefficient of thermal expansion, in./°F/in. $\times$ 10^{-6}				
at −58–68°F	12		12.1	12.1
at 68–212°F	12.9	13.2	13.0	13.0
at 68–392°F	13.5		13.5	13.6
at 68–572°F	13.9		14.1	14.2

Table 6-127 Allowable Design Stress for Aluminum Alloys, Welded Construction

Operating temperature (°F/°C)	Allowable stress (psi)				
	1060	3003	5052	6061	6063
100/38	1650	3350	6250	6000	4250
150/66	1650	3150	6250	5900	4200
200/93	1600	2900	6200	5700	4000
250/121	1450	2700	6000	5400	3800
300/149	1250	2400	5400	5000	3600
350/177	1200	2100	4650	4200	2750
400/204	1050	1800	3500	3200	1900

Table 6-128 Allowable Design Stress for Aluminum Alloys, Seamless Construction

Operating temperature (°F/°C)	Allowable stress (psi)	
	3003-0	6061-T-6
−452/−269	3350	12650
100/38	3350	12650
150/66	3350	12650
200/93	3350	12650
250/121	3250	12250
300/149	2400	10500
350/177	2100	
400/204	1800	

Table 6-129 Maximum Allowable Operating Pressure of Welded Aluminum Alloy 1060 Pipe

Nominal pipe size (in.)	Maximum allowable operating pressure (psi) at °F/°C											
	Schedule 40S						Schedule 80S					
	100/38	200/93	250/121	300/149	350/177	400/204	100/38	200/93	250/121	300/149	350/177	400/204
½	412	400	362	312	300	262	580	560	510	436	419	366
¾	336	326	295	255	244	214	472	468	415	358	343	300
1	314	305	276	238	229	200	434	421	382	329	316	276
1½	233	226	205	176	169	149	328	318	288	248	238	209
2	196	190	172	149	143	125	283	275	249	215	206	180
3	186	181	164	151	135	119	263	255	231	200	192	168
4	156	153	139	120	115	100	228	221	201	173	166	145
6	126	122	110	95	91	80	197	191	173	150	144	126
8	111	107	97	89	81	70	174	169	153	132	127	111
10	100	97	88	76	73	64	139	135	122	105	201	88
12	87	84	76	66	63	55	116	113	102	88	85	74

Table 6-130 Maximum Allowable Operating Pressure of Welded Aluminum Alloy 3003 Pipe

Nominal pipe size (in.)	Maximum allowable operating pressure (psi) at °F/°C											
	Schedule 40S						Schedule 80S					
	100/38	200/93	250/121	300/149	350/177	400/204	100/38	200/93	250/121	300/149	350/177	400/204
½	840	730	680	600	530	450	1170	1020	950	840	740	630
¾	690	600	550	489	428	367	960	830	780	690	610	520
1	640	560	520	458	401	343	890	770	720	640	560	474
1½	473	410	381	339	297	254	670	580	540	477	418	358
2	399	345	322	286	250	215	580	498	464	413	361	310
3	379	328	305	271	238	204	540	463	431	384	336	288
4	321	278	259	230	201	173	464	402	374	332	291	249
6	256	221	206	183	161	138	401	347	323	288	252	216
8	225	195	182	162	141	121	355	307	286	254	222	191
10	205	177	165	147	128	110	282	244	228	202	177	152
12	177	153	143	127	111	95	237	205	191	170	149	128
14	161	139	130	115	101	87	215	186	174	154	135	116
16	140	121	113	101	88	76	188	163	151	135	118	101
18	124	108	100	89	78	67	167	144	134	120	105	90
20	112	97	90	80	70	60	150	130	121	107	94	81
24	93	81	75	67	58	50	124	108	100	89	78	67
30	74	64	60	53	46	40	99	86	80	71	62	53
36	62	53	50	44	39	33	82	71	66	59	52	44

Table 6-131 Maximum Allowable Operating Pressure of Welded Aluminum Alloy 3003-0 Pipe

Maximum allowable operating pressure (psi) at °F/°C

Nominal pipe size (in.)	Schedule 40S					Schedule 80S				
	-452/-269 to 200/93	250/121	300/149	350/177	400/204	-452/-269 to 200/93	250/121	300/149	350/177	400/204
½	836	810	599	524	449	1169	1134	837	732	678
¾	682	661	488	427	366	958	929	686	600	514
1	638	619	457	400	342	882	855	631	552	473
1¼	525	509	376	329	282	733	711	525	459	394
1½	472	458	338	296	253	666	646	477	417	357
2	398	386	285	249	213	575	557	411	360	308
2½	435	422	311	272	233	603	585	432	378	324
3	378	366	270	237	203	534	518	382	335	287
4	320	311	229	200	172	463	449	331	290	248
6	255	247	182	160	137	400	388	286	251	215
8	224	218	161	140	120	354	343	253	222	190
10	203	197	146	127	109	281	273	201	176	151
12	176	170	126	110	94	236	229	169	148	127
14	160	155	114	100	85	214	208	153	134	115
16	139	135	100	87	75	187	189	134	117	100
18	123	120	88	77	66	166	161	118	104	89
20	111	108	79	69	59	149	144	106	93	80
24	92	89	66	58	49	123	120	88	77	66

Table 6-132 Maximum Allowable Operating Pressure of Welded Aluminum Alloy 5052 Pipe

| Nominal pipe size (in.) | Maximum allowable operating pressure (psi) at °F/°C | | | | | | | | | | | |
| | Schedule 40S | | | | | | Schedule 80S | | | | | |
	100/38	200/93	250/121	300/149	350/177	400/204	100/38	200/93	250/121	300/149	350/177	400/204
½	1570	1550	1500	1350	1170	880	2190	2170	2100	1890	1630	1230
¾	1280	1270	1230	1100	950	720	1790	1780	1720	1550	1330	1010
1	1200	1190	1150	1030	890	670	1650	1640	1580	1430	1230	930
1½	890	880	850	770	660	494	1250	1240	1200	1080	930	700
2	750	740	720	650	560	416	1080	1070	1030	930	800	610
3	710	700	680	610	530	395	1000	990	960	870	750	560
4	600	600	580	520	445	335	870	860	830	750	650	484
6	475	472	457	411	354	267	750	750	720	650	360	419
8	419	416	403	362	312	235	670	660	640	580	492	370
10	381	378	365	329	283	213	530	530	510	454	391	294
12	329	326	316	284	245	184	441	438	424	381	328	247
14	299	296	287	258	222	167	401	397	385	346	298	224
16	261	259	250	225	194	146	349	347	335	302	260	196
18	231	229	222	200	172	130	310	308	298	268	231	174
20	207	206	199	176	154	116	278	276	267	240	207	156
24	172	171	166	149	128	97	231	229	222	200	172	130
30	138	137	133	119	103	77	184	183	177	159	137	103
36	115	114	110	99	86	64	154	153	148	133	114	86

Table 6-133 Maximum Allowable Operating Pressure of Welded Aluminum Alloy 6061 Pipe

Nominal pipe size (in.)	Maximum allowable operating pressure (psi) at °F/°C											
	Schedule 40S						Schedule 80S					
	100/38	200/93	250/121	300/149	350/177	400/204	100/38	200/93	250/121	300/149	350/177	400/204
½	1500	1430	1350	1250	1050	800	2100	1990	1890	1750	1470	1120
¾	1230	1170	1100	1020	860	660	1720	1640	1550	1430	1210	920
1	1150	1090	1030	960	810	610	1580	1510	1430	1320	1110	850
1½	850	810	770	710	600	452	1200	1140	1080	1000	840	640
2	720	680	650	600	500	381	1040	980	930	860	730	550
3	680	650	610	570	475	362	960	910	870	800	680	520
4	580	555	520	479	402	307	830	790	750	700	590	443
6	458	435	412	381	321	244	720	690	650	600	510	383
8	403	333	363	336	282	215	640	610	580	530	444	339
10	366	348	329	305	256	195	510	480	455	421	354	270
12	316	300	285	263	221	169	424	403	382	353	297	226

Table 6-134 Maximum Allowable Operating Pressure of Seamless Aluminum Alloy 6061-T6 Pipe

Nominal pipe size (in.)	Maximum allowable operating pressure (psi) at °F/°C					
	Schedule 40S			Schedule 80S		
	−452/−259 to 200/93	250/121	300/149	−452/−269 to 200/93	250/121	300/149
½	3159	3059	2622	4414	4275	3664
¾	2576	2495	2138	3618	3503	3003
1	2409	2333	2000	3330	3225	2764
1¼	1984	1921	1646	2770	2682	2299
1½	1784	1728	1481	2515	2436	2088
2	1503	1456	1248	2171	2102	1802
3	1427	1382	1185	2018	1954	1675
4	1210	1172	1004	1749	1694	1452
6	964	933	800	1512	1464	1255
8	848	821	704	1337	1295	1110
10	769	745	639	1064	1030	883
12	664	643	551	892	864	740
14	604	585	501	810	785	673
16	527	510	437	810	684	587
18	468	453	388	627	607	520
20	420	407	349	563	545	467
24	349	338	290	468	453	388

Table 6-135 Maximum Allowable Operating Pressure of Welded Aluminum Alloy 6063 Pipe

Nominal pipe size (in.)	Schedule 40S								Schedule 80S							
	100/38	200/93	250/121	300/149	350/177	400/204			100/38	200/93	250/121	300/149	350/177	400/204		
½	1070	1000	950	900	690	475			1490	1400	1330	1260	960	670		
¾	870	820	780	740	570	388			1220	1150	1090	1030	790	550		
1	810	770	730	690	530	362			1120	1060	1010	950	730	510		
1½	600	570	540	510	389	269			850	800	760	720	550	378		
2	510	476	452	429	327	226			730	690	660	620	473	327		
3	480	452	430	407	311	215			680	640	610	580	439	304		
4	407	383	364	345	264	182			590	560	530	498	381	263		
6	324	305	290	275	210	145			510	479	455	431	329	228		
8	286	269	255	242	185	128			450	423	402	381	291	201		
10	259	244	232	220	168	116			358	337	320	303	232	160		
12	224	211	200	190	145	100			301	283	269	255	195	135		

Maximum allowable operating pressure (psi) at °F/°C

Table 6-136 Maximum Nonshock Ratings of Forged Aluminum Flanges at Different Primary Service Ratings

Operating temperature (°F/°C)	Maximum nonshock rating (psi)											
	150 psi		300 psi		600 psi		900 psi		1500 psi		2500 psi	
	3003	6061	3003	6061	3003	6061	3003 H112	6061[a] T-6	3003 H112	6061[a] T-6	3003 H112	6061[a] T-6
−20 to 100/−29 to 38							310	2160	515	3600	860	6000
100/38	40	275	105	720	210	1440						
150/66	40	270	100	710	200	1420						
200/93	35	265	95	700	190	1400	290	2100	485	3500	805	5825
250/121	35	260	95	675	190	1350						
300/149	35	215	85	565	170	1130	260	1690	430	2820	720	4700
350/177	30	155	80	410	160	820						
400/204	25	100	60	265	120	530	185	800	310	1345	515	2225

[a]The ratings of slip-on and socket welding types of flanges are two thirds of the tabulated values, since welding these types of flanges to pipe decreases their strength.

Table 6-137 Support Spacing for Aluminum Alloy 6063 Pipe

Nominal pipe size (in.)	Support spacing (ft)		
	5S	10S	40S
½	5.0	5.5	6.0
¾	5.5	6.0	6.5
1	6.0	7.0	7.5
1½	6.5	7.5	8.5
2	6.5	8.0	9.0
3	7.5	8.5	10.0
4	8.0	9.0	11.5
6	9.0	10.0	13.0
8	9.5	11.0	14.0
10	10.0	11.5	15.0
12	10.5	13.0	15.5

Support spacing is based on the pipe being uninsulated, operating at a maximum temperature of 400°F (204°C), and conveying a fluid having a specific gravity of 1.35. No allowance has been made for concentrated loads on the pipe such as valves, meters, etc. These must be supported independently.

Table 6-138 Support Spacing for Aluminum Alloy 6061 Pipe

Nominal pipe size (in.)	Support spacing (ft)		
	5S	10S	40S
½	6.5	7.0	7.5
¾	7.0	7.5	8.4
1	7.5	9.0	9.7
1½	8.4	9.7	11.0
2	8.4	10.4	11.5
3	9.7	11.0	13.0
4	10.4	11.5	15.0
6	11.5	13.0	16.8
8	12.3	14.25	18.0
10	13.0	15.0	19.0
12	13.6	16.8	20.0

Support spacing is based on the pipe being uninsulated, operating at a maximum temperature of 400°F (204°C), and conveying a fluid having a specific gravity of 1.35. No allowance has been made for concentrated loads on the pipe such as valves, meters, etc. These must be supported independently.

Table 6-139 Support Spacing for Aluminum Alloy 5052 Pipe

Nominal pipe size (in.)	Support spacing (ft)		
	5S	10S	40S
½	6.75	7.4	7.0
¾	7.40	7.0	8.8
1	8.00	9.5	10.0
1½	8.80	10.0	11.5
2	8.80	10.8	12.2
3	10.00	11.5	13.5
4	10.80	12.2	15.6
6	12.00	13.5	17.6
8	12.80	14.9	19.0
10	13.50	15.6	20.3
12	14.25	17.6	21.0

Support spacing is based on the pipe being uninsulated, operating at a maximum temperature of 400°F (204°C), and conveying a fluid having a maximum specific gravity of 1.35. No allowance has been made for concentrated loads such as valves, meters, etc. These must be supported independently.

Table 6-140 Support Spacing for Aluminum Alloy 3003 Pipe

Nominal pipe size (in.)	Support spacing (ft)		
	5S	10S	40S
½	4.8	5.3	5.8
¾	5.3	5.8	6.3
1	5.8	6.8	7.3
1½	6.3	7.3	8.25
2	6.3	7.7	8.7
3	7.3	8.2	9.7
4	7.7	8.75	11.0
6	8.7	9.7	12.6
8	9.25	10.7	13.6
10	9.7	11.0	14.6
12	10.2	12.6	15.0

Support spacing is based on the pipe being uninsulated, operating at a maximum temperature of 400°F (204°C), and conveying a fluid having a maximum specific gravity of 1.35. No allowance has been made for concentrated loads. Items such as valves, meters, etc. installed in the piping system must be installed independently.

Table 6-141 Support Spacing for Aluminum
Alloy 1060 Pipe

Nominal pipe size (in.)	Support spacing (ft)		
	5S	10S	40S
½	3.7	4.0	4.4
¾	4.0	4.4	4.8
1	4.4	5.2	5.5
1½	4.8	5.5	6.3
2	4.8	5.9	6.6
3	5.5	6.3	7.4
4	4.9	6.6	8.5
6	6.6	7.4	9.6
8	7.0	8.0	10.4
10	7.4	8.5	11.0
12	7.8	9.6	11.5

Support spacing is based on the pipe being uninsulated,
operating at a maximum temperature of 400°F (204°C),
and conveying a fluid having a specific gravity of 1.35.
No allowance has been made for concentrated loads such
as valves, meters, etc. These must be supported independently.

Table 6-142 Compatibility of Aluminum Alloys with Selected Corrodents
The chemicals listed are in the pure state or in a saturated solution unless otherwise indicated. Compatibility is shown to the maximum allowable temperature for which data are available. Incompatibility is shown by an x. A blank space indicates that data are unavailable. When compatible, corrosion rate is <20 mpy.

Chemical	Maximum temp.		Chemical	Maximum temp.	
	°F	°C		°F	°C
Acetaldehyde	360	182	Barium carbonate	x	x
Acetamide	340	171	Barium chloride 30%	180	82
Acetic acid 10%	110	43	Barium hydroxide	x	x
Acetic acid 50%	130	54	Barium sulfate	210	99
Acetic acid 80%	90	32	Barium sulfide	x	x
Acetic acid, glacial	210	99	Benzaldehyde	120	49
Acetic anhydride	350	177	Benzene	210	99
Acetone	500	260	Benzene sulfonic acid 10%	x	x
Acetyl chloride	x	x	Benzoic acid 10%	400	204
Acrylic acid			Benzyl alcohol	110	43
Acrylonitrile	210	99	Benzyl chloride	x	x
Adipic acid	210	99	Borax	x	x
Allyl alcohol	150	66	Boric acid	100	38
Allyl chloride	x	x	Bromine gas, dry	60	16
Alum	110	43	Bromine gas, moist	x	x
Aluminum acetate	60	16	Bromine liquid	210	99
Aluminum chloride, aqueous	x	x	Butadiene	110	43
Aluminum chloride, dry	60	16	Butyl acetate	110	43
Aluminum fluoride	120	49	Butyl alcohol	210	99
Aluminum hydroxide	80	27	n-Butylamine	90	32
Aluminum nitrate	110	43	Butyl phthalate	x	x
Aluminum oxychloride			Butyric acid	180	82
Aluminum sulfate	x	x	Calcium bisulfide		
Ammonia gas	x	x	Calcium bisulfite	x	x
Ammonium bifluoride			Calcium carbonate	x	x
Ammonium carbonate	350	177	Calcium chlorate	140	60
Ammonium chloride 10%	x	x	Calcium chloride 20%	100	38
Ammonium chloride 50%	x	x	Calcium hydroxide 10%	x	x
Ammonium chloride, sat.	x	x	Calcium hydroxide, sat.	x	x
Ammonium fluoride 10%	x	x	Calcium hypochlorite	x	x
Ammonium fluoride 25%	x	x	Calcium nitrate	170	77
Ammonium hydroxide 25%	350	177	Calcium oxide	90	32
Ammonium hydroxide, sat.	350	177	Calcium sulfate[a]	210	99
Ammonium nitrate	350	177	Caprylic acid	300	149
Ammonium persulfate	350	177	Carbon bisulfide	210	99
Ammonium phosphate	x	x	Carbon dioxide, dry	570	299
Ammonium sulfate 10–40%	x	x	Carbon dioxide, wet	170	77
Ammonium sulfide	170	77	Carbon disulfide	210	99
Ammonium sulfite	x	x	Carbon monoxide	570	299
Amyl acetate	350	177	Carbon tetrachloride	x	x
Amyl alcohol	170	77	Carbonic acid	80	27
Amyl chloride	90	32	Cellosolve	210	99
Aniline[a]	350	177	Chloracetic acid, 50% water	x	x
Antimony trichloride	x	x	Chloracetic acid	x	x
Aqua regia 3 : 1	x	x	Chlorine gas, dry	210	99

365

Table 6-142 *Continued*

Chemical	°F	°C	Chemical	°F	°C
Chlorine gas, wet	x	x	Magnesium chloride	x	x
Chlorine, liquid			Malic acid	210	99
Chlorobenzene	150	66	Manganese chloride		
Chloroform, dry	170	77	Methyl chloride	x	x
Chlorosulfonic acid, dry	170	77	Methyl ethyl ketone	150	66
Chromic acid 10%	200	93	Methyl isobutyl ketone	150	66
Chromic acid 50%	100	38	Muriatic acid	x	x
Chromyl chloride	210	99	Nitric acid 5%	x	x
Citric acid 15%	210	99	Nitric acid 20%	x	x
Citric acid, concentrated	70	21	Nitric acid 70%	x	x
Copper acetate	x	x	Nitric acid, anhydrous	90	32
Copper carbonate	x	x	Nitrous acid, concentrated	x	x
Copper chloride	x	x	Oleum	100	38
Copper cyanide	x	x	Perchloric acid 10%	x	x
Copper sulfate	x	x	Perchloric acid 70%	x	x
Cresol	150	66	Phenol	210	99
Cupric chloride 5%	x	x	Phosphoric acid 50–80%	x	x
Cupric chloride 50%			Picric acid	210	99
Cyclohexane	180	82	Potassium bromide 30%[a]	80	27
Cyclohexanol	x	x	Salicylic acid	130	54
Dichloroacetic acid			Silver bromide 10%	x	x
Dichloroethane (ethylene di-chloride)	110	43	Sodium carbonate	x	x
			Sodium chloride	x	x
Ethylene glycol	100	38	Sodium hydroxide 10%	x	x
Ferric chloride	x	x	Sodium hydroxide 50%	x	x
Ferric chloride 50% in water	x	x	Sodium hydroxide, con-centrated	x	x
Ferric nitrate 10–50%	x	x	Sodium hypochlorite 20%	80	27
Ferrous chloride	x	x	Sodium hypochlorite, con-centrated	x	x
Ferrous nitrate					
Fluorine gas, dry	470	243	Sodium sulfide to 50%	x	x
Fluorine gas, moist	x	x	Stannic chloride	x	x
Hydrobromic acid, dilute	x	x	Stannous chloride, dry	x	x
Hydrobromic acid 20%	x	x	Sulfuric acid 10%	x	x
Hydrobromic acid 50%	x	x	Sulfuric acid 50%	x	x
Hydrochloric acid 20%	x	x	Sulfuric acid 70%	x	x
Hydrochloric acid 38%	x	x	Sulfuric acid 90%	x	x
Hydrocyanic acid 10%	100	38	Sulfuric acid 98%	x	x
Hydrofluoric acid 30%	x	x	Sulfuric acid 100%	x	x
Hydrofluoric acid 70%	x	x	Sulfuric acid, fuming	90	32
Hydrofluoric acid 100%	x	x	Sulfurous acid	370	188
Hypochlorous acid	x	x	Thionyl chloride	x	x
Iodine solution 10%	x	x	Toluene	210	99
Ketones, general	100	38	Trichloroacetic acid	x	x
Lactic acid 25%	80	27	White liquor	100	38
Lactic acid, concentrated[b]	100	38	Zinc chloride	x	x

[a]Material subject to pitting.
[b]Material subject to intergranular corrosion.
Source: Schweitzer, Philip A. (1991). *Corrosion Resistance Tables*, Marcel Dekker, Inc., New York, Vols. 1 and 2.

Table 6-143 Mechanical and Physical Properties of Copper Pipe

Property	Annealed	Hard-drawn
Modulus of elasticity $\times 10^6$, psi	17	17
Tensile strength $\times 10^3$, psi	33	45
Yield strength 0.2% offset $\times 10^3$, psi	10	40
Elongation in 2 in., %	45	10
Hardness, Rockwell	F-45	B-40
Density, lb/in.^{3}	0.323	0.323
Specific gravity	8.91	8.91
Specific heat, Btu/hr °F	.092	.092
Thermal conductivity at 68°F, Btu/hr ft^2 °F	2364	2364
Coefficient of thermal expansion at 77–572 °F, in./in. °F $\times 10^{-6}$	9.8	9.8

Table 6-144 Dimensions of Thin-Wall Copper Pipe Not Suitable for Threading

Nominal pipe size (in.)	Outside diameter (in.)	Inside diameter (in.)	Wall thickness (in.)
¼	0.540	0.410	0.065
⅜	0.675	0.545	0.065
½	0.840	0.710	0.065
¾	1.050	0.920	0.065
1	1.315	1.185	0.065
1¼	1.660	1.530	0.065
1½	1.900	1.770	0.065
2	2.375	2.245	0.065
2½	2.875	2.745	0.065
3	3.500	3.334	0.083
3½	4.000	3.810	0.095
4	4.500	4.286	0.107
5	5.562	5.298	0.132
6	6.625	6.309	0.158
8	8.625	8.215	0.205
10	10.750	10.238	0.256
12	12.750	12.124	0.313

Table 6-145 Dimensions of Standard Copper Pipe Suitable for Threading

Nominal pipe size (in.)	Outside diameter	Regular (in.)		Extra strong (in.)		Double extra strong (in.)	
		Inside diameter	Wall thickness	Inside diameter	Wall thickness	Inside diameter	Wall thickness
⅛	0.405	0.281	0.062	0.205	0.100		
¼	0.540	0.376	0.082	0.294	0.123		
⅜	0.675	0.495	0.090	0.421	0.127		
½	0.840	0.626	0.107	0.542	0.149	0.252	0.294
¾	1.050	0.822	0.114	0.736	0.157	0.434	0.308
1	1.315	1.063	0.126	0.951	0.182	0.599	0.358
1¼	1.660	1.368	0.146	1.272	0.194	0.896	0.382
1½	1.900	1.600	0.150	1.494	0.203	1.100	0.400
2	2.375	2.063	0.156	1.933	0.221	1.503	0.436
2½	2.875	2.501	0.187	2.315	0.280	1.771	0.552
3	3.500	3.062	0.219	2.892	0.304	2.300	0.600
3½	4.000	3.500	0.250	3.358	0.321	2.728	0.636
4	4.500	4.000	0.250	3.818	0.341	3.152	0.674
5	5.562	5.062	0.250	4.812	0.375	4.062	0.750
6	6.625	6.125	0.250	5.751	0.437	4.897	0.864
8	8.625	8.001	0.312	7.625	0.500	6.875	0.875
10	10.750	10.020	0.365	9.750	0.500		
12	12.750	12.000	0.375				

Table 6-146 Allowable Design Stress for Copper Pipe

Operating temperature (°F/°C)	Allowable design stress (psi)		
	Annealed	Hard-drawn[a]	Light-drawn[b]
≤150/66	6000	11,300	9000
250/121	5800	10,500	8300
300/149	5000	8000	8000
350/177	3800	5000	5000
400/204	2500	2500	2500
450/332	1500	15000	1500
500/260	750	750	750

[a]Size range ½ inch to 2 inches inclusive.
[b]Size range 2 to 12 inches inclusive.

Table 6-147 Maximum Allowable Operating Pressure of Threaded Copper Pipe

Nominal pipe size (in.)	Maximum allowable operating pressure (psi) at °F/°C									
	Regular					Extra strong				
	150/66	250/121	300/149	350/177	400/204	150/66	250/121	300/149	350/177	400/204
⅛	280	270	220	180	110	1470	1420	1210	930	600
¼	650	630	540	410	270	1660	1600	1370	1050	680
⅜	670	650	540	420	270	1380	1330	1140	870	570
½	670	650	550	420	270	1320	1280	1100	840	550
¾	610	590	500	390	250	1130	1090	940	720	470
1	480	460	340	300	190	1010	980	830	640	410
1¼	520	500	430	330	210	880	850	720	560	360
1½	480	460	390	300	190	820	790	680	520	340
2	410	400	330	260	160	750	720	610	470	300
2½	340	330	270	220	130	730	710	600	460	300
3	380	370	310	240	150	680	660	560	430	280
3½	430	420	350	270	120	650	630	530	410	260
4	380	370	310	240	150	630	610	510	400	250
5	310	300	250	200	120	580	560	470	370	280
6	260	250	210	160	100	600	580	490	380	240
8	270	260	220	170	110	530	510	440	340	220
10	270	260	220	170	110	420	410	350	270	170
12	240	230	210	150	100					

Table 6-148 Maximum Allowable Operating Pressure of Plain End Copper Pipe

| Nominal pipe size (in.) | Maximum allowable operating pressure (psi) at °F/°C | | | | | | | | | |
| | Regular | | | | | Extra strong | | | | |
	150/66	250/121	300/149	350/177	400/204	150/66	250/121	300/149	350/177	400/204
⅛	1980	1910	1650	1250	820	3470	3350	2890	2200	1450
¼	1960	1890	1630	1240	862	3150	3040	2630	1990	1320
⅜	1710	1650	1420	1080	720	2520	2440	2090	1600	1050
½	1620	1570	1350	1030	670	2350	2270	1960	1490	980
¾	1350	1300	1130	850	570	1920	1860	1600	1220	800
1	1190	1150	990	750	500	1770	1710	1470	1120	740
1¼	1080	1040	900	680	450	1460	1410	1220	920	610
1½	960	930	800	610	400	1330	1290	1100	840	560
2	790	760	660	500	330	1140	1100	960	720	470
2½	780	750	650	490	320	1200	1160	1000	760	500
3	750	720	630	470	320	1070	1030	880	680	440
3½	750	720	630	470	320	980	950	820	620	410
4	660	640	550	420	270	920	890	770	580	380
5	540	520	440	340	220	810	780	680	510	340
6	450	440	380	280	190	800	770	660	510	330
8	420	410	350	270	170	680	660	570	930	280
10	390	380	330	250	170	540	520	440	340	220
12	340	330	280	220	140					

Table 6-149 Maximum Allowable Operating Pressure of Thin-Wall Copper Pipe

Nominal pipe size (in.)	Maximum allowable operating pressure (psi) at °F/°C				
	150/66	250/121	300/149	350/177	400/204
¼	1500	1450	1250	950	620
⅜	1170	1130	980	740	480
½	920	890	770	580	380
¾	730	710	610	460	310
1	580	560	480	370	240
1¼	450	440	380	280	190
1½	400	390	330	250	170
2	300	290	250	190	120
2½	250	240	210	160	100
3	260	250	220	160	110
3½	270	260	220	170	120
4	270	260	220	170	110
5	270	260	220	170	110
6	270	260	230	170	120
8	270	260	230	170	120
10	270	260	230	170	120
12	280	270	230	180	120

Table 6-150 Maximum Allowable Operating Pressure of Joints Made Using Copper Pipe and Solder-Type Fittings

Solder or brazing alloy	Operating temperature (°F/°C)	Maximum allowable operating pressure (psi) for copper pipe (inches)				
		¼–1	1¼–2	2½–4	5–8	10–12
50-50 Tin-Lead	100/38	200	175	150	130	100
	150/66	150	125	100	90	70
	200/93	100	90	75	70	50
	250/121	85	75	50	50	40
95-5 Tin-Antimony	100/38	500	400	300	150	150
	150/66	400	350	275	150	150
	200/93	300	250	200	150	140
	250/121	200	175	150	140	110
Brazing alloys melting at or above 1000°F (538°C)	250/121	300	210	170	150	150
	350/177	270	190	150	150	150
95-5 Tin-Silver	100/38	525	330	235		
	150/66	365	245	235		
	200/93	275	170	170		
	250/121	200	120	120		

Table 6-151 Support Spacing for Copper Pipe

Nominal pipe size (in.)	Support spacing (ft)	
	Regular schedule	Heavy schedule
½	6.5	6.5
1	8.0	8.0
1½	9.5	9.5
2	10.5	11.0
3	12.5	13.0
4	13.5	15.0
6	15.5	17.5
8	17.0	20.0
10	20.0	21.5

Spacing is based on lines being uninsulated, operating at a maximum temperature of 300°F (149°C), and carrying a fluid with a specific gravity of 1.35.

Table 6-152 Compatibility of Copper, Aluminum Bronze, and Red Brass with Selected Corrodents

The chemicals listed are in the pure state or in a saturated solution unless otherwise indicated. Compatibility is shown to the maximum allowable temperature for which data are available. Incompatibility is shown by an x. A blank space indicates that data are unavailable. When compatible, corrosion rate is <20 mpy.

| Chemical | Maximum temperature (°F/°C) | | |
	Copper	Aluminum bronze	Red brass
Acetaldehyde	x	x	x
Acetamide		60/16	
Acetic acid 10%	100/38		x
Acetic acid 50%	x	x	x
Acetic acid 80%	x	x	x
Acetic acid, glacial	x	x	x
Acetic anhydride	80/27	90/32	x
Acetone	140/60	90/32	220/104
Acetyl chloride	x	60/16	x
Acrylic acid			
Acrylonitrile	80/27	90/32	210/99
Adipic acid	80/27		
Allyl alcohol	90/32	90/32	90/32
Allyl chloride			
Alum	90/32	60/16	80/27
Aluminum acetate	60/16		
Aluminum chloride, aqueous	x	x	x
Aluminum chloride, dry	60/16		
Aluminum fluoride	x	90/32	
Aluminum hydroxide	90/32	x	80/27
Aluminum nitrate		x	x
Aluminum oxychloride			
Aluminum sulfate	80/27	x	x
Ammonia gas	x	90/32	x
Ammonium bifluoride	x		x
Ammonium carbonate	x		x
Ammonium chloride 10%	x	x	
Ammonium chloride 50%	x	x	
Ammonium chloride, sat.	x	x	x
Ammonium fluoride 10%	x		x
Ammonium fluoride 25%	x	x	x
Ammonium hydroxide 25%	x	x	x
Ammonium hydroxide, sat.	x	x	x
Ammonium nitrate	x	x	x
Ammonium persulfate	90/32	x	x
Ammonium phosphate	x	90/32	x
Ammonium sulfate 10–40%	x	x	x
Ammonium sulfide	x	x	x
Ammonium sulfite	x		x
Amyl acetate	90/32	x	400/204
Amyl alcohol	80/27	90/32	90/32

Table 6-152 Continued

Chemical	Maximum temperature (°F/°C)		
	Copper	Aluminum bronze	Red brass
Amyl chloride	80/27	90/32	80/27
Aniline	x	90/32	x
Antimony trichloride	80/27	x	x
Aqua regia 3:1	x	x	x
Barium carbonate	80/27	90/32	90/32
Barium chloride	80/27	80/27	80/27
Barium hydroxide	80/27	x	80/27
Barium sulfate	80/27	60/16	210/99
Barium sulfide	x	x	x
Benzaldehyde	80/27	90/32	210/99
Benzene	100/38	80/27	210/99
Benzene sulfonic acid 10%			90/32
Benzoic acid 10%	80/27	90/32	210/99
Benzyl alcohol	80/27	90/32	210/99
Benzyl chloride	x		x
Borax	80/27	90/32	80/27
Boric acid	100/38	90/32	x
Bromine gas, dry	60/16	x	
Bromine gas, moist	x	x	
Bromine liquid		x	
Butadiene	80/27		80/27
Butyl acetate	80/27		300/149
Butyl alcohol	80/27	90/32	90/32
n-Butylamine			
Butyl phthalate	80/27		210/99
Butyric acid	60/16	90/32	80/27
Calcium bisulfide			
Calcium bisulfite	80/27	x	x
Calcium carbonate	80/27	x	80/27
Calcium chlorate	x	x	x
Calcium chloride	210/99	x	80/27
Calcium hydroxide 10%	210/99	80/27	
Calcium hydroxide, sat.	210/99	60/16	210/99
Calcium hypochlorite	x	x	x
Calcium nitrate		x	x
Calcium oxide			
Calcium sulfate	80/27	x	80/27
Caprylic acid	x		x
Carbon bisulfide	80/27		x
Carbon dioxide, dry	90/32	90/32	570/299
Carbon dioxide, wet	90/32	90/32	x
Carbon disulfide	80/27		x
Carbon monoxide		60/16	570/299
Carbon tetrachloride	210/99	90/32	180/82
Carbonic acid	80/27	x	210/99
Cellosolve	80/27	60/16	210/99

Table 6-152 Continued

Chemical	Maximum temperature (°F/°C)		
	Copper	Aluminum bronze	Red brass
Chloracetic acid, 50% water	x		
Chloracetic acid	x	80/27	x
Chlorine gas, dry	210/99	90/32	570/299
Chlorine gas, wet	x	x	x
Chlorine, liquid			
Chlorobenzene	90/32	60/16	210/99
Chloroform	80/27	90/32	80/27
Chlorosulfonic acid	x	x	x
Chromic acid 10%	x	x	x
Chromic acid 50%	x	x	x
Chromyl chloride			
Citric acid 15%	210/99	90/32	x
Citric acid, concentrated	x	x	x
Copper acetate	90/32	x	x
Copper carbonate	90/32		
Copper chloride	x	x	x
Copper cyanide	x	x	x
Copper sulfate	x	x	x
Cresol			
Cupric chloride 5%	x		
Cupric chloride 50%			
Cyclohexane	80/27	80/27	80/27
Cyclohexanol	80/27		80/27
Dichloroacetic acid			
Dichloroethane			210/99
Ethylene glycol	100/38	80/27	80/27
Ferric chloride	80/27	x	x
Ferric chloride 50% in water	x	x	x
Ferric nitrate 10–50%	x	x	x
Ferrous chloride		x	x
Ferrous nitrate			
Fluorine gas, dry	x	x	x
Fluorine gas, moist	x		
Hydrobromic acid, dilute	x	x	x
Hydrobromic acid 20%	x	x	x
Hydrobromic acid 50%	x	x	x
Hydrochloric acid 20%	x	x	x
Hydrochloric acid 38%	x	x	x
Hydrocyanic acid 10%	x	x	x
Hydrofluoric acid 30%	x	x	x
Hydrofluoric acid 70%	x	x	x
Hydrofluoric acid 100%	x	x	x
Hypochlorous acid	x		x
Iodine solution 10%			
Ketones, general		90/32	100/38
Lactic acid 25%		x	90/32
Lactic acid, concentrated	90/32	90/32	90/32

Table 6-152   Continued

Chemical	Maximum temperature (°F/°C)		
	Copper	Aluminum bronze	Red brass
Magnesium chloride	300/149	90/32	x
Malic acid	x		
Manganese chloride	x		
Methyl chloride	90/32	x	210/99
Methyl ethyl ketone	80/27	60/16	210/99
Methyl isobutyl ketone	90/32		210/99
Muriatic acid	x		
Nitric acid 5%	x	x	x
Nitric acid 20%	x	x	x
Nitric acid 70%	x	x	x
Nitric acid, anhydrous	x	x	x
Nitrous acid, concentrated	80/27	x	
Oleum		x	x
Perchloric acid 10%			x
Perchloric acid 70%			x
Phenol	x	x	570/299
Phosphoric acid 50–80%	x	x	x
Picric acid	x	x	x
Potassium bromide 30%	80/27		
Salicylic acid	90/32		210/99
Silver bromide 10%	x		
Sodium carbonate	120/49	60/16	90/32
Sodium chloride to 30%	210/99	60/16	210/99
Sodium hydroxide 10%	210/99	60/16	210/99
Sodium hydroxide 50%	x	x	x
Sodium hydroxide, concentrated	x	x	x
Sodium hypochlorite 20%	x	x	80/27
Sodium hypochlorite, concentrated	x	x	x
Sodium sulfide to 50%	x	x	x
Stannic chloride	x	x	x
Stannous chloride	x	x	x
Sulfuric acid 10%	x	x	200/93
Sulfuric acid 50%	x	x	x
Sulfuric acid 70%	x	x	x
Sulfuric acid 90%	x	x	x
Sulfuric acid 98%	x	x	x
Sulfuric acid 100%	x	x	x
Sulfuric acid, fuming	x	x	x
Sulfurous acid	x	x	90/32
Thionyl chloride			
Toluene	210/99	90/32	210/99
Trichloroacetic acid	80/27	x	80/27
White liquor			
Zinc chloride	x	x	x

Source: Schweitzer, Philip A. (1991). *Corrosion Resistance Tables*, Marcel Dekker, Inc., New York, Vols. 1 and 2.

Table 6-153   Mechanical and Physical Properties of 90-10 Copper-Nickel Alloy Pipe

Modulus of elasticity $\times$ 10^6, psi	18
Tensile strength $\times$ 10^3, psi	
4½ in. O.D.	40
4½ in. O.D.	38
Yield strength at 0.5% extension under load $\times$ 10^3, psi	
4½ in. O.D.	15
4½ in. O.D.	13
Elongation in 2 in., %	25
Density, lb/in.^{3}	0.323
Specific heat, Btu/lb °F	0.09
Thermal conductivity, Btu/hr/ft/at ft.2/°F	26
Coefficient of thermal expansion at 68–572°F, in./in. °F $\times$ 10^{-6}	9.5

Table 6-154   Support Spacing for Alloy 706 Pipe

Nominal pipe size (in.)	Maximum distance between supports (ft)
1	7
2	9
4	12
6	15
8	17
≥10	20

Table 6-155   Mechanical and Physical Properties of Red Brass Pipe

Modulus of elasticity $\times$ 10^6, psi	17
Tensile strength $\times$ 10^3, psi	40
Yield strength 0.2% offset $\times$ 10^3, psi	15
Elongation in 2 in., %	50
Hardness, Brinell	50
Density, lb/in.^{3}	0.316
Specific gravity	8.75
Specific heat, Btu/lb °F	8.75
Specific heat, Btu/lb °F	0.09
Thermal conductivity at 32–212°F, Btu/ft^2/hr/°F/in.	1100
Coefficient of thermal expansion at 31–212°F, in./in. °F $\times$ 10^{-6}	9.8

Table 6-156   Allowable Design Stress for Red Brass Pipe

Operating temperature (°F/°C)	Allowable design stress (psi)
≥300°F/149°C	8000
350/177	6000
400/204	3000
450/232	2000

Table 6-157 Maximum Allowable Operating Pressure of Threaded Red Brass Pipe

Nominal pipe size (in.)	Maximum allowable operating pressure (psi)							
	Regular				Extra strong			
	<300/149	350/177	400/204	450/232	<300/149	350/177	400/204	450/232
⅛	370	280	140	90	1960	1470	740	490
¼	870	650	330	220	2210	1660	830	550
⅜	890	670	346	220	1840	1380	690	460
½	900	670	340	220	1760	1320	660	440
¾	810	610	310	200	1510	1130	570	380
1	630	480	240	160	1340	1010	510	340
1¼	690	520	260	170	1160	880	440	290
1½	630	480	240	160	1090	820	410	270
2	540	410	210	140	1000	750	380	250
2½	450	340	170	110	970	730	370	240
3	510	380	190	130	910	680	340	230
3½	570	430	220	140	860	650	330	220
4	510	380	190	130	840	630	320	210
5	410	310	160	100	770	580	290	190
6	340	260	130	80	800	600	300	200
8	360	270	140	90	710	530	270	180
10	360	270	140	90	550	420	210	140
12	320	240	120	80				

Table 6-158 Maximum Allowable Operating Pressure of Threaded Red Brass Pipe

Nominal pipe size (in.)	Maximum allowable operating pressure (psi)								
	Regular				Extra strong				
	<300/149	350/177	400/204	450/232	<300/149	350/177	400/204	450/232	
⅛	2640	1980	990	660	4630	3470	1740	1160	
¼	2610	1960	980	650	4200	3150	1580	1050	
⅜	2280	1710	860	570	3360	2520	1260	840	
½	2160	1620	810	540	3130	2350	1180	780	
¾	1800	1350	680	450	2650	1920	960	640	
1	1580	1190	600	400	2360	1770	890	590	
1¼	1440	1080	540	360	1950	1460	730	490	
1½	1280	960	480	320	1770	1330	670	440	
2	1050	790	400	260	1520	1140	570	380	
2½	1040	780	390	260	1600	1200	600	400	
3	1000	750	380	250	1420	1070	540	360	
3½	1000	750	380	250	1300	980	490	320	
4	880	660	330	220	1230	920	460	310	
5	710	540	270	180	1080	810	410	270	
6	600	450	230	150	1060	800	400	260	
8	560	420	210	140	910	680	340	230	
10	520	390	200	130	710	540	270	180	
12	450	340	170	110					

Table 6-159 Support Spacing for Red Brass Pipe

Nominal pipe size (in.)	Support spacing (ft)	
	Regular schedule	Heavy schedule
½	5.0	5.0
1	6.0	6.0
1½	6.5	7.5
2	7.0	8.5
3	9.0	10.0
4	10.0	11.5
6	11.0	14.0
8	11.5	15.5
10	15.5	16.5

Support spacing is based on uninsulated pipe, operating at a maximum temperature of 400°F (204°C) and conveying a fluid with a maximum specific gravity of 1.35.

Table 6-160 Mechanical and Physical Properties of Titanium Pipe (Grade 2)

Modulus of elasticity $\times 10^6$, psi	14.9
Tensile strength $\times 10^3$, psi	50
Yield strength 0.2% offset $\times 10^3$, psi	40
Elongation in 2 in., %	20
Density, lb/in.^{3}	0.163
Specific gravity	4.48
Specific heat at 75°F, Btu/lb °F	0.125
Thermal conductivity at 75°F, Btu/ft^2/hr/°F/in.	114
Coefficient of thermal expansion at 32–600°F, in./in. °F $\times 10^{-6}$	5.1

Table 6-161 Maximum Allowable Operating Pressure of Seamless Titanium Pipe at 650°F/343°C

Nominal pipe size (in.)	Maximum allowable operating pressure (psi)			
	5S	10S	40S	80S
⅛		1250	1815	2680
¼		1250	1750	2480
⅜		985	1420	2020
½	1010	1010	1360	1900
¾	615	800	1110	1560
1	490	820	1090	1430
1¼	384	655	855	1195
1½	335	570	770	1080
2	266	453	695	935
2½	282	410	710	980
3	230	335	615	855
3½	200	292	560	805
4	178	260	520	735
5	190	234	457	675
6	159	196	416	650

Table 6-162 Support Spacing for Titanium Pipe

Nominal pipe size (in.)	Support spacing (ft)		
	5S	10S	40S
½	8.0	8.5	9.0
¾	9.0	9.5	10.0
1	9.5	10.5	11.0
1¼	10.0	11.5	12.0
1½	10.0	11.5	12.5
2	10.5	12.5	13.5
2½	11.0	13.0	15.0
3	12.0	13.5	16.5
3½	12.0	14.0	17.0
4	12.5	14.5	17.5
5	14.0	15.5	19.0
6	14.5	15.5	20.0

Table is based on a maximum operating temperature of 650°F (343°C), conveying a fluid with a specific gravity of 1.35, with the lines being uninsulated.

Table 6-163 Compatibility of Titanium, Zirconium, and Tantalum with Selected Corrodents
The chemicals listed are in the pure state or in a saturated solution unless otherwise indicated. Compatibility is shown to the maximum allowable temperature for which data are available. Incompatibility is shown by an x. A blank space indicates that data are unavailable. When compatible, corrosion rate is <20 mpy.

Chemical	Maximum temperature (°F/°C)		
	Titanium	Zirconium	Tantalum
Acetaldehyde	300/104	250/121	90/32
Acetamide			
Acetic acid 10%	260/127	220/104	302/150
Acetic acid 50%	260/127	230/110	302/150
Acetic acid 80%	260/127	230/110	302/150
Acetic acid, glacial	260/127	230/110	302/150
Acetic anhydride	280/138	250/121	302/150
Acetone	190/88	190/88	302/150
Acetyl chloride		80/27	80/27
Acrylic acid			
Acrylonitrile	210/93	210/93	210/93
Adipic acid	450/232		210/93
Allyl alcohol	200/93	200/93	300/149
Allyl chloride		200/93	
Alum	200/93	210/99	90/32
Aluminum acetate			
Aluminum chloride, aqueous	10%	40%	302/150
	310/154	200/93	
Aluminum chloride, dry		37%	
	200/93	210/99	302/150

Table 6-163 Continued

Chemical	Maximum temperature (°F/°C)		
	Titanium	Zirconium	Tantalum
Aluminum fluoride	80/27	x	x
Aluminum hydroxide	190/88	200/93	100/38
Aluminum nitrate	200/93		80/27
Aluminum oxychloride			
Aluminum sulfate	210/99	210/99	302/150
Ammonia gas		100/38	
Ammonium bifluoride			
Ammonium carbonate	200/93		200/93
Ammonium chloride 10%	210/99	210/99	302/150
Ammonium chloride 50%	190/88	220/104	302/150
Ammonium chloride, sat.	200/93		302/150
Ammonium fluoride 10%	90/32	x	x
Ammonium fluoride 25%	80/27	x	x
Ammonium hydroxide 25%	80/27	210/99	302/150
Ammonium hydroxide, sat.	210/99	210/99	302/150
Ammonium nitrate	210/99	210/99	210/99
Ammonium persulfate	80/27	220/104	90/32
Ammonium phosphate 10%	210/99	210/99	302/150
Ammonium sulfate 10–40%	210/99	210/99	302/150
Ammonium sulfide			90/32
Ammonium sulfite			210/99
Amyl acetate	210/99	210/99	302/150
Amyl alcohol	200/93	200/93	320/160
Amyl chloride		210/99	302/150
Aniline	210/99	210/99	210/99
Antimony trichloride	110/43		210/99
Aqua regia 3:1	80/27	x	302/150
Barium carbonate	80/27	210/99	90/32
Barium chloride 25%	210/99	210/99	210/99
Barium hydroxide	210/99	200/93	302/150
Barium sulfate	210/99	210/99	210/99
Barium sulfide	90/32	90/32	90/32
Benzaldehyde	100/38	210/99	210/99
Benzene	230/110	230/110	230/110
Benzene sulfonic acid 10%		210/99	210/99
Benzoic acid	400/204	400/204	210/99
Benzyl alcohol	210/99	210/99	210/99
Benzyl chloride			230/110
Borax	190/88		x
Boric acid	210/99	210/99	300/149
Bromine gas, dry	x	x	302/150
Bromine gas, moist	190/88	60/16	302/150
Bromine liquid	x	60/16	570/299
Butadiene			80/27
Butyl acetate	210/99	210/99	80/27
Butyl alcohol	200/93	200/93	80/27

Table 6-163 Continued

Chemical	Maximum temperature (°F/°C)		
	Titanium	Zirconium	Tantalum
n-Butylamine	210/99		
Butyl phthalate	210/99	210/99	210/99
Butyric acid	210/99	210/99	302/150
Calcium bisulfide			
Calcium bisulfite	210/99	90/32	80/27
Calcium carbonate	230/110	230/110	230/110
Calcium chlorate	140/60		210/99
Calcium chloride	310/154	210/99	302/150
Calcium hydroxide 10%	210/99	210/99	302/150
Calcium hydroxide, sat.	230/110	210/99	302/150
Calcium hypochlorite	200/93	200/93	302/150
Calcium nitrate	210/99		80/27
Calcium oxide			
Calcium sulfate	210/99	210/99	210/99
Caprylic acid	210/99	210/99	300/149
Carbon bisulfide	210/99		210/99
Carbon dioxide, dry	90/32	410/210	310/154
Carbon dioxide, wet	80/27		300/149
Carbon disulfide	210/99		210/99
Carbon monoxide	300/149		
Carbon tetrachloride			302/150
Carbonic acid	210/99	210/99	300/149
Cellosolve	210/99	210/99	210/99
Chloracetic acid, 50% water	210/99	210/99	210/99
Chloracetic acid	210/99	210/99	302/150
Chlorine gas, dry	x	90/32	460/238
Chlorine gas, wet	390/199	x	570/299
Chlorine, liquid		x	300/149
Chlorobenzene	200/93	200/93	300/149
Chloroform	210/99	210/99	210/99
Chlorosulfonic acid	210/99		210/99
Chromic acid 10%	210/99	210/99	302/150
Chromic acid 50%	210/99	210/99	302/150
Chromyl chloride	60/16		210/99
Citric acid 15%	210/99	210/99	302/150
Citric acid, concentrated	180/82	180/82	302/150
Copper acetate		200/93	300/149
Copper carbonate	80/27		300/149
Copper chloride	200/93	x	300/149
Copper cyanide	90/32	x	300/149
Copper sulfate	210/99	210/99	300/149
Cresol	210/99		
Cupric chloride 5%	210/99	x	300/149
Cupric chloride 50%	210/99	190/88	90/32
Cyclohexane			
Cyclohexanol			

Table 6-163 Continued

Chemical	Maximum temperature (°F/°C)		
	Titanium	Zirconium	Tantalum
Dichloroacetic acid	280/138	350/177	260/127
Dichloroethane			
Ethylene glycol	210/99	210/99	90/32
Ferric chloride	300/149	x	302/150
Ferric chloride 50% in water	210/99	x	302/150
Ferric nitrate 10–50%	90/32		210/99
Ferrous chloride	210/99	210/99	210/99
Ferrous nitrate			
Fluorine gas, dry	x	x	x
Fluorine gas, moist	x	x	x
Hydrobromic acid, dilute	90/32	80/27	302/150
Hydrobromic acid 20%	200/93	x	302/150
Hydrobromic acid 50%	200/93	x	302/150
Hydrochloric acid 20%	x	300/149	302/150
Hydrochloric acid 38%	x	140/60	302/150
Hydrocyanic acid 10%			
Hydrofluoric acid 30%	x	x	x
Hydrofluoric acid 70%	x	x	x
Hydrofluoric acid 100%	x	x	x
Hypochlorous acid	100/38		302/150
Iodine solution 10%	90/32		
Ketones, general	90/32		
Lactic acid 25%	210/99	300/149	302/150
Lactic acid, concentrated	300/149	300/149	300/149
Magnesium chloride	300/149		302/150
Malic acid	210/99	210/99	210/99
Manganese chloride 5–20%	210/99	210/99	210/99
Methyl chloride	210/99		210/99
Methyl ethyl ketone	210/99	210/99	210/99
Methyl isobutyl ketone	200/93	200/93	210/99
Muriatic acid	x		302/150
Nitric acid 5%	360/182	500/260	302/150
Nitric acid 20%	400/204	500/260	302/150
Nitric acid 70%	390/199	500/260	302/150
Nitric acid, anhydrous	210/99	90/32	302/150
Nitrous acid, concentrated			300/149
Oleum			x
Perchloric acid 10%	x		302/150
Perchloric acid 70%	x	210/99	302/150
Phenol	90/32	210/99	302/150
Phosphoric acid 50–80%	x	180/82	302/150
Picric acid	90/32		200/93
Potassium bromide 30%	200/93	200/93	90/32
Salicylic acid	90/32		210/99
Silver bromide 10%			90/32
Sodium carbonate	210/99	210/99	210/99

Table 6-163 Continued

Chemical	Maximum temperature (°F/°C)		
	Titanium	Zirconium	Tantalum
Sodium chloride	210/99	250/121	302/150
Sodium hydroxide 10%	210/99	210/99	x
Sodium hydroxide 50%	200/93	200/93	x
Sodium hydroxide, concentrated	200/93	210/99	x
Sodium hypochlorite 20%	200/93	100/38	302/150
Sodium hypochlorite, concentrated			302/150
Sodium sulfide to 10%	210/99	x	210/99
Stannic chloride 20%	210/99	210/99	300/149
Stannous chloride	90/32		210/99
Sulfuric acid 10%	x	300/149	302/150
Sulfuric acid 50%	x	300/149	302/150
Sulfuric acid 70%	x	210/99	302/150
Sulfuric acid 90%			302/150
Sulfuric acid 98%	x	x	302/150
Sulfuric acid 100%	x	x	300/149
Sulfuric acid, fuming	x		x
Sulfurous acid	170/77	370/188	300/149
Thionyl chloride			300/149
Toluene	210/99	80/27	300/149
Trichloroacetic acid	x	x	300/149
White liquor		250/121	
Zinc chloride			210/99

Source: Schweitzer, Philip A. (1991). *Corrosion Resistance Tables*, Marcel Dekker, Inc., New York, Vols. 1 and 2.

Table 6-164 Mechanical and Physical Properties of Zirconium Pipe, Grade 2

Property	Grade	
	702	705
Modulus of elasticity $\times 10^6$, psi	14.4	14.0
Tensile strength $\times 10^3$, psi	65	85
Yield strength 0.2% offset $\times 10^3$, psi	45	65
Elongation in 2 in., %	25	20
Hardness, Rockwell	B-90	
Density, lb/in.^{3}	0.235	0.240
Specific gravity	6.51	6.64
Specific heat, Btu/lb °F	0.068	0.067
Thermal conductivity at 32–212 °F, Btu/ft^2/hr/°F/in.	95	95
Coefficient of thermal expansion, in./in. °F $\times 10^{-6}$		
at 60°F	3.2	3.5
at 60–750 °F	3.7	

Table 6-165 Dimensions of Welded and Seamless Zirconium Pipe

Nominal pipe size (in.)	Outside diameter (in.)	Nominal wall thickness (in.)			
		5S	10S	40S	80S
1/8	0.405		0.409	0.068	0.095
1/4	0.505		0.065	0.068	0.119
3/8	0.675		0.065	0.091	0.126
1/2	0.840	0.065	0.083	0.109	0.147
3/4	1.050	0.065	0.083	0.113	0.154
1	1.315	0.065	0.109	0.133	0.179
1 1/4	1.660	0.065	0.109	0.140	0.191
1 1/2	1.900	0.065	0.109	0.145	0.200
2	2.375	0.065	0.109	0.154	0.218
2 1/2	2.875	0.083	0.120	0.203	0.276
3	3.500	0.083	0.120	0.216	0.300
3 1/2	4.000	0.083	0.120	0.226	0.318
4	4.500	0.083	0.120	0.237	0.337
5	5.563	0.109	0.134	0.258	0.375
6	6.625	0.109	0.134	0.280	0.432
8	8.625	0.109	0.148	0.322	0.500
10	10.750	0.134	0.165	0.365	0.500
12	12.750	0.156	0.180	0.375	0.500

Table 6-166 Maximum Allowable Stress Values for Zirconium Pipe, Grades 702 and 705

Operating temperature (°F/°C)	Maximum allowable stress value (psi)			
	Grade 702		Grade 705	
	Seamless	Welded[a]	Seamless	Welded[a]
100/38	13000	11100	20000	17000
200/93	11000	9400	16600	14100
300/149	9300	7900	14200	12000
400/204	7000	6000	12500	10600
500/260	6100	5200	11300	9600
600/316	6000	5100	10400	8500
700/371	4800	4100	9900	7600

[a]85% joint efficiency has been used in determining the allowable stress values for welded pipe.

Table 6-167 Maximum Allowable Operating Pressure of Seamless Schedule 5 Zirconium Pipe Grade 702

Nominal pipe size (in.)	Maximum allowable operating pressure (psi) at °F/°C						
	100/38	200/93	300/149	400/204	500/260	600/316	700/371
½	1787	1512	1394	966	840	821	661
¾	1415	1198	1103	764	665	651	523
1	1121	949	874	605	527	515	414
1¼	1401	746	688	476	414	406	326
1½	768	650	599	415	361	353	284
2	612	518	477	330	287	281	226
2½	670	570	523	362	315	308	248
3	548	466	428	296	257	252	203
3½	479	407	373	258	225	222	177
4	425	361	331	225	199	195	157
5	452	384	299	244	212	207	167
6	378	321	251	204	178	174	140
8	290	246	192	156	136	133	107
10	285	243	223	154	134	131	105
12	280	238	220	151	131	129	103

Table 6-168 Maximum Allowable Operating Pressure of Seamless Schedule 10 Zirconium Pipe Grade 702

Nominal pipe size (in.)	Maximum allowable operating pressure (psi) at °F/°C						
	100/38	200/93	300/149	400/204	500/260	600/316	700/371
⅛	2788	2369	1994	1501	1308	1286	1029
¼	3421	2907	2447	1842	1605	1578	1263
⅜	2536	2156	1814	1365	1190	1170	936
½	2321	1973	1660	1250	1089	1071	857
¾	1801	1562	1315	989	862	848	678
1	2105	1789	1506	1133	987	977	777
1¼	1646	1399	1177	886	772	761	607
1½	1429	1214	1022	769	670	659	527
2	1133	963	800	610	531	523	418
2½	1026	872	734	552	481	473	379
3	838	712	600	451	393	387	309
3½	716	621	523	393	343	337	270
4	648	551	463	349	304	299	239
5	571	485	408	307	267	263	210
6	564	480	403	304	264	260	208
8	396	337	283	213	186	183	146
10	352	299	252	189	165	162	130
12	324	275	232	174	152	149	119

Table 6-169 Maximum Allowable Operating Pressure of Seamless Schedule 40 Zirconium Pipe Grade 702

Nominal pipe size (in.)	Maximum allowable operating pressure (psi) at °F/°C						
	100/38	200/93	300/149	400/204	500/260	600/316	700/371
⅛	4349	3718	3143	2366	2061	2028	1622
¼	4684	3963	3351	2522	2198	2162	1729
⅜	3382	2861	2419	1821	1585	1560	1248
½	3260	2757	2331	1754	1529	1504	1203
¾	2647	2239	1893	1425	1242	1221	977
1	2475	2095	1771	1333	1161	1148	914
1¼	2039	1725	1458	1097	956	941	752
1½	1833	1551	1311	987	860	846	677
2	1545	1307	1105	831	724	713	570
2½	1689	1429	1208	909	792	779	623
3	1467	1241	1049	790	688	677	541
3½	1338	1132	957	720	627	617	494
4	1243	1062	889	669	583	574	459
5	1090	922	780	587	511	503	402
6	990	838	708	533	464	457	365
8	872	740	627	469	409	402	321
10	791	669	569	426	371	365	292
12	683	578	488	367	320	315	252

Table 6-170 Maximum Allowable Operating Pressure of Seamless Schedule 80 Zirconium Pipe Grade 702

Nominal pipe size (in.)	Maximum allowable operating pressure (psi) at °F/°C						
	100/38	200/93	300/149	400/204	500/260	600/316	700/371
⅛	6381	5426	4594	3445	2999	2935	2361
¼	6435	5469	4633	3474	3024	2960	2380
⅜	4884	4152	3517	2637	2295	2247	1807
½	4537	3856	3266	2450	2132	2087	1678
¾	3718	3160	2677	2007	1747	1710	1375
1	3422	2909	2464	1848	1608	1574	1266
1¼	2846	2416	2047	1535	1336	1307	1051
1½	2585	2197	1861	1396	1215	1189	956
2	2231	1896	1605	1205	1048	1026	825
2½	2341	1996	1685	1264	1100	1077	866
3	2074	1763	1493	1120	975	954	767
3½	1943	1652	1399	1044	913	894	719
4	1797	1526	1242	969	843	825	664
5	1609	1368	1158	869	756	740	595
6	1554	1321	1119	839	730	715	575
8	1374	1169	989	742	646	632	508
10	1093	929	787	590	504	509	404
12	917	779	660	495	431	421	339

Table 6-171   Mechanical and Physical Properties of Tantalum Pipe

Modulus of elasticity $\times$ 10^6, psi	27
Tensile strength $\times$ 10^3, psi	
Grade VM	30
Grade PM	40
Yield strength 0.2% offset $\times$ 10^3, psi	
Grade VM	20
Grade PM	30
Elongation in 2 in., %	30+
Hardness, Rockwell	
Grade VM	B-55
Grade PM	B-65
Density, lb/in.^{3}	0.6
Specific gravity	16.6
Specific heat, Btu/lb °F	0.036
Thermal conductivity at 68°F, Btu/hr./ft^2/°F/in.	377
Coefficient of thermal expansion, in./in./°F $\times$ 10^{-6}	3.6

Table 6-172   Wall Thicknesses of Welded Tantalum Tubing

O.D. range (in.)	Minimum wall (in.)	Maximum wall (in.)
½–1	0.005	0.020
1–3	0.010	0.049
3–9	0.020	0.125

Table 6-173   Mechanical and Physical Properties of High-Silicon Iron Pipe

Property	Duriron	Durichlor
Modulus of elasticity $\times$ 10^6, psi	23	23
Tensile strength $\times$ 10^3, psi	16	16
Elongation in 2 in., %	nil	nil
Hardness, Brinell	520	520
Density, lb/in.^{3}	0.255	0.255
Specific gravity	7.0	7.0
Specific heat at 32–212°F, Btu/lb °F	0.13	0.13
Coefficient of thermal expansion $\times$ 10^{-6}, Btu/ft^2 hr/°F/in.		
at 32–212°F		7.2
at 68–392°F	7.4	

Table 6-174 Maximum Lengths of Duriron and Durichlor Pipe

Nominal pipe size (in.)	Maximum length available (ft)
1	3
1¼	3
1½	3
2	4
2½	5
3	5
4	5
5	5
6	5
8	5
10	5
12	5

Table 6-175 Compatibility of High-Silicon Iron[a] with Selected Corrodents

The chemicals listed are in the pure state or in a saturated solution unless otherwise indicated. Compatibility is shown to the maximum allowable temperature for which data are available. Incompatibility is shown by an x. A blank space indicates that data are unavailable. When compatible, corrosion rate is <20 mpy.

Chemical	Maximum temp. °F	Maximum temp. °C	Chemical	Maximum temp. °F	Maximum temp. °C
Acetaldehyde	90	32	Barium carbonate	80	27
Acetamide			Barium chloride	80	27
Acetic acid 10%	200	93	Barium hydroxide		
Acetic acid 50%	200	93	Barium sulfate	80	27
Acetic acid 80%	260	127	Barium sulfide	80	27
Acetic acid, glacial	230	110	Benzaldehyde	120	49
Acetic anhydride	120	49	Benzene	210	99
Acetone	80	27	Benzene sulfonic acid 10%	90	32
Acetyl chloride	80	27	Benzoic acid	90	32
Acrylic acid			Benzyl alcohol	80	27
Acrylonitrile	80	27	Benzyl chloride	90	32
Adipic acid	80	27	Borax	90	32
Allyl alcohol	80	27	Boric acid	80	27
Allyl chloride	90	32	Bromine gas, dry	x	x
Alum	240	116	Bromine gas, moist	80	27
Aluminum acetate	200	93	Bromine liquid		
Aluminum chloride, aqueous			Butadiene		
Aluminum chloride, dry			Butyl acetate		
Aluminum fluoride	x	x	Butyl alcohol	80	27
Aluminum hydroxide	80	27	n-Butylamine		
Aluminum nitrate	80	27	Butyl phthalate	80	27
Aluminum oxychloride			Butyric acid	80	27
Aluminum sulfate	80	27	Calcium bisulfide		
Ammonia gas			Calcium bisulfite	x	x
Ammonium bifluoride	x	x	Calcium carbonate	90	32
Ammonium carbonate	200	93	Calcium chlorate	80	27
Ammonium chloride 10%			Calcium chloride	210	99
Ammonium chloride 50%	200	93	Calcium hydroxide 10%		
Ammonium chloride, sat.			Calcium hydroxide, sat.	200	93
Ammonium fluoride 10%	x	x	Calcium hypochlorite	80	27
Ammonium fluoride 25%	x	x	Calcium nitrate		
Ammonium hydroxide 25%	210	99	Calcium oxide		
Ammonium hydroxide, sat.			Calcium sulfate	80	27
Ammonium nitrate	90	32	Caprylic acid	90	32
Ammonium persulfate	80	27	Carbon bisulfide	210	99
Ammonium phosphate	90	32	Carbon dioxide, dry	570	299
Ammonium sulfate 10–40%	80	27	Carbon dioxide, wet	80	27
Ammonium sulfide			Carbon disulfide		
Ammonium sulfite			Carbon monoxide		
Amyl acetate	90	32	Carbon tetrachloride	210	99
Amyl alcohol	90	32	Carbonic acid	80	27
Amyl chloride	90	32	Cellosolve	90	32
Aniline	250	121	Chloracetic acid, 50% water	80	27
Antimony trichloride	80	27	Chloracetic acid	90	32
Aqua regia 3:1	x	x	Chlorine gas, dry		

392

Table 6-175 *Continued*

Chemical	°F	°C	Chemical	°F	°C
Chlorine gas, wet			Magnesium chloride 30%	250	121
Chlorine, liquid			Malic acid	90	32
Chlorobenzene	80	27	Manganese chloride		
Chloroform	90	32	Methyl chloride		
Chlorosulfonic acid, dry			Methyl ethyl ketone	80	27
Chromic acid 10%	200	93	Methyl isobutyl ketone	80	27
Chromic acid 50%	200	93	Muriatic acid		
Chromyl chloride	210	99	Nitric acid 5%	180	82
Citric acid 15%			Nitric acid 20%	180	82
Citric acid, concentrated	200	93	Nitric acid 70%	186	86
Copper acetate			Nitric acid, anhydrous	150	66
Copper carbonate			Nitrous acid, concentrated	80	27
Copper chloride	x	x	Oleum	x	x
Copper cyanide	80	27	Perchloric acid 10%	80	27
Copper sulfate	100	38	Perchloric acid 70%	80	27
Cresol			Phenol	100	38
Cupric chloride 5%			Phosphoric acid 50–80%	210	99
Cupric chloride 50%			Picric acid	80	27
Cyclohexane	80	27	Potassium bromide 30%	100	38
Cyclohexanol	80	27	Salicylic acid	80	27
Dichloroacetic acid			Silver bromide 10%		
Dichloroethane (ethylene dichloride)	80	27	Sodium carbonate		
			Sodium chloride to 30%	150	66
Ethylene glycol	210	99	Sodium hydroxide 10%	170	77
Ferric chloride	x	x	Sodium hydroxide 50%	x	x
Ferric chloride 50% in water			Sodium hydroxide, concentrated	x	x
Ferric nitrate 10–50%	90	32	Sodium hypochlorite 20%	60	16
Ferrous chloride	100	38	Sodium hypochlorite, concentrated		
Ferrous nitrate					
Fluorine gas, dry	x	x	Sodium sulfide to 50%	90	32
Fluorine gas, moist			Stannic chloride	x	x
Hydrobromic acid, dilute	x	x	Stannous chloride	x	x
Hydrobromic acid 20%			Sulfuric acid 10%	212	100
Hydrobromic acid 50%	x	x	Sulfuric acid 50%	295	146
Hydrochloric acid 20%[b]	80	27	Sulfuric acid 70%	386	197
Hydrochloric acid 38%			Sulfuric acid 90%	485	252
Hydrocyanic acid 10%	x	x	Sulfuric acid 98%	538	281
Hydrofluoric acid 30%	x	x	Sulfuric acid 100%	644	340
Hydrofluoric acid 70%	x	x	Sulfuric acid, fuming		
Hydrofluoric acid 100%	x	x	Sulfurous acid	x	x
Hypochlorous acid			Thionyl chloride		
Iodine solution 10%			Toluene		
Ketones, general	90	32	Trichloroacetic acid	80	27
Lactic acid 25%	90	32	White liquor		
Lactic acid, concentrated	90	32	Zinc chloride		

[a]Resistance applies to Duriron unless otherwise noted.
[b]Resistance applies only to Durichlor.
Source: Schweitzer, Philip A. (1991). *Corrosion Resistance Tables*, Marcel Dekker, Inc., New York, Vols. 1 and 2.

7

Miscellaneous Piping Systems

In addition to plastic, metallic, and lined piping systems, there are a few miscellaneous piping systems. Each of these systems have more or less been designed for specific purposes and serve a need in the area of corrosion resistant piping. Although they may have special fields of application, they are generally widely used and for the most part are readily available.

7.1. BOROSILICATE GLASS PIPE

Of the many glass compositions available, the one most commonly used for piping systems is borosilicate glass. This particular composition has been selected because of its wide range of corrosion resistance, relatively high operating temperature, good heat resistance due to low thermal expansion, transparency to ultraviolet light, and ability to be prestressed. The physical properties of borosilicate glass are given in Table 7-1.

7.1.1 Thermal Shock

Glass piping systems must be guarded against thermal shock. Thermal shock can be defined as an almost instant temperature change, such as by low-pressure steam, followed by an immediate flush of cold water. It can also be caused by water falling on the outside of a hot uninsulated or unarmored pipe. Thermal shock can also be caused by a sudden increase in temperature.

Some of the borosilicate piping systems are supplied with an external armor, which does provide some protection against external thermal shock. All piping systems operating at a temperature of 250°F (121°C) or above must be insulated to guard against external thermal shock. The maximum allowable sudden temperature differentials to prevent shock for the various types of borosilicate glass pipe systems are shown in Table 7-2. If the temperature change is expected to be greater than the value shown in Table 7-2, a uniform difference of approximately 50°F (28°C) per minute is within safe limits for any of the borosilicate glass systems through 6 inches in diameter.

7.1.2 Electrostatic Grounding

When a piping system is conveying nonconductive materials, it is possible for static charges of electricity to build up. In order for dangerous voltage to develop the resistivity of the transfer fluid must be of a value of 10 inches ohm-cm. The existence of high voltages within the pipe will not be subject to spark hazard unless the transfer fluid carries highly oxidizable or unstable compounds. These materials could explode in the absence of oxygen. On this basis it is usually sufficient to ground glass piping systems externally.

External grounding can be accomplished by looping a grounding wire around the pipe on 12-inch centers, grounding it every 20 feet to ensure continuity even if there is a break in the wire. Make three revolutions around each joint. Use a wire that will be resistant to atmospheric corrosion that may be present.

If internal grounding is required, as inert wire must be threaded through the system. Use a 0.0020 maximum diameter wire. Bring it out through the joint by sandwiching it between two donut gaskets, grounding it every 20 feet to ensure continuity even though there may be a break in the line. Make sure that the wire will be resistant to the fluid being transported. Periodic continuity checks should be made to ensure that no breaks have occurred in the system.

7.1.3 Piping Systems

There are five basic borosilicate glass piping systems, differing from each other in terms of allowable operating pressures and temperatures. Table 7-3 lists the allowable operating pressures and temperature for each system.

The general rules to be followed for installing glass pipe are basically the same for all five of the glass piping systems. It is important to keep in mind that glass is a nonductile material with a compressive strength of approximately 2000,000 psi, which is greater than steel. The tensile strength of glass is variable depending upon the surface condition. A scratch or other surface defect will cause stress concentration under load conditions. Care should be taken during installation to protect the glass from surface damage. Because of these factors, all installation techniques attempt to take advantage of the high compressive strength of glass and keep tensile stresses at a minimum.

Pipe lengths should never be sprung into or out of alignment. This is especially true of glass pipe, since it is a nonductile material. Excessive bending load on a glass pipe can cause breakage. The break may not occur during installation, but at some later date. The greater the bending load, the sooner the pipe will leak. Leaky joints are symptoms of misaligned pipe. If a joint leaks after having been tightened in a normal manner, the joint should be disassembled and the natural alignment of the pipe ends observed. If there is any movement of the pipe ends away from each other, in any direction, the system must be realigned.

Proper length of each pipe section is just as critical as proper alignment. If a length of pipe is installed that is slightly shorter than required and the joints made up, severe tensile stresses will have been induced into the glass by attempting to "stretch" the glass to fit. This will result in eventual breakage of the pipe length. Use the proper length of pipe.

In general, when installing glass pipe, the following rules should be adhered to:

1. All lines should be free to move laterally. When padded clevis hangers are used or when the pipe is permitted to rest on a flat, padded piping bridge, little or no bending load will be induced into the piping system. Padding is also recommended to protect the external surface of the glass from scratches. Notched supports or U-bolts should not be used, since they all restrain the lateral movement of the pipe.

2. Every straight run of pipe should have one longitudinal anchor. It is important that only one such anchor be installed in each longitudinal run, unless there is an expansion joint between the anchors.

Connection of the pipe to a fixed piece of equipment constitutes an anchor point. If a pipe is being run between two fixed points in a straight line, an expansion joint or flexible joint of some kind is required. A PTFE flexible coupling as described on page 26 is adequate for this purpose.

3. All vertical lines should be supported from the bottom. The first support on the horizontal run from the top of the riser should be 8–10 feet from the riser. By so doing, the glass is kept in compression.

A loose lateral guide should be installed near the top of every vertical run and additional guides at 10-foot intervals. Every 30 feet in a vertical run, a flexible swing joint or bellows should be installed in conjunction with a vertical support anchor.

4. Ordinary manual valves do not require individual supports in lines above 1 inch in diameter. Two clevis hangers installed within 3 inches of the valve face will supply adequate support. If a valve is to be used as the means of supplying the anchor point in the run, care should be taken that the support and means of anchoring do not impose a bending load on the pipe. All actuated valves should be rigidly mounted. Flexible couplings should be installed on either side of the valve.

5. All other heavy equipment such as meters, strainers, filters, etc. should be supported independently of the piping system. These units then may become anchor points in the system.

6. Flexible PTFE bellows should be installed on pumps or other equipment to prevent any vibration from being transmitted to the pipeline.

Conical Pipe System

This is the original glass piping system. It has the lowest allowable operating pressure of any of the process glass piping systems. It is available in diameters of ½ inch through 18 inches. Allowable operating pressures of the ½- through 6-inch diameters are given in Table 7-3. The allowable operating pressures of the larger diameters are as follows:

9-inch diameter—15 psi
12-inch diameter—11 psi
18-inch diameter—8 psi

All sizes have a maximum allowable operating temperature of 450°F (232°C).

Joining of conical pipe is by means of flanges. The ends of the pipe are flared to take one of three special metal flanges, which are cushioned from the glass by means of a filled neoprene insert. The insert distributes bolting forces uniformly over the conical pipe ends. An interface gasket is placed between the grooved pipe ends. The style 1 flange is a triangular-shaped aluminum flange used for joining conical pipe and fittings. The bolt circle of this flange does not match the bolt circle of the equivalent size USASI flange. This is the most economical flange and is usually installed in noncorrosive atmospheres and when the piping system is being used to convey a material such as distilled water.

One of the most commonly used flanges on conical pipe for joining the glass pipe both to itself and to USASI flanges is the style 2 flange. This cast iron flange has the same bolt circle and number of bolt holes as an equivalent size UASAI flange, although the diameter of the bolt is smaller. By using centering washers on the USASI flange, this flange can be used for joining conical pipe to USASI flanges. This permits connections to be made to flanged vessel nozzles, pumps, valves, and other pieces of processing equipment.

Style 3 flanges are used for joining conical pipe to USASI flanges. Dimensionally this flange is identical to a standard equivalent USASI flange. These flanges are not used for joining conical pipe to itself.

Flanges used for joining the 9-, 12-, and 18-inch-diameter pipes are dimensionally different from USASI flanges. In order to join these sizes to USASI flanges, special adapters must be employed.

Conical pipe can be field-fabricated to provide odd lengths using equipment available from the manufacturer. Support spacing is given in Table 7-4.

Low-Pressure Beaded System

This piping system is available in nominal diameters of 1½ through 6 inches and is rated for 15 psi (all sizes) at a maximum temperature of 250°F (121°C). It can also be used under full vacuum. The main areas of application of this system are in laboratory drain lines and vent lines. The pipe is joined by means of one of two single bolt couplings with a PTFE liner. For joining factory ends, a bead-to-bead coupling is used, while a bead-to-plain end coupling is used for joining a cut length to a factory end. Factory lengths of pipe and fittings are furnished with a bead on either end. Field fabrication is simple, consisting only of cutting the pipe to length, smoothing the cut end, and installing the bead-to-plain coupling. Two cut lengths cannot be joined to each other. Hanger supports are required every 8–10 feet.

This pipe is suitable for burial when furnished with special outside coverings supplied by the manufacturer.

Beaded Pressure System

This piping system is available in nominal diameters of ½ inch through 6 inches, with a maximum allowable operating temperature of 350°F (177°C). Refer to Table 7-3 for the maximum allowable operating pressure for each size of pipe. This system is furnished from the factory with beads on the straight lengths of pipe and fittings. By using a field-fabrication kit furnished by the manufacturer, odd lengths can be cut and field-beaded.

The pipe is joined by means of a single bolt coupling with a PTFE liner. Only glass or PTFE contacts the fluid being handled. Pipe supports are required every 8–10 feet and preferably located in the vicinity of a coupling.

Beaded Armored System

This system is basically the same as beaded pressure pipe with the addition of an outer laminate of fiberglass cloth impregnated with a modified polyester resin. Operating temperatures and pressures are the same as for the beaded pressure pipe (see Table 7-3). The outer laminate is translucent but will tend to darken with prolonged use at temperatures above 275°F (135°C) and serves only one purpose, that of holding the system together in the event of a pipe failure. The product will weep or drip through the laminate, but the pipe system will be operable. It will also provide some protection against external thermal shock. The outer laminates are made with resins that are so heavily loaded with fiberglass that the result is a self-extinguishing outer pipe.

This piping system can be field-fabricated using equipment supplied by the manufacturer. Straight lengths can be cut, ends beaded, and rewrapped with armor. Joining is by means of a single bolt coupling with a PTFE liner. Factory lengths of pipe and fittings are furnished with beaded ends. Pipe supports should be spaced every 8–10 feet and located preferably in the vicinity of a coupling.

Corgard System*

This glass piping system has the highest pressure rating of any of the glass piping systems. Refer to Table 7-3 for the pressure ratings. The borosilicate glass pipe is armored with an outer laminate of filament-wound fiberglass reinforcing and modified polyester resin. The outer laminate serves one purpose only: if a failure to the inner pipe should occur for any accidental reason, the outer pipe holds the system together. The pipe will weep or drip, perhaps even spray a little, but the plant stays on stream until an orderly maintenance program can be scheduled. The outer laminate is self-extinguishing.

Connections are made using either the ball coupling, which offers a 3° angular flexibility at each joint, or style 4 split flanges. Pipe supports should be spaced every 8–10 feet and located in the vicinity of a coupling. Corgard can be field-fabricated using equipment supplied by the manufacturer.

7.1.4   Corrosion Resistance

The chemical stability of borosilicate glass is one of the most comprehensive of any known construction material. It is highly resistant to water, acids, salt solutions, organic substances, and even halogens like chlorine and bromine.

Only hydrofluoric acid, phosphoric acid with fluorides, or strong alkalis at temperatures above 102°F (49°C) can visibly affect the glass surface. Refer to Table 7-5 for the compatibility of borosilicate glass with selected corrodents.

7.2.   VITRIFIED CLAY PIPE

Vitrified clay pipe is virtually impervious to every chemical except hydrofluoric acid. It was and still is used for the handling of sewerage. With the advent of the problems of cleaning up of toxic dumps and landfills, new applications have arisen. The primary reasons for using clay pipe in these new applications are that it:

1. Is chemically inert, unaffected by sewer gases and acids
2. Is rigid, will not flatten out or sag
3. Is rustproof
4. Is unaffected by harsh household cleaning compounds and solvents
5. Withstands the extra stresses of heavy backfill loads
6. Will not soften or swell under any condition
7. Is durable, will not roughen, erode, or wear out
8. Is unaffected by gases and acids generated by ground garbage
9. Is made impervious through vitrification

---

*Trademark of Corning.

Vitrified clay pipe is presently being used to conduct the leachate from the Love Canal to holding tanks for processing.

Pipe sections and fittings are joined by means of bell and spigot joints, which are sealed by means of an O-ring. For lines that are handling corrosive spent acids and solvents a Furathane/O-ring combination joint is used. A fillet and bead of Furathane mortar at the pipe junction provides the necessary resistance against attack from nonoxidizing acids, chlorinated organic solvents, and detergents, to a maximum operating temperature of 140°F (60°C). Furathane is a thermosetting furan resin–based corrosion-resistant mortar containing a 100% carbon filler. Set and cured, Furathane exhibits the chemical and thermal resistance of furan mortar with the high bond strength of epoxy mortar.

Dimensions of vitrified clay pipe are shown in Table 7-6. Vitrified clay pipe is covered by ASTM specification C700 containing data for both standard and extra strength. For normal sanitary wastes the standard strength is usually adequate, unless live loading requires the use of a heavier section. When used to transport contaminated storm drainage or industrial waste, the extra strength should be specified.

Though salt-glazed pipe may be used for sanitary sewers, unglazed pipe should be employed when handling industrial wastes. Clay pipe is normally applied for gravity flow lines. However, they should be designed for heads of 5–10 feet to guard against the possibility of blockage or of the sudden or abrupt introduction of material into the line.

Clay pipe is brittle, and care must be exercised during installation if a liquid-tight line is to be obtained. It is recommended that all pipe be continuously supported by a concrete pad or saddles, especially if there is any chance of movement of the soil. Concrete cover or other adequate protection must be provided to protect the pipe from crushing and vibration at highway or rail crossings. Care must also be exercised during backfilling that no stone be included within 12 inches of the pipe body or bells. Tamping must be done carefully to guard against movement or damage to the pipe or joints. Table 7-7 provides the compatibility of vitrified clay pipe with selected corrodents.

7.3 WOOD PIPE

Wood stave pipe is available in two types: machine-banded pipe and continuous-stave pipe. Machine-banded pipe is banded with wire and is made with wood or metal collars or with inserted joints. Continuous-stave pipe is manufactured in units consisting of staves, bands, and shoes shipped in knocked-down form and constructed in the trench. In building this type of pipe, the staves are laid so as to break joints, with the completed pipe being without joints. Continuous-stave pipe is banded with individual bands, ranging in size from ⅜ to 1 inch, depending upon the size of the pipe. A safety factor of 4 is maintained in the band based on an ultimate strength of 60,000 psi of cross section. The maximum pressure to which a continuous-stave pipe may be subjected depends upon the size of the pipe. The head for small pipes may run as high as 400 feet, while in the largest sizes the head would be less than 200 feet.

Machine-banded pipe is made for pressures of 50–400 feet. For underground installations machine-banded wood stave pipe is generally suitable. The staves have double tongues and grooves on lateral edges and are wound with 1-inch-wide steel strips of adequate gage and spacing to provide pressure classes of 43, 86, 130, and 172 psi. It is recommended that working pressures be limited to 60% of class pressure, which provides working pressures of 26, 52, 78, and 103 psi. Fittings for wood pipe are made of cast iron or wood. Table 7-8 provides the dimensions of machine-banded wood pipe.

The most common woods used for wood pipe are Douglas fir, white pine, redwood, and cyprus. All woods are affected adversely by acids, particularly the strong oxidizing acids, but they are regularly used in dilute hydrochloric acid solutions at ambient temperature. Improved corrosion resistance can be imparted to wood by impregnating the wood under pressure conditions with certain resin solutions that include asphalt, phenolic, and furan. This greatly extends the area of application of woods in corrosion services. Strong alkaline solutions, particularly caustic ones, generally cause disintegration and cannot be used with impregnated wood. Weak alkaline solutions can be used with wood with reasonably good service life. Wood stave pipe is used to a large extent for municipal water supply, outfall sewers, mining, irrigation, and various other uses providing for the transportation of water. Wooden pipe is most often built in the West where it is close to the material lumber market.

Table 7-1 Physical Properties of Borosilicate Glass

Thermal conductivity:	0.73 Btu/hr ft^2 °F/ft
	0.0035 Cal/sec-cm^2 °C/cm
Specific heat:	0.20 Btu/lb °F
	0.20 Cal/g °F
Dielectric constant at 23°C and 1 MHz per ASTM method D150 4.6 ± 0.2	
Density:	1.39 to 1.40 lb/ft^3
	2.23 g/cm^2
Young's modulus per ASTM method C215:	in the range of 9–10 × 10^6
Linear coefficient of expansion:	
32–572°F	18.1 × 10^{-7} in./in. °F
(0–300°C) per ASTM method E228	32.5 × 10^{-7} cm/cm/°C

Table 7-2 Maximum Allowable Sudden Temperature Differential to Prevent Thermal Shock

Nominal pipe size (in.)	Corgard[a]	Temperature (°F/°C)			
		Beaded armored system	Conical system	Beaded pressure system	Low-pressure beaded system
1	200/111	200/111	200/111	200/111	200/111
1½	200/111	200/111	200/111	200/111	200/111
2	200/111	200/111	200/111	200/111	200/111
3	180/100	180/100	180/111	180/100	200/111
4	140/78	140/78	140/78	140/78	175/97
6	122/68		122/68		160/89
9			100/55		
12			100/55		
18			90/50		

[a]Trademark of Corning Incorporated.

Table 7-3 Allowable Operating Pressures and Temperatures for Borosilicate Glass Piping Systems

Nominal pipe size (in.)	Maximum allowable operating pressure (psi)				
	Conical system	Low-pressure beaded system	Beaded pressure system	Beaded armored system	Corgard[a]
½	100		100	100	
¾	100		100	100	
1	100		100	100	150
1½	60	15	75	75	150
2	50	15	75	75	150
3	40	15	50 or 75[b]	50 or 75[b]	100
4	35	15	50	50	75
6	20	15	30	30	60
Allowable operating temp. range all sizes	Cryogenic to 450°F cryogenic to 232°C	0–250°F		−18°C to 121°C	

[a]Trademark of Corning Incorporated.
[b]Allowable operating pressure dependent upon manufacturer.

Table 7-4 Support Spacing for Glass Pipe Carrying Fluids of Different Specific Gravities

Nominal pipe size	Support spacing (ft) for specific gravity:	
	<1.3	>1.3
½	8	6
¾	8	6
1	8	7
1½	9	7
2	9	8
3	9	8
4	10	8
6	10	8
9	5	5
12	5	5
18	5	5

Table 7-5 Compatibility of Borosilicate Glass with Selected Corrodents

The chemicals listed are in the pure state or in a saturated solution unless otherwise indicated. Compatibility is shown to the maximum allowable temperature for which data are available. Incompatibility is shown by an x. A blank space indicates that data are unavailable.

Chemical	Maximum temp. °F	Maximum temp. °C	Chemical	Maximum temp. °F	Maximum temp. °C
Acetaldehyde	450	232	Barium carbonate	250	121
Acetamide	270	132	Barium chloride	250	121
Acetic acid 10%	400	204	Barium hydroxide	250	121
Acetic acid 50%	400	204	Barium sulfate	250	121
Acetic acid 80%	400	204	Barium sulfide	250	121
Acetic acid, glacial	400	204	Benzaldehyde	200	93
Acetic anhydride	250	121	Benzene	200	93
Acetone	250	121	Benzene sulfonic acid 10%	200	93
Acetyl chloride			Benzoic acid	200	93
Acrylic acid			Benzyl alcohol	200	93
Acrylonitrile			Benzyl chloride	200	93
Adipic acid	210	99	Borax	250	121
Allyl alcohol	120	49	Boric acid	300	149
Allyl chloride	250	121	Bromine gas, dry		
Alum	250	121	Bromine gas, moist	250	121
Aluminum acetate			Bromine liquid	90	32
Aluminum chloride, aqueous	250	121	Butadiene	90	32
Aluminum chloride, dry	180	82	Butyl acetate	250	121
Aluminum fluoride	x	x	Butyl alcohol	200	93
Aluminum hydroxide	250	121	n-Butylamine		
Aluminum nitrate	100	38	Butyl phthalate		
Aluminum oxychloride	190	88	Butyric acid	200	93
Aluminum sulfate	250	121	Calcium bisulfide		
Ammonia gas			Calcium bisulfite	250	121
Ammonium bifluoride	x	x	Calcium carbonate	250	121
Ammonium carbonate	250	121	Calcium chlorate	200	93
Ammonium chloride 10%	250	121	Calcium chloride	200	93
Ammonium chloride 50%	250	121	Calcium hydroxide 10%	250	121
Ammoinum chloride, sat.	250	121	Calcium hydroxide, sat.	x	x
Ammonium fluoride 10%	x	x	Calcium hypochlorite	200	93
Ammonium fluoride 25%	x	x	Calcium nitrate	100	38
Ammonium hydroxide 25%	250	121	Calcium oxide		
Ammonium hydroxide, sat.	250	121	Calcium sulfate		
Ammonium nitrate	200	93	Caprylic acid		
Ammonium persulfate	200	93	Carbon bisulfide	250	121
Ammonium phosphate	90	32	Carbon dioxide, dry	160	71
Ammonium sulfate 10–40%	200	93	Carbon dioxide, wet	160	71
Ammonium sulfide			Carbon disulfide	250	121
Ammonium sulfite			Carbon monoxide	450	232
Amyl acetate	200	93	Carbon tetrachloride	200	93
Amyl alcohol	250	121	Carbonic acid	200	93
Amyl chloride	250	121	Cellosolve	160	71
Aniline	200	93	Chloracetic acid, 50% water	250	121
Antimony trichloride	250	121	Chloracetic acid	250	121
Aqua regia 3:1	200	93	Chlorine gas, dry	450	232

Table 7-5 *Continued*

Chemical	Maximum temp. °F	Maximum temp. °C	Chemical	Maximum temp. °F	Maximum temp. °C
Chlorine gas, wet	400	204	Malic acid	160	72
Chlorine, liquid	140	60	Manganese chloride		
Chlorobenzene	200	93	Methyl chloride	200	93
Chloroform	200	93	Methyl ethyl ketone	200	93
Chlorosulfonic acid	200	93	Methyl isobutyl ketone	200	93
Chromic acid 10%	200	93	Muriatic acid		
Chromic acid 50%	200	93	Nitric acid 5%	400	204
Chromyl chloride			Nitric acid 20%	400	204
Citric acid 15%	200	93	Nitric acid 70%	400	204
Citric acid, concentrated	200	93	Nitric acid, anhydrous	250	121
Copper acetate			Nitrous acid, concentrated		
Copper carbonate			Oleum	400	204
Copper chloride	250	121	Perchloric acid 10%	200	93
Copper cyanide			Perchloric acid 70%	200	93
Copper sulfate	200	93	Phenol	200	93
Cresol	200	93	Phosphoric acid 50–80%	300	149
Cupric chloride 5%	160	71	Picric acid	200	93
Cupric chloride 50%	160	71	Potassium bromide 30%	250	121
Cyclohexane	200	93	Salicylic acid		
Cyclohexanol			Silver bromide 10%		
Dichloroacetic acid	310	154	Sodium carbonate	250	121
Dichloroethane (ethylene di-chloride)	250	121	Sodium chloride	250	121
			Sodium hydroxide 10%	x	x
Ethylene glycol	210	99	Sodium hydroxide 50%	x	x
Ferric chloride	290	143	Sodium hydroxide, con-centrated	x	x
Ferric chloride 50% in water	280	138			
Ferric nitrate 10–50%	180	82	Sodium hypochlorite 20%	150	66
Ferrous chloride	200	93	Sodium hypochlorite, con-centrated	150	66
Ferrous nitrate					
Fluorine gas, dry	300	149	Sodium sulfide to 50%	x	x
Fluorine gas, moist	x	x	Stannic chloride	210	99
Hydrobromic acid, dilute	200	93	Stannous chloride	210	99
Hydrobromic acid 20%	200	93	Sulfuric acid 10%	400	204
Hydrobromic acid 50%	200	93	Sulfuric acid 50%	400	204
Hydrochloric acid 20%	200	93	Sulfuric acid 70%	400	204
Hydrochloric acid 38%	200	93	Sulfuric acid 90%	400	204
Hydrocyanic acid 10%	200	93	Sulfuric acid 98%	400	204
Hydrofluoric acid 30%	x	x	Sulfuric acid 100%	400	204
Hydrofluoric acid 70%	x	x	Sulfuric acid, fuming		
Hydrofluoric acid 100%	x	x	Sulfurous acid	210	99
Hypochlorous acid	190	88	Thionyl chloride	210	99
Iodine solution 10%	200	93	Toluene	250	121
Ketones, general	200	93	Trichloroacetic acid	210	99
Lactic acid 25%	200	93	White liquor	210	99
Lactic acid, concentrated	200	93	Zinc chloride	210	99
Magnesium chloride	250	121			

Source: Schweitzer, Philip A. (1991). *Corrosion Resistance Tables*, Marcel Dekker, Inc., New York, Vols. 1 and 2.

Table 7-6 Dimensions of Vitrified Clay Pipe

Nominal pipe size (in.)	Minimum laying length (ft)	Minimum outside diameter of barrel	Minimum wall thickness (in.)	
			Standard strength	Extra strength
4	2	$4\frac{7}{8}$	$\frac{7}{16}$	
6	2	$7\frac{1}{16}$	$\frac{1}{2}$	$\frac{9}{16}$
8	2	$9\frac{1}{4}$	$\frac{9}{16}$	$\frac{3}{4}$
10	2	$11\frac{1}{2}$	$\frac{11}{16}$	$\frac{7}{8}$
12	2	$13\frac{1}{4}$	$\frac{13}{16}$	$1\frac{1}{16}$
16	3	$17\frac{3}{16}$	$\frac{15}{16}$	$1\frac{3}{8}$
18	3	$20\frac{3}{8}$	$1\frac{1}{8}$	$1\frac{3}{4}$
21	3	$24\frac{1}{8}$	$1\frac{5}{16}$	2
24	3	$27\frac{1}{2}$	$1\frac{1}{2}$	$2\frac{1}{4}$
27	3	31	$1\frac{11}{16}$	$2\frac{1}{2}$
30	3	$34\frac{3}{8}$	$1\frac{7}{8}$	$2\frac{3}{4}$
33	3	$37\frac{5}{8}$	2	3
36	3	$40\frac{3}{4}$	$2\frac{1}{16}$	$3\frac{1}{4}$

Table 7-7 Compatibility of Vitrified Clay Pipe with Selected Corrodents

The chemicals listed are in the pure state or in a saturated solution unless otherwise indicated. Compatibility is shown to the maximum allowable temperature for which data are available. Incompatibility is shown by an x. A blank space indicates that data are not available.

Chemical	Maximum temperature (°F/°C)
Acetic acid 5%	150/66
Acetone	73/23
Aluminum chloride	x
Aluminum sulfate 5%	150/66
Ammonium chloride 5%	150/66
Ammonium chloride 10%	x
Ammonium chloride 25%	x
Ammonium hydroxide 5%	73/23
Ammonium hydroxide 10%	73/23
Aniline	73/23
Benzene	73/23
Borax 3%	150/66
Carbon tetrachloride	73/23
Chromic acid 40%	150/66
Citric acid 10%	150/66
Copper sulfate 3%	150/66
Ferric chloride 1%	150/66
Hydrochloric acid 10%	120/49
Hydrofluoric acid 30%	x
Hydrofluoric acid 70%	x
Hydrofluoric acid 100%	x
Nitric acid 1%	150/66
Nitric acid 10%	150/66
Nitric acid 20%	150/66
Sodium carbonate 20%	150/66
Sodium chloride 30%	150/66
Sodium hydroxide 10%	150/66
Sulfuric acid 20%	150/66
Sulfuric acid 30%	150/66
Toluene	120/49

Table 7-8 Dimensions of Machine-Banded Wood Pipe

Inside diameter (in.)	Outside diameter (in.)	Wood Wall thickness (in.)
2	5	1½
3	6	1½
4	7¼	1⅝
6	9½	1¾
8	11½	1¾
10	13½	1¾
12	15½	1¾
14	17½	1¾
16	19½	1¾
18	21½	1¾
20	23½	1¾
24	27½	1¾
24	29½	2¾
30	35½	2¾
36	41½	2¾
48	53½	2¾

8

Double Containment Piping Systems

Double containment piping systems are not a new concept to the process piping industry. Dual arrangements of metallic pipe utilizing a carrier pipe with a secondary containment pipe have been used for many years. Applications have included systems designed and installed in the nuclear, gas, petroleum-refining, and chemical process industries where hazardous or highly toxic chemicals and wastes have been transported. Using the same techniques learned from the manufacturers of double pipe heat exchangers, double containment piping systems were developed.

In the late 1970s and early 1980s, interest in dual arrangements using thermoplastic pipe was sparked, primarily by large semi-conductor manufacturers to transport their chemical wastes to treatment areas. They were concerned about potential leakage of the hazardous chemicals being handled.

The initial attempts to produce double containment piping systems using thermoplastic piping materials were not very successful as the systems were designed piecemeal with combinations of piping and materials whose geometries were not readily compatible. These systems were limited to having containment piping serve merely as a temporary holding chamber without any pressure capability. Problems such as sagging and deformation of the inner pipe due to a lack of support, distortion of the inner pipe and/or outer pipe due to uncontrolled thermal expansion, failure of fabricated welds, as well as many others were prevalent.

As more and more companies became interested in using double-containment piping as a standard practice, manufacturers of thermoplastic and thermoset piping undertook the task of developing reliable double-containment piping systems. This interest on the part of the user companies was the result of rising insurance premiums and awareness of company image and a concern for public and employee safety.

Simultaneously, public pressure at local, state, and federal levels resulted in the passage of strict laws to protect the environment. There have been many concerns about

409

groundwater contamination. In response to these concerns and the public pressure generated, the EPA has issued regulations governing the transportation of hazardous materials. Noncompliance with these laws can result in severe penalties for owners, managers, and employees responsible for environmental damage resulting from spills and leaks. For example, the Clean Water Act permits fines of $25,000–$50,000 per day of violation and prison terms of from 1 to 3 years.

One of the regulations requires the use of double containment piping when transporting toxic waste from processing site to treatment facilities or settling pond. The EPA selected this approach because it is one of the safest and most effective means of transporting hazardous materials and providing a safeguard against contamination of groundwater.

Secondary containment piping systems are now being used in tank farms, underground storage tanks, and similar applications throughout the chemical process and allied industries. The most prevalent application is in underground piping applications, where monitoring difficulties are significant and where environmental consequences of leaks can be severe.

However, because of the inherent protection against the leakage of material from the carrier pipe reaching external areas, other applications are also being found. Secondary containment piping systems provide an ideal means of transporting any hazardous material overhead through workplaces. The basic design of these systems provide safety for workers in these areas.

One such installation is at the American Airlines maintenance and engineering center in Tulsa, Oklahoma. In this facility dangerous waste materials are collected from a sump and pumped overhead to a waste treatment facility. The construction of the building was such that it would have been impractical to lay an underground pipe. The double-walled piping installation passes through an area where aircraft engines and landing gears are overhauled, safely transporting the hazardous material overhead.

Other potential installations can be designed to handle corrosive materials such as acids, caustics, or other hazardous materials, not just industrial wastes.

8.1 SELECTION OF A DOUBLE CONTAINMENT PIPING SYSTEM

The selection of a double containment piping system requires a considerable amount of thought in order to specify the best system for a specific application. Step one in the process is to determine the best material of construction for the carrier pipe. This is the pipe that will be continually exposed to the corrodent being handled. It must be completely resistant to this corrodent under all conditions of temperature and pressure to which it will be subjected. Thought should also be given to the possibility of concentration changes resulting from process upsets or other abnormal conditions. Additionally it must meet the physical and mechanical requirement of the system, since most leaks in a piping system are the result of mechanical failure of the pipe or of leaky joints. This latter problem will be discussed separately.

Step two is to select the best material of construction for the containment pipe. This pipe must be able to contain a leak should there be a failure of the carrier pipe. It must therefore be resistant to the corrodent being handled, but it must also meet other requirements since it has an additional job to perform. The containment pipe must be resistant to weathering, withstand ambient temperature variations (possibly severe depending upon geographic location), resist impact by outside forces, and withstand the compressive loads of underground installation if necessary. It must be kept in mind that the secondary containment piping has the dual function of enhancing leak detection and protecting the environment against hazardous chemical spills.

Step three is to consider the combination of materials selected. They must be able to withstand thermal expansion and contraction of the pipe as well as changes in temperature and humidity of the annular space. Consideration must also be given to the methods used in joining the pipe and fittings. Remember that once the containment pipe is installed and joined, access to the carrier pipe has been eliminated or at least greatly reduced. Therefore it is important that the carrier pipe be free of flanges, unions, and other mechanical joints, since these are areas where leaks are most apt to occur.

Engineering plastics provide one of the most cost-effective means of supplying a double containment piping system, providing the mechanical, physical, and corrosion-resistant properties are compatible with the requirements. The most reliable method of joining pipe to pipe or pipe to fittings when using polypropylene, polyvinylidene fluoride, and polyethylene is by means of heat fusion. There are three techniques whereby this may be done: butt fusion or socket fusion in carrier pipe and butt fusion or electrofusion in secondary containment pipe.

Butt fusion is accomplished by heating the squared piping ends and holding them together with a constant force while they cool. A high degree of care is required in preparing the joint and in the fusion operation to maintain proper alignment and joint strength.

Socket fusion of the carrier pipe requires the use of a heated nonstick female bushing to meet the outside of the pipe and a heated male bushing to heat the inside of the corresponding size fitting. When both are partially melted, the pipe is pushed into the fitting. The resulting joint is considerably stronger than the pipe itself because of the large contact area. By providing an interference fit between the cold piping's outside diameter and the fitting's inside diameter, the possibility of accidently leaving the joint unfused is avoided.

Electrofusion of secondary containment piping involves the use of a specially designed coupling that incorporates integral resistance wires for heating and a portable, fully automated power supply. The couplings are prestressed during manufacture. When assembled in the field, the coupling freely slides over the pipe. However, once heat is applied, the prestressing is relieved and the couplings shrink to form a tight leak-proof joint.

When fiberblass-reinforced thermoset piping is used to form a double containment piping system, the carrier pipe is joined by means of a straight socket adhesive bond. Care should be taken that the adhesive used is resistant to the materials being handled. The outer containment pipe is also joined by means of straight socket, adhesive bonded, with couplings for joint-to-joint connections.

Step four is to select a suitable leak-detection system.

8.2 LEAK-DETECTION SYSTEMS

Three types of leak-detection systems are used in secondary containment piping systems:

Cable detection systems
Liquid-sensing systems
Visual systems

They may be used singly or in conjunction with each other. The particular system to be selected will depend upon several factors, including location of the pipe line and the material being transported.

Cable detection systems are relatively expensive and sophisticated. They are highly accurate in their ability to detect leaks and locate the leaks precisely. It should be remembered that the more sophisticated the electronic system, the greater the chance of having a problem such as power failure, false alarms, or a corroded cable. However, there are instances where such a system is required.

Other less sophisticated methods range from the use of a slight glass drip tube or detector probes sensing humidity, conductivity, pH, etc.

8.2.1   Cable Detection Systems

The basis of operation of leak-detecting cables is the continuous measurement of their electrical properties. When these properties change as the result of contact with a corrodent, an alarm is given, which signals the plant operators that there is a leak. Most of the commercially available systems are also capable of reporting the location of the leak.

For a cable detection system to operate properly, certain design requirements of the double containment piping system must be met.

1. Frequent access ports must be provided in the piping system. As a general rule, access ports should be provided after every two changes in direction (preferably after every change in direction), at tee branches and Y-lateral connections, where connections or splices are being used, at least in every 100 feet of straight run, and at the beginning and end of each line. This access must be provided for installation and maintenance purposes.

2. The alarm module/control box must be powered for continuous operation and monitor the sensing cable that has been installed. Under these conditions, if a leak occurs the alarm will sound and the location of the leak given. Otherwise the alarm will report an average location when it is turned back on.

3. The containment pipe must be thoroughly clean of contaminants and water before the cable is installed. If the containment pipe is filled with conductive media, the cable will sense it.

4. There must be a minimum annular clearance of ¾ inch throughout the entire system. This will permit easy installation of the cable between the 100-foot spaced access ports.

8.2.2   Sensing Probes

Sensing probes can be used when the double containment system is divided into separate isolated leak detection compartments. A probe is installed in each of these compartments. Types of probes that can be used are varied. They can be designed to sense conductivity (resistivity), pH, moisture, pressure change, or a combination of these or other characteristics. Any type of probe that is capable of detecting the presence of the material being transported in the carrier pipe can be used.

Conductivity sensing involves the measurement of a change in conductivity from a fixed stable condition. This type of measurement can be used with a wide range of conductive fluids such as acids, bases, and many organic chemicals. When the leaking fluid comes into contact with the probe, an alarm is sounded. No indication is given as to the location of the leak other than the compartment in which the probe is installed.

Use of pH-sensing probes is applicable only for acids or bases. The probe measures a change in acidity from a fixed neutral point. This approach follows the same principle as conductivity measurement.

Moisture-sensing devices follow the same type of approach as the preceding systems. However, application is limited to sensing leakage of fluid, primarily water, containing trace amounts of hazardous organics or other corrodents.

Pressure change in an isolated compartment can be sensed by installing a pressure sensor. If a leak occurs in the carrier pipe, the pressure in the annular space will increase, thus setting off an alarm.

8.2.3   Visual Detection Methods

There are several visual detection methods that may be employed, including installing sight glasses directly into the containment piping, the use of side branches underneath the bottom of the containment piping with a sight glass, or installing short sections of flanged, clear piping into the containment pipe.

The system of detection to be selected will depend upon the location of the line, the fluids being handled, the amount of safety needed for the system, and the cost. In some instances dual systems may be used as a means of providing back-up for safety reasons.

8.3   SIZING OF THE PIPING SYSTEM

Sizing of the carrier pipe is identical to the sizing of any single pipe. Sizing of the outer containment pipe is generally determined according to the fabrication (assembly) needs and the leak-detection system to be used. Most systems are provided with outer containment piping that is designed to drain fluids to a collection vessel. The drainage piping can be pressurized or nonpressurized. On occasion the containment piping is designed to act as a holding tank to contain fluid until the leak can be located and repaired. When it is necessary to calculate friction loss through the annular portion of the pipe, the following method may be used.

Since the width of the annular opening is small relative to the length of the piping involved, the hydraulic radius is equal to approximately one-half the width of the passage, or:

$$R_H = \frac{W}{2}$$

where:

R_H = hydraulic radius
W = width of annular space

The conventional equations for fluid flow in pipes may then be used, substituting $4R_H$ for the diameter in the equations.

The friction losses due to flow across pipe support clips and through semi-annular dogbone fittings should be included as well. Estimates for these values will have to be made since data are not established. Reasonable comparisons should be made to well-known frictional losses across devices having similar geometry. A safety factor should be included.

The values thus calculated become important when determining the amount and pressure of the flushing water to be used during cleaning of the annular space prior to a repair situation.

8.3.1   Sizing of the Drainage System

An assumption must be made when designing the pipe system that at some time a leak will develop and that repairs will have to be made. This means that provisions must be made to drain and vent the outer containment piping. When the outer containment piping is designed to contain separate compartments (in order to establish separate zones of containment), the compartment may have to be drained prior to repair operations if a leak should develop in the carrier pipe.

The drain should be placed at the low point of the piping compartment. In addition, an atmospheric vent must be installed at the high point of the compartment. Both of the connections must be valved to remain closed during normal operations. The size of the drain is determined by the length of time required to drain the annular space of the compartment.

Flow from the annular space discharging liquid to an atmospheric drainline through an opening in the containment pipe is affected by both the area and shape of the opening. Total head at the orifice is converted into kinetic energy by the following equation:

$$V_o = C_v \sqrt{2gh}$$

where:

V_o = velocity (ft/sec)
C_v = coefficient of velocity
   =   0.82 for short tube, no separation of fluid from walls
   = 0.98 for a short tube with rounded entrance
g = gravitational acceleration
   = 32(lb m)(ft)/lb-ft sec^2
h = fluid height (ft)

The discharge from the orifice is:

$$Q_o = (C_cA_o)V_o = C_cA_o \sqrt{2gh}$$

where:

Q_o = flow rate (gpm)
C_c = coefficient of contraction
   = 1.00 for short tube
   = 0.99 for short tube with rounded entrance
A_o = area of the opening (ft^2)
C_d = C_vC_c dimensionless

The head loss due to turbulence at the orifice is:

$$h_f = \left[\frac{1}{CV^2} - 1\right] \frac{V_o^2}{2g}$$

To this must be added the head loss of downstream piping and any valves included in the drain piping. If the liquid in the annular space compartment is not constantly being replenished, the static head forcing discharge through the orifice will decrease. For a tank with varying cross sections, the following basic relationship holds:

$$Q_{dt} = -A_a dh$$

where:

t = time (sec)
A_a = area of the annular space (ft²)

An expression for the area of the annular space, A_a as a function of h must be determined. Then the time to empty the annular area from height h_1 and h_2 is:

$$t = \int_{h_1}^{h_2} \frac{A_a dh}{C_d A_o \sqrt{2gh}}$$

It is assumed in these equations that the outer containment piping is vented to the atmosphere. If the fluid is being discharged in a pressurized manner, the total head will be increased by the gauge pressure converted to head of fluid by means of the following equation:

$$h_p = \frac{P}{\rho}$$

where:

h_p = head (ft)
P = pressure (lb/ft²)
ρ = fluid density (lb/ft³)

During flushing operations, the annular space will be constantly fed and simultaneously emptied in a pressurized manner. Since the flushing water will be fed to the annular space at a rate greater than the discharge from the annular space, the time equation is modified to determine the time to fill the annular space:

$$t = \int_{h_1}^{h_2} \frac{A_a dh}{[C_d A_o \sqrt{2gh} - V_{in}]}$$

where:

V_{in} = volumetric flow rate of flushing water (ft³/sec)
A_{ia} = area of annular space

The size of vent opening required can be determined using the standard equations for the flow of compressible fluids through orifices:

$$q = YCA \frac{\sqrt{2g(144)\Delta P}}{[1 - \beta^4]}$$

where:

q = volumetric air flow rate (ft³/sec)
Y = expansion factor (dependent upon the specific heat ratio, the ratio of orifice diameter to inlet diameter, and the ratio of downstream to upstream absolute pressure)
C = flow coefficient of orifice

A = cross-sectional area of the orifice (ft²)
g = gravitational acceleration (32 lb m ft/lb/ft sec²)
P = differential pressure, equal to inlet pressure of venting to the atmosphere (lb/in²)
ρ = density (lb/ft³)
β = ratio of throat diameter to pipe diameter, dimensionless.

If the outer containment piping is designed with separate compartments that will maintain pressure, then a pressure-relief valve should be installed on each compartment in addition to the vent valve.

8.4 THERMAL EXPANSION

Double containment piping systems present a unique problem of thermal expansion, since the inner (carrier) pipe may be subject to different temperature changes than the outer (containment) pipe. This problem of differential thermal expansion can be subdivided into four distinct design cases:

1. Greater magnitude outer piping temperature change:
    above-ground systems
    below-ground systems
2. Greater magnitude inner piping temperature change:
    above-ground systems
    below-ground systems

In above-ground systems when the outer piping has a larger temperature change than the inner pipe, compensation can be provided by the addition of an expansion joint in the outer piping (see Fig. 8-1). This condition is usually encountered when the piping is installed outside. The temperature changes in the external pipe are the result of ambient temperature changes, while the inner pipe temperature remains relatively stable since it is conducting a constant temperature fluid. If PTFE expansion joints are required, the cost can become prohibitive. An alternate method is to add heat tracing and insulation to the outer pipe. This design limits the temperature changes to minimal positive temperature

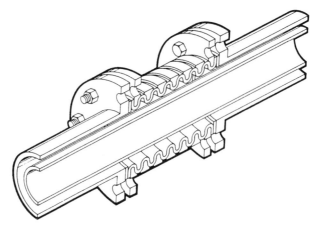

Figure 8-1 Typical expansion joint arrangement in containment pipe. (Courtesy of Asahi/America.)

changes from the set temperature of the heat tracing. This will result in small magnitudes of expansion, which will not create a problem.

Normally when the piping is installed underground, temperature changes in the outer piping are negative, resulting from a high installation temperatures. Since the pipes are buried, the use of expansion joints, loops, or offsets is not possible.

In most cases the soil will impart friction to prevent movement of the outer pipe. If extreme temperature changes are expected, it may be necessary to install concrete piping vaults or trenches to allow movement underground.

In above-ground piping systems, a greater magnitude of temperature change will be experienced in the inner pipe when a hot fluid enters. The resulting expansion can be compensated for by installing expansion loops, offsets, or in changes of direction (see Figs. 8-2, 8-3, and 8-4) or by a totally restrained design.

Relieving thermal stress and preventing fitting failure through a totally restrained system can be accomplished by placing anchors at the end of straight runs of pipe and before or after fittings of complex fitting arrangements. By so doing, the thermal stress is transferred to the pipe itself. In plastic pipe, the pipe reacts by relieving itself through some permanent deformation and relaxation of the walls. This could eventually lead to failure of the pipe. The amount of stress applied, the number and duration of the cycles, and the internal operating pressure of the pipe will determine the life of the pipe. There are means available to determine whether or not a restrained design may be used without danger of pipe or fitting failure.

An analysis of the system, taking into account the combined effect of all stresses involved, must be made. This analysis must include the thermal stress, internal hydrostatic stress, and any other possible stresses. The actual situation is a very complex one and should be analyzed or modeled on a computer by an expert. However, an approximation can be made by taking into account only the thermal and hydrostatic stresses. From these results it would be possible to rule out the use of a totally restrained system, but not necessarily to allow such an installation.

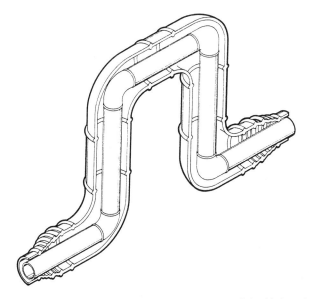

Figure 8-2 Typical expansion loop configuration. (Courtesy of Asahi/America.)

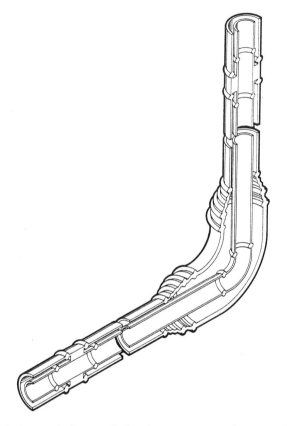

Figure 8-3 Typical use of change of direction to accommodate expansion. (Courtesy of Asahi/America.)

The aforementioned stresses can be approximately combined for the carrier pipe as follows:

$$J_c = \sqrt{J_T^2 + J_P^2} \tag{1}$$

where:

J_c = combined stress (psi)
J_T = thermal stress (psi)
J_P = internal hydrostatic stress (psi)

$$J_T = E \propto \Delta T \tag{2}$$

where:

E = long-term modulus of elasticity (psi)
$\propto$ = coefficient of thermal expansion (°F^{-1})
ΔT = maximum temperature differential from installation temperature (°F)

$$J_P = \frac{P(D - t)}{2t}$$

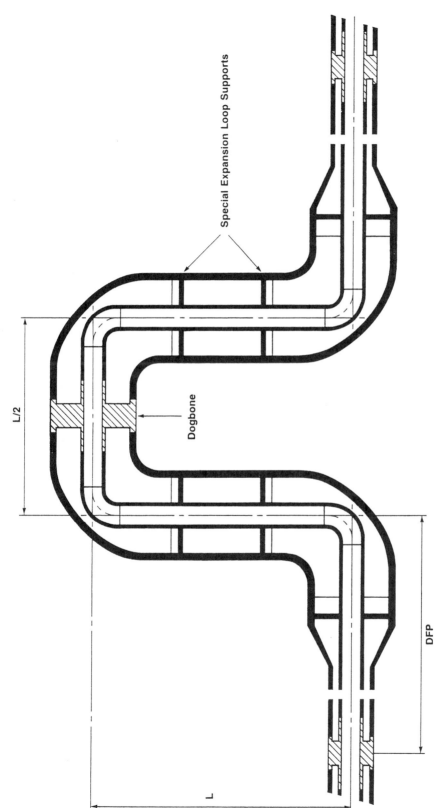

Figure 8-4 Typical expansion loop assembly. (Courtesy of Asahi/America.)

where:

  P = internal pressure (psi)
  D = outside diameter (in.)
  t = wall thickness (in.)

The estimated value for the combined pressure (J_c) can be compared to the design stress rating S of the material. If the value of J_c is less than S, then a restrained design can be utilized.

    If it is not possible to use expansion loops, offsets, directional changes, or expansion joints because of space or other limitations, then the system can be designed with closely spaced guides (pipe clips). However, it is necessary to provide a flexibility analysis based upon piping stress calculations or be able to demonstrate that the principles of elastic behavior can be met. The maximum guide spacing can be determined from the following equation:

$$L_s = \frac{f \; 48 \; E \; I}{4 \; SM \; J_c}$$

where:

  L_s = maximum distance between supports (in.)
   f = allowable sag (= 0.098 inches)
   E = long term E modulus (psi)
   I = moment of inertia (in.^{4})
 SM = section modulus (in.^{3})
  J_c = combined stress (psi)

L_s is a maximum spacing and should be compared to the recommended maximum hanger spacing at the maximum operating temperature and the smaller value used. Expansion joints are not recommended unless the expansion joint can be placed in a tank or other access device to permit maintenance.

    In order to provide proper movement of the inner elbows, the outer elbows must be increased in size in the area of the expansion loop or offset. Normally the outer piping in the expansion loop is kept at the same size as the elbows (see Fig. 8-4). The same procedure would be followed at a change in direction of the pipe (see Fig. 8-5).

    For underground piping systems where the inner pipe experiences greater temperature changes than the outer pipe, the methods used to control the resulting expansion are the same as those for above-ground piping having the same type of temperature differential. In all cases the manufacturer should be consulted and recommendations followed as to the proper means to protect the system from thermal changes.

8.5 INSTALLATION AND SUPPORT

Installation and support of double containment piping systems follows the same general rules as for other piping systems. Support spacing of above-ground piping should be based on the size of the outer pipe. Refer to the appropriate piping system to determine the proper support spacing.

    When a double containment piping system is to be buried, all designs should be based on the containment pipe rather than the carrier pipe.

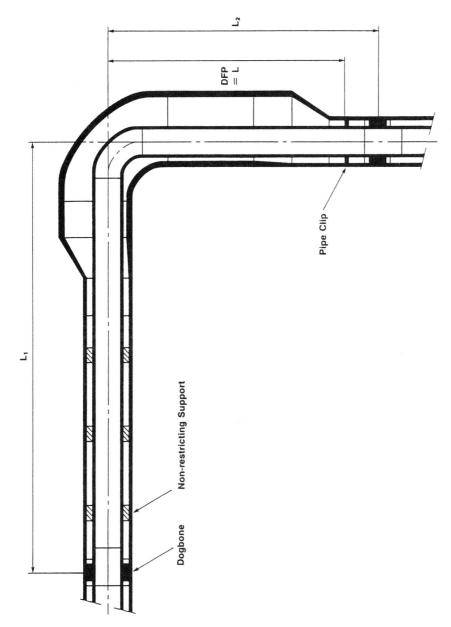

Figure 8-5 Typical change of direction assembly. (Courtesy of Asahi/America.)

8.6 THERMOPLASTIC SYSTEMS

Because of the complexities involved in the design and fabrication of a double containment piping system, it is best to use a system provided by a competent piping manufacturer. There are several standard combinations of different thermoplastics available from various manufacturers.

Asahi/America produces a double containment piping system under the tradename of Duo-Pro from copolymer polypropylene and PVDF. The polypropylene piping is available in two pressure ratings, designated as Pro 45 and Pro 150. These are both designed according to standard dimension ratio. The Pro 45 is rated at 45 psi at 73°F (23°C), while the Pro 150 is rated at 150 psi at 73°F (23°C). As the operating temperature increases, the pressure ratings are reduced. Refer to Table 8-1 for the pressure corrections at elevated temperatures.

The PVDF piping system is designed as Super-Pro 230 and Super-Pro 150 and is also available in two pressure ratings, depending upon pipe diameter. Pipe sizes of ⅜ inch through 2½ inches are available in the Super-Pro 230 system, which is rated at 230 psi at 70°F (21°C). Pipe sizes of 3 through 12 inches are available in the Super-Pro system, which is rated at 150 psi at 70°F (21°C). As the operating temperature increases, the pressure ratings are reduced. Refer to Table 8-2 for the correction factors at elevated temperatures.

Standard size combinations range from ½-inch carrier pipe to 24-inch containment pipe. There are six combinations of these piping materials available to form a double containment piping system. The Pro and Super-Pro systems may be combined as follows:

Carrier pipe	Containment pipe
Pro 150	Pro 45
Pro 150	Pro 150
Pro 45	Pro 45
Super-Pro 230	Super-Pro 150
Super-Pro 230	Pro 150
Super-Pro 150	Pro 45

Refer to Tables 2-48 and 2-68 for compatibility of these materials with selected corrodents.

The Plexico Division of Chevron Chemical Company supplies dual containment piping systems produced from high-density polyethylene as a standard in two pressure ratings: LP, which is basically designed for gravity flow and has a test pressure rating of 15 psi, and HP, which has a system operating pressure of 160 psi. Both are designed primarily for underground service. Carrier pipe is available in sizes of 1 inch through 8 inches, and containment pipes are available in sizes of 4 through 14 inches. Also available as carrier or containment pipes are high-temperature polyethylene and high-impact polypropylene copolymer. Refer to Table 2-55 for the compatibility of high-density polyethylene with selected corrodents.

The R. G. Sloan Company supplies a secondary containment piping system produced from a clear, unpigmented PVC. This is the containment pipe and accessories only. It is designed to be installed over any conventional piping system. The pipe and fittings are manufactured split horizontally. These fittings and/or pipe sections are placed over any existing piping system and joined through an injection bonding process. This is achieved by injecting the bonding adhesive into the predrilled injection ports that are located on each fitting and on the seams of the split pipe. The adhesive then flows through the adhesive channels forming a bonded leak-free joint.

The clear PVC allows visual inspection of an above-ground system to check for leaking carrier pipe. Refer to Table 8-3 for the compatibility of unpigmented PVC with selected corrodents.

The Perm Alert Company furnishes a system in which the carrier and containment pipe are manufactured from type 1 homopolymer polypropylene. The hompolymer polypropylene has increased tensile strength and stiffness over that of the copolymer material, particularly at elevated temperatures. The carrier and containment pipes are manufactured to an SDR so that the pressure rating of the pipe is consistent for all pipe sizes. Carrier pipes are available with the following pressure ratings:

SDR 11—150 psi
SDR 17.6—90 psi

while containment pipes are available in the following pressure ratings:

SDR 17.6—90 psi
SDR 32.5—45 psi

These pressure ratings are at 73.4°F (23°C).

Carrier/containment pipe combinations are available as follows:

Carrier pipe size	Containment pipe size
1	4
1.5	4
2	4
2	6
2.5	6
3	6
4	8
6	10
8	12
10	14
12	16

The companies mentioned here are not the only sources for thermoplastic double containment piping systems. They have been selected because they provide examples of the choices available for these piping systems.

8.7 THERMOSET SYSTEMS

Double containment piping systems are also produced from the various thermoset resins with appropriate reinforcing materials. Among the resins used are epoxies, vinyl esters, and furan resins. The most common reinforcing material used is glass fiber, although other materials are available.

The Conley Corporation produces systems utilizing aromatic amine-cured epoxy,

Novalac-based vinyl ester, bisphenol A–based vinyl ester, and furan resin systems. Carrier pipes are available in schedule 40 ratings, while the containment pipes are produced in schedule 20 ratings. Tables 8-4 and 8-5 show the pressure ratings of these two schedules of pipe, while Table 8-6 provides the sizes of the double containment systems.

The carrier pipe and containment pipe need not have the same resin formulation. Different resins may be used for each. The sizes and pressure ratings shown are the same for all resin formulations.

Thermal expansion of the containment pipe can cause end loads on anchors. These loads will basically be independent of the temperature of the carrier pipe. Table 8-7 tabulates these loads based on the change in temperature of the containment pipe, plus 10% of the maximum load from the carrier pipe.

Joining of pipe and fittings is by means of standard socket adhesive connection in both the carrier and containment pipe. Installation follows the same general procedures and principles as for single thermoset piping systems. Support spacing for the different size systems at various carrier pipe operating temperatures are given in Table 8-8. Table 8-9 provides the corrosion resistance of the various resins in contact with selected corrodents.

Fibercast is another producer of a thermoset double containment piping system, using either epoxy or vinyl ester resins with fiberglass reinforcing. Carrier pipe is available in nominal pipe sizes of 1 inch through 12 inches, with containment pipes available in nominal pipe sizes of 3 through 16 inches.

Standard pipe combinations are as follows:

Dualcast	Nominal pipe size (in.)	
size designation	Carrier pipe	Containment pipe
1N3	1	3
1½N3	1½	3
1N4	1	4
1½N4	1½	4
2N4	2	4
2N6	2	6
3N6	3	6
4N6	4	6
4N8	4	8
6N8	6	8
6N10	6	10
8N12	8	12
10N14	10	14
12N16	12	16

The maximum ratings of the epoxy and vinyl ester pipes are shown in Tables 8-10 and 8-11. However, because of the lower allowable pressure ratings of fittings, the total system operating pressures of the carrier pipes are somewhat reduced. The system operating pressures of the carrier pipes are shown in Table 8-12. Joining of the pipe is by means of adhesive joints with integral female sockets on all fittings. Support spacing for both epoxy and vinyl ester piping is given in Table 8-13.

These are just two examples of double containment piping systems produced from thermoset resins and are illustrative of the types and sizes of systems available.

8.8 METALLIC SYSTEMS

For applications where temperature, pressure, or other considerations do not permit the use of thermoset or thermoplastic double containment piping systems, it may be necessary to use a metallic system. The most commonly available systems are constructed of carbon steel or stainless steel. In many instances multiple carrier pipes will be installed within a single containment pipe.

When carbon steel is used as the containment pipe in an underground system, some form of external coating is usually applied to prevent corrosion. Among the coatings used are coal tar urethanes, asphalt, and epoxies. Cathodic protection is also used. The specific coating to be used will depend upon the ground conditions.

The containment pipe, when installed above ground, may be galvanized carbon steel for protection against atmospheric corrosion. Stainless steel systems usually contain multiple carrier pipes to reduce the cost of containment piping. The specific grade of stainless steel to be used will depend upon the material being transported.

Metallic systems besides carbon steel and stainless steel are also available. Systems may be obtained in practically any metal or metallic alloy.

8.9 MIXED SYSTEMS

In many instances the most effective double containment pipe system from the viewpoint of economics, corrosion resistance, and mechanical requirements will be a combination of types of materials. When high temperatures or pressures are required for the carrier pipe, a metallic material may be dictated. The containment pipe may be an epoxy or vinyl ester material for burial. The possible combinations between thermoplastic, thermoset, and metallic systems is virtually unlimited. Several manufacturers offer as standard configurations such mixed systems, while others offer mixed systems as options.

Table 8-1   Proline Pressure Rating Correction Chart

Temperature (°F/°C)	Correction factor
73/22.7	1.00
100/37.7	0.64
140/60.0	0.40
180/82.2	0.28
200/93.3	0.10

Source: Courtesy of Asahi/America Inc.

Table 8-2   Pressure Rating Correction Chart for Super Proline

Temperature (°F/°C)	Correction factor
70/21	1.00
80/27	0.95
90/32	0.87
100/38	0.80
120/49	0.68
140/60	0.58
160/71	0.49
180/82	0.42
200/93	0.36
240/115	0.25
280/138	0.18

Source: Courtesy of Asahi/America Inc.

Table 8-3 Compatibility of Unpigmented PVC with Selected Corrodents

The chemicals listed are in the pure state or in a saturated solution unless otherwise indicated. Compatibility is shown to the maximum allowable temperature for which data are available. Incompatibility is shown by an x. A blank space indicates that data are unavailable.

Chemical	Maximum temp.		Chemical	Maximum temp.	
	°F	°C		°F	°C
Acetaldehyde	x	x	Barium carbonate	120	49
Acetamide			Barium chloride	120	49
Acetic acid 10%	120	49	Barium hydroxide	120	49
Acetic acid 50%	72	22	Barium sulfate	120	49
Acetic acid 80%	72	22	Barium sulfide	120	49
Acetic acid, glacial	72	22	Benzaldehyde	x	x
Acetic anhydride	x	x	Benzene	x	x
Acetone	x	x	Benzene sulfonic acid 10%	120	49
Acetyl chloride	x	x	Benzoic acid	120	49
Acrylic acid			Benzyl alcohol	x	x
Acrylonitrile			Benzyl chloride		
Adipic acid			Borax	120	49
Allyl alcohol	x	x	Boric acid	120	49
Allyl chloride	120	49	Bromine gas, dry		
Alum			Bromine gas, moist		
Aluminum acetate	120	49	Bromine liquid	x	x
Aluminum chloride, aqueous	120	49	Butadiene	120	49
Aluminum chloride, dry			Butyl acetate	x	x
Aluminum fluoride	120	49	Butyl alcohol	120	49
Aluminum hydroxide	120	49	*n*-Butylamine		
Aluminum nitrate	120	49	Butyl phthalate	72	22
Aluminum oxychloride	120	49	Butyric acid	x	x
Aluminum sulfate	120	49	Calcium bisulfide	120	49
Ammonia gas	120	49	Calcium bisulfite	120	49
Ammonium bifluoride	120	49	Calcium carbonate	120	49
Ammonium carbonate	120	49	Calcium chlorate	120	49
Ammonium chloride 10%	120	49	Calcium chloride	120	49
Ammonium chloride 50%	120	49	Calcium hydroxide 10%	120	49
Ammonium chloride, sat.	120	49	Calcium hydroxide, sat.	120	49
Ammonium fluoride 10%	120	49	Calcium hypochlorite	72	22
Ammonium fluoride 25%	72	22	Calcium nitrate	120	49
Ammonium hydroxide 25%	120	49	Calcium oxide		
Ammonium hydroxide, sat.	120	49	Calcium sulfate	120	49
Ammonium nitrate	120	49	Caprylic acid		
Ammonium persulfate	120	49	Carbon bisulfide	x	x
Ammonium phosphate	120	49	Carbon dioxide, dry	120	49
Ammonium sulfate 10–40%	120	49	Carbon dioxide, wet	120	49
Ammonium sulfide	120	49	Carbon disulfide	x	x
Ammonium sulfite			Carbon monoxide	120	49
Amyl acetate	x	x	Carbon tetrachloride	x	x
Amyl alcohol	120	49	Carbonic acid	120	49
Amyl chloride	x	x	Cellosolve	72	22
Aniline	x	x	Chloracetic acid, 50% water	x	x
Antimony trichloride	72	22	Chloracetic acid	x	x
Aqua regia 3:1	x	x	Chlorine gas, dry	x	x

Table 8-3 *Continued*

Chemical	Maximum temp. °F	Maximum temp. °C	Chemical	Maximum temp. °F	Maximum temp. °C
Chlorine gas, wet	x	x	Malic acid	120	49
Chlorine, liquid	x	x	Manganese chloride	72	22
Chlorobenzene	x	x	Methyl chloride	x	x
Chloroform	x	x	Methyl ethyl ketone	x	x
Chlorosulfonic acid	72	22	Methyl isobutyl ketone	x	x
Chromic acid 10%	72	22	Muriatic acid	120	49
Chromic acid 50%	x	x	Nitric acid 5%	72	22
Chromyl chloride			Nitric acid 20%	72	22
Citric acid 15%	120	49	Nitric acid 70%	72	22
Citric acid, concentrated	120	49	Nitric acid, anhydrous	x	x
Copper acetate			Nitrous acid, concentrated	x	x
Copper carbonate	120	49	Oleum	x	x
Copper chloride	120	49	Perchloric acid 10%	120	49
Copper cyanide	120	49	Perchloric acid 70%	72	22
Copper sulfate	120	49	Phenol	72	22
Cresol	x	x	Phosphoric acid 50–80%	120	49
Cupric chloride 5%			Picric acid		
Cupric chloride 50%			Potassium bromide 30%	120	49
Cyclohexane	120	49	Salicylic acid		
Cyclohexanol	x	x	Silver bromide 10%		
Dichloroacetic acid			Sodium carbonate	120	49
Dichloroethane (ethylene di-chloride)	x	x	Sodium chloride	120	49
			Sodium hydroxide 10%	120	49
Ethylene glycol	120	49	Sodium hydroxide 50%	120	49
Ferric chloride	120	49	Sodium hydroxide, con-centrated	120	49
Ferric chloride 50% in water	120	49			
Ferric nitrate 10–50%	120	49	Sodium hypochlorite 20%	120	49
Ferrous chloride	120	49	Sodium hypochlorite, con-centrated	120	49
Ferrous nitrate	120	49			
Fluorine gas, dry			Sodium sulfide to 50%	120	49
Fluorine gas, moist	120	49	Stannic chloride	120	49
Hydrobromic acid, dilute	120	49	Stannous chloride	120	49
Hydrobromic acid 20%	120	49	Sulfuric acid 10%	120	49
Hydrobromic acid 50%	x	x	Sulfuric acid 50%	120	49
Hydrochloric acid 20%	120	49	Sulfuric acid 70%	120	49
Hydrochloric acid 38%	120	49	Sulfuric acid 90%	120	49
Hydrocyanic acid 10%	120	49	Sulfuric acid 98%	x	x
Hydrofluoric acid 30%	72	22	Sulfuric acid 100%	x	x
Hydrofluoric acid 70%			Sulfuric acid, fuming		
Hydrofluoric acid 100%			Sulfurous acid	120	49
Hypochlorous acid	120	49	Thionyl chloride	x	x
Iodine solution 10%	x	x	Toluene	x	x
Ketones, general	x	x	Trichloroacetic acid	72	22
Lactic acid 25%	120	49	White liquor	120	49
Lactic acid, concentrated			Zinc chloride	120	49
Magnesium chloride	120	49			

Source: Schweitzer, Philip A. (1991). *Corrosion Resistance Tables,* Marcel Dekker, Inc., New York, Vols. 1 and 2.

Table 8-4 Pressure Rating of Schedule 40 Conley Pipe

Nominal pipe O.D. (in.)	Operating pressure (psi)	
	Internal	Vacuum
2.375	150	459.1
2.875	150	259.6
3.520	150	148.0
4.500	150	67.7
6.570	150	37.3
8.600	150	26.2
10.655	150	15.9
13.000	100	12.3

Source: Courtesy of The Conley Corp.

Table 8-5 Pressure Rating of Schedule 20 Conley Pipe

Nominal pipe O.D. (in.)	Operating pressure (psi)	
	Internal	Vacuum
3.25	150	74.8
4.27	150	42.9
6.31	150	22.1
8.31	150	9.7
10.37	150	8.3
12.57	100	9.0
14.57	100	5.8

Source: Courtesy of Conley Corp.

Table 8-6 Sizes of Conley Double Containment Piping Systems

Nominal size (in.)	Annulus clearance (in.)	Annulus pressure (psi)	Annulus vacuum (psi)
2/4	0.81	150	42.88
2/6	1.81	150	22.07
3/6	1.24	150	22.07
3/8	2.24	150	9.67
4/6	0.75	150	22.07
4/8	1.75	150	9.67
6/8	0.72	150	0.67
6/10	1.72	150	8.29
8/10	0.70	150	8.29
8/12	1.76	100	9.00
10/12	0.74	100	9.00
10/14	1.74	100	5.78
12/14	0.56	100	5.78

Source: Courtesy of Conley Corp.

Table 8-7 Anchor Loads Due to Thermal Expansion with Temperature Changes in Containment Pipe

Nominal Size (in.)	Load (psi) with temperature change (°F/°C)		
	25/14	50/28	75/42
2/4	1309	2142	2975
2/6	1918	3361	4803
3/6	2201	3644	5086
3/8	2670	4581	6492
4/6	2394	3836	5279
4/8	2862	4773	6684
6/8	3515	5426	7337
6/10	4385	7165	9946
8/10	5010	7791	10572
8/12	6357	10485	14613
10/12	7156	11284	15412
10/14	7825	12621	17417
12/14	9035	13832	18628

Coefficient of thermal expansion 9.5×10^{-6}.
Based on uninsulated containment pipe.
Source: Courtesy of the Conley Corp.

Table 8-8 Support Spans for Conley Double Containment Piping Systems at Different Carrier Pipe Temperatures

Nominal size (in.)	Support span (ft) at °F/°C							
	75/24	100/38	125/52	150/66	175/79	200/93	225/107	250/121
2/4	14.7	14.7	14.7	14.6	14.6	14.6	14.6	14.6
2/6	18.8	18.8	18.7	18.7	18.7	18.7	18.7	18.7
3/6	17.2	17.2	17.2	17.2	17.1	17.1	17.1	17.1
3/8	20.3	20.3	20.3	20.2	20.2	20.2	20.2	20.2
4/6	16.3	16.3	16.3	16.2	16.2	16.2	16.1	16.1
4/8	19.0	19.0	19.0	19.0	19.0	18.9	18.9	18.9
6/8	17.8	17.7	17.6	17.6	17.5	17.5	17.4	17.4
6/10	20.3	20.3	20.3	20.2	20.2	20.2	20.1	20.1
8/10	19.5	19.5	19.4	19.3	19.3	19.2	19.1	19.1
8/12	22.1	22.0	22.0	22.0	21.9	21.9	21.8	21.8
10/12	21.3	21.2	21.1	21.1	21.0	20.9	20.8	20.8
10/14	22.8	22.8	22.7	22.7	22.6	22.6	22.5	22.5
12/14	22.3	22.2	22.1	22.0	22.0	21.9	21.8	21.7

These spans are valid for containment pipe operating in an ambient temperature environment.
Source: Courtesy of Conley Corp.

Table 8-9 Compatibility of Conley Piping Systems with Selected Corrodents
The chemicals listed are in the pure state or in a saturated solution unless otherwise
indicated. Compatibility is shown to the maximum allowable temperature for which
data are available. Incompatibility is shown by an x. A blank space indicates that data
are unavailable.

| | Maximum temperature (°F/°C) | | | |
| | | Vinyl esters | | |
Chemical	Epoxy	411	470	Furan
Acetaldehyde		x		225/107
Acetamide				
Acetic acid 10%	150/66	200/93	210/99	225/107
Acetic acid 50%	125/52	180/82	180/82	180/82
Acetic acid 80%				
Acetic acid, glacial				
Acetic anhydride		x	100/38	225/107
Acetone	180/82	x	x	70/21
Acetyl chloride				180/82
Acrylic acid 25%		100/38	100/38	
Acrylonitrile				
Adipic acid				
Allyl alcohol				
Allyl chloride				
Alum				
Aluminum acetate				
Aluminum chloride, aqueous	300/149	200/93	210/99	250/121
Aluminum chloride, dry				
Aluminum fluoride	150/66	80/27	80/27	225/107
Aluminum hydroxide	150/66	80/27	80/27	225/107
Aluminum nitrate	250/121	160/71	160/71	
Aluminum oxychloride				
Aluminum sulfate	300/149	200/93	210/99	250/121
Ammonia gas	150/66	80/27	100/38	250/121
Ammonium bifluoride				
Ammonium carbonate	200/93	150/66	150/66	225/107
Ammonium chloride 10%	200/93	200/93	210/99	220/104
Ammonium chloride 50%	200/93	200/93	210/99	220/104
Ammonium chloride, sat.	200/93	200/93	210/99	220/104
Ammonium fluoride 10%		150/66	150/66	225/107
Ammonium fluoride 25%	150/66	150/66	150/66	225/107
Ammonium hydroxide 25%	100/38	100/38	100/38	180/82
Ammonium hydroxide, sat.				
Ammonium nitrate	250/121	200/93	250/121	220/104
Ammonium persulfate		180/82	180/82	180/82
Ammonium phosphate	150/66	200/93	210/99	
Ammonium sulfate 10–40%	300/149	200/93	210/99	220/104
Ammonium sulfide		120/49	120/49	250/121
Ammonium sulfite				
Amyl acetate	75/24		120/49	200/93
Amyl alcohol				
Amyl chloride				

Table 8-9 Continued

Chemical		Maximum temperature (°F/°C)		
			Vinyl esters	
	Epoxy	411	470	Furan
Aniline	75/24	x	70/21	250/121
Antimony trichloride				
Aqua regia 3 : 1				
Barium carbonate	250/121	200/93	210/99	
Barium chloride	250/121	200/93	210/99	200/93
Barium hydroxide				150/66
Barium sulfate	250/121	200/93	210/99	
Barium sulfide	300/149	150/66	180/82	150/66
Benzaldehyde		x	70/21	200/93
Benzene	100/38	x	100/38	
Benzene sulfonic acid 10%	x	150/66	150/66	200/93
Benzoic acid	210/99	200/93	210/99	250/121
Benzyl alcohol				
Benzyl chloride		x	80/27	200/93
Borax				
Boric acid	200/93	200/93	210/99	200/93
Bromine gas, dry				
Bromine gas, moist				
Bromine liquid		x	x	
Butadiene	100/38			
Butyl acetate	70/21	x	80/27	220/104
Butyl alcohol				
n-Butylamine				
Butyl phthalate				
Butyric acid		80/27	120/49	150/66
Calcium bisulfide				
Calcium bisulfite	200/93	180/82	180/82	225/107
Calcium carbonate	300/149	180/82	180/82	
Calcium chlorate	200/93	200/93	210/99	
Calcium chloride	300/149	200/93	250/121	250/121
Calcium hydroxide 10%		180/82	180/82	225/107
Calcium hydroxide, sat.	200/93	200/93	210/99	225/107
Calcium hypochlorite	x	120/49	160/71	
Calcium nitrate	250/121	200/93	210/99	220/104
Calcium oxide				
Calcium sulfate	250/121	200/93	210/99	250/121
Caprylic acid				
Carbon bisulfide				80/27
Carbon dioxide, dry	250/121	200/93	210/99	
Carbon dioxide, wet	250/121	200/93	210/99	
Carbon disulfide				80/27
Carbon monoxide				
Carbon tetrachloride	100/38	180/82	200/93	225/107
Carbonic acid				
Cellosolve				
Chloracetic acid, 50% water	200/93	100/38	100/38	

Table 8-9 Continued

| Chemical | Epoxy | Vinyl esters | | Furan |
		411	470	
Chloracetic acid		x	x	
Chlorine gas, dry		200/93	210/99	225/107
Chlorine gas, wet				
Chlorine, liquid				
Chlorobenzene	100/38	x	100/38	250/121
Chloroform	100/38	x	x	80/27
Chlorosulfonic acid				225/107
Chromic acid 10%	120/49		150/66	
Chromic acid 50%				
Chromyl chloride				
Citric acid 15%				
Citric acid, concentrated	250/121	200/93	210/99	190/88
Copper acetate		160/71	160/71	225/107
Copper carbonate				
Copper chloride	250/121	200/93	210/99	250/121
Copper cyanide				
Copper sulfate	250/121	200/93	210/99	250/121
Cresol				250/121
Cupric chloride 5%				
Cupric chloride 50%				
Cyclohexane				
Cyclohexanol				
Dichloroacetic acid				
Dichloroethane				
Ethylene glycol	200/93	200/93	210/99	250/121
Ferric chloride	300/149	200/93	210/99	250/121
Ferric chloride 50% in water				
Ferric nitrate 10–50%	250/121	200/93	210/99	250/121
Ferrous chloride	250/121	200/93	210/99	
Ferrous nitrate				
Fluorine gas, dry				
Fluorine gas, moist				
Hydrobromic acid, dilute	150/66	180/82	180/82	
Hydrobromic acid 20%	150/66	180/82	180/82	
Hydrobromic acid 50%	150/66	150/38	100/38	
Hydrochloric acid 20%	200/93	180/82	230/110	150/66
Hydrochloric acid 38%	200/93	150/66	180/82	150/66
Hydrocyanic acid 10%		100/38	100/38	
Hydrofluoric acid 30%				
Hydrofluoric acid 70%				
Hydrofluoric acid 100%				
Hypochlorous acid	200/93		180/82	
Iodine solution 10%				
Ketones, general				
Lactic acid 25%				
Lactic acid, concentrated	200/93	200/93	210/99	225/107

Table 8-9 Continued

| | | Maximum temperature (°F/°C) | | |
| | | Vinyl esters | | |
Chemical	Epoxy	411	470	Furan
Magnesium chloride		200/93	210/99	225/107
Malic acid				
Manganese chloride				
Methyl chloride				
Methyl ethyl ketone	100/38	x	70/21	150/66
Methyl isobutyl ketone				
Muriatic acid				
Nitric acid 5%		150/66	180/82	
Nitric acid 20%		120/49	150/66	
Nitric acid 70%				
Nitric acid, anhydrous				
Nitrous acid, concentrated				
Oleum				
Perchloric acid 10%		150/66	150/66	
Perchloric acid 70%				
Phenol				
Phosphoric acid 50–80%	220/104	200/93	210/99	250/121
Picric acid		x	100/38	165/74
Potassium bromide 30%	200/93	100/38	120/49	200/93
Salicylic acid	250/121	140/60		250/121
Silver bromide 10%				
Sodium carbonate	300/149	160/71	180/82	225/107
Sodium chloride	300/149	200/93	210/99	250/121
Sodium hydroxide 10%	200/93	150/66	150/66	212/100
Sodium hydroxide 50%	200/93	150/66	150/66	212/100
Sodium hydroxide, concentrated				
Sodium hypochlorite 20%				
Sodium hypochlorite, conc				
Sodium sulfide to 50%	200/93	120/49	210/99	220/104
Stannic chloride	200/93	200/93	210/99	225/107
Stannous chloride				
Sulfuric acid 10%				
Sulfuric acid 50%	150/66	200/93	210/99	225/107
Sulfuric acid 70%		180/82	180/82	
Sulfuric acid 90%				
Sulfuric acid 98%				
Sulfuric acid 100%				
Sulfuric acid, fuming				
Sulfurous acid	200/93	80/27	100/38	200/93
Thionyl chloride				
Toluene	150/66	80/27	120/49	225/107
Trichloroacetic acid		200/93	210/99	
White liquor				
Zinc chloride	250/121	200/93	210/99	

Source: Schweitzer, Philip A. (1991). *Corrosion Resistance Tables,* Marcel Dekker, Inc., New York, Vols. 1 and 2.

Table 8-10 Maximum Allowable Operating Pressure of Centricast Epoxy Pipes

Nominal pipe size (in.)	Maximum allowable operating pressure (psi)			
	RB 2530[a]		111 EP[a]	
	Internal pressure at 250°F (121°C)	External pressure at 75°F (24°C)	Internal pressure at 250°F (121°C)	External pressure at 75°F (24°C)
1	300	2125		
1½	1275	2065	300	865
2	1000	1825	200	265
3	650	733	200	120
4	500	360	175	43
6	500	180	150	21
8	375	70	150	14
10	300	35	150	11
12	250	23	150	7

[a]Carrier pipes.

Nominal pipe size (in.)	Maximum allowable operating pressure (psi) for 111 EP[a]	
	Internal at 225°F (107°C)	External at 75°F (24°C)
3	200	120
4	175	43
6	150	21
8	150	14
10	150	11
12	150	7
14	150	7
16	100	7

[a]Containment pipe.

Table 8-11 Maximum Allowable Operating Pressure of Centricast Vinyl Ester Pipes

	Maximum allowable operating pressure (psi)			
	Cl 2030[a]		111 VE[a]	
Nominal pipe size (in.)	Internal pressure at 175°F (79°C)	External pressure at 75°F (24°C)	Internal pressure at 175°F (79°C)	External pressure at 75°F (24°C)
1			300	1975
1½	875	983	300	615
2	800	890	200	225
3	525	525	200	68
4	500	360	175	40
6	500	160	150	15
8	375	65	150	11
10	300	33	150	8
12	250	23	150	6

[a]Carrier pipes.

	Maximum allowable operating pressure (psi) for 111 VE[a]	
Nominal pipe size (in.)	Internal at 175°F (79°C)	External at 75°F (24°C)
3	200	68
4	175	40
6	150	15
8	150	11
10	150	8
12	150	6
14	150	6
16	100	6

[a]Containment pipe.

Table 8-12 Allowable Operating Pressure of Dualcast Piping Systems (Carrier Pipe)

Nominal pipe size (in.)	Epoxy resin[a] pressure rating (psi) at 225°F (107°C)	Vinyl Ester Resin[b] pressure rating (psi) at 175°F (79°C)
1N3	300	300
1½N3	450	300
1N4	300	300
1½N4	450	300
2N4	450	275
2N6	450	275
3N6	300	200
4N6	225	150
4N8	225	150
6N8	225	150
6N10	225	150
8N12	225	150
10N14	225	150
12N16	225	150

[a]Based on Weldfast 440 adhesive.
[b]Based on Weldfast Cl 200 and Cl 200QS adhesives.

Table 8-13 Support Spacing for Fibercast Double Containment Piping

Nominal pipe size (in)	Maximum unsupported span (ft)	
	Epoxy	Vinyl
1N3	16.2	15.7
1½N3	15.1	14.7
1N4	18.8	18.2
1½N4	17.5	17.0
2N4	16.7	16.2
2N6	20.0	20.0
3N6	20.0	19.7
4N6	18.8	18.4
4N8	20.0	20.0
6N8	20.0	20.0
6N10	20.0	20.0
8N12	20.0	20.0
10N14	20.0	20.0
12N16	20.0	20.0

Based on 125°F (52°C) ambient temperature and 175°F (79°C) fluid temperature, with maximum rated internal pressure and deflection considerations for pipe full of fluid, specific gravity 1.00, and ½-inch maximum deflection. For specific gravities other than 1.00, the following multipliers may be used:

Specific gravity	Multiplier
1.25	0.96
1.50	0.93
2.00	0.87
3.00	0.80
air	1.12

Index